'With food security at the top of the international agenda, this book fills an important niche in its broad framing of the challenges and solutions; its recognition that food insecurity has household to global dimensions; and its coverage of the multiple causes, ranging from climate change to land grabs to the ways supermarkets are accessing food. It is an important read for academics and practitioners grappling with one of the greatest challenges facing humankind in the twenty-first century.'
Professor Bruce Campbell, Director,
CGIAR Research Program on Climate Change,
Agriculture and Food Security

'Ian Christoplos and Adam Pain have got it spot on. They have brought together a great team of co-authors to craft a highly relevant and topical reader on the different faces of food insecurity today. Reaching far beyond the familiar paradigms of food unavailability or entitlement problems, the book analyses how food insecurity is the complex outcome of a range of structural yet contextual processes. It rightly seeks the way out of food crises in the realm of politics and a rethinking of the nature of the social contract between states and their citizens.'
Dorothea Hilhorst, Professor of Humanitarian Aid
and Reconstruction, Wageningen University

'This timely book highlights the resurgent interest in food security as a global policy imperative. Ranging over topics from agricultural innovation to property rights in land and natural resources, humanitarian emergencies, nutrition, human rights, risk and resilience, Christoplos and Pain and their co-authors lay out new directions for research and policy at multiple levels, from household to global. This book is essential reading for food security scholars, practitioners and policy makers alike.'
Professor Dan Maxwell, Feinstein International Center,
Tufts University

NEW CHALLENGES TO FOOD SECURITY

Food security, always high on the political agenda, has risen to even greater prominence following the food price and financial crisis of 2008. Fears about societal insecurity due to food price increases and hunger, grave scenarios regarding the effects of climate change and general uncertainty about the impact of investment in biofuels and 'land grabbing' on food prices and availability have meant that food security is now recognised as a global challenge. Recent crises in the Horn of Africa, chronic conflicts in countries such as Afghanistan and natural disasters such as the Haiti earthquake have demonstrated that food security has many dimensions. This book is the first to bring together analyses of the various factors that have an impact on food security.

With contributions from leading experts from around the world, *New Challenges to Food Security* benefits from global perspectives and case studies. The topics explored include market volatility, climate change and state fragility. Analyses of responses to food security crises and risk cover rural and urban contexts, arenas of national policy formation and global food regimes, and investment in land and productive technologies.

This book is unique in two respects. First, it takes a step back from the normative literature focused on specific factors of, for example, climate change, agricultural production or market volatility to look instead at the dynamic interplay between these new challenges. It helps readers to understand that food security is not one discourse, but is related to how the different factors generate multiple risks and opportunities. Second, through case studies, the book highlights how these factors come together at local levels as farmers, entrepreneurs, consumers, government officials and others are making key decisions about what will be done to address food security and whose food security will be given priority. The book examines how food production and consumption are embedded in powerful political and market forces and how these influence local actions.

Exploring how different actors respond to these challenges, *New Challenges to Food Security* provides a new and timely perspective on what the search for enhanced food security means today, and where research, development and humanitarian efforts may need to focus tomorrow.

Ian Christoplos is a researcher in Natural Resources and Poverty at the Danish Institute for International Studies working in development cooperation and humanitarian assistance. He has worked as a researcher and consultant for over 25 years. His research focuses on how local actors manage market and environmental risks and conflict.

Adam Pain is a Visiting Professor in Rural Development at the Swedish University of Agricultural Sciences in Uppsala. Previously a lecturer in Natural Resources at the School of Development Studies, University of East Anglia, UK, he has worked in natural resource management in Africa and Asia. He is currently also a researcher at the Danish Institute for International Studies.

NEW CHALLENGES TO FOOD SECURITY

From climate change to fragile states

Ian Christoplos and Adam Pain

Routledge
Taylor & Francis Group

LONDON AND NEW YORK

First published 2015
by Routledge
2 Park Square, Milton Park, Abingdon, Oxon OX14 4RN

and by Routledge
711 Third Avenue, New York, NY 10017

Routledge is an imprint of the Taylor & Francis Group, an informa business

© 2015 Ian Christoplos and Adam Pain

The right of the editor to be identified as the author of the editorial material, and of the authors for their individual chapters, has been asserted in accordance with sections 77 and 78 of the Copyright, Designs and Patents Act 1988.

British Library Cataloguing in Publication Data
A catalogue record for this book is available from the British Library

Library of Congress Cataloging-in-Publication Data
New challenges to food security : from climate change to fragile states/
    edited by Ian Christoplos, Adam Pain.
        pages cm
    Includes bibliographical references and index.
    1. Food supply—Developing countries.   2. Agriculture—Economic
aspects—Developing countries.   3. Food industry and trade—Security
measures.   I. Christoplos, Ian.   II. Pain, Adam.
    HD9018.D44N49 2014
    338.1'91724—dc23
    2014025232

ISBN: 978-0-415-82255-8 (hbk)
ISBN: 978-0-415-82256-5 (pbk)
ISBN: 978-0-203-37117-6 (ebk)

Typeset in ApexBembo
by Apex CoVantage, LLC

CONTENTS

FIGURES

TABLES

CONTRIBUTORS

Jagannath Adhikari Independent researcher and consultant in Nepal, and visiting scholar at the Australian National University, Canberra, Australia

Michael Brüntrup German Development Institute (DIE), Bonn, Germany

Ian Christoplos Senior Project Researcher, Danish Institute for International Studies (DIIS), Copenhagen, Denmark

Lisa Denney Researcher, Secure Livelihoods Research Consortium, Overseas Development Institute, London, UK

Wenche Barth Eide Department of Nutrition, Faculty of Medicine, University of Oslo, Norway

Sandrine Fréguin-Gresh Researcher in Agricultural Economics and Geography, CIRAD/ART-Dev Instituto Nitlapan, Universidad Centroamericana, Managua, Nicaragua

Mikkel Funder Danish Institute for International Studies (DIIS), Copenhagen, Denmark

Arthur H. Grigsby Independent researcher

Susanne Jaspars PhD candidate, School of Sociology, Politics and International Studies, Bristol University, UK

Niek Koning Centre for Sustainable Development and Food Security, Wageningen University, the Netherlands

Richard Mallett Researcher, Secure Livelihoods Research Consortium, Overseas Development Institute, London, UK

Amit Mitra Formerly Associate Director, Centre for Science and Environment; now independent researcher based in New Delhi, interested in development issues ranging from water to food security

Anders Riel Müller Danish Institute for International Studies (DIIS) and Department of Society and Globalisation, Roskilde University, Denmark

Carol Mweemba PhD student, Integrated Water Resources Management Centre, University of Zambia

Le Duc Ngoan Director, Centre for Climate Change Studies in Central Viet Nam

Imasiku Nyambe Professor, Integrated Water Resources Management Centre, University of Zambia

Hemant R. Ojha Senior Fellow, Melbourne School of Land and Environment, University of Melbourne, Australia, and Research Director and Board Chair, Southasia Institute of Advanced Studies, Kathmandu, Nepal

Adam Pain Danish Institute for International Studies (DIIS), Copenhagen, Denmark

Francisco J. Perez Instituto Nicaragüense de Investigaciones Económicas y Sociales, Universidad Nacional Autónoma de Nicaragua

Helle Munk Ravnborg Senior researcher, Danish Institute for International Studies (DIIS), Copenhagen, Denmark

Le Thi Hoa Sen Senior researcher, Centre for Climate Change Studies in Central Viet Nam

Winnie Wangari Wairimu PhD researcher, Special Chair Humanitarian Aid and Reconstruction, Wageningen University, the Netherlands

Dirk Willenbockel Institute of Development Studies at the University of Sussex, UK

ACKNOWLEDGEMENTS

The authors wish to thank the research programme on Climate Change, Agriculture and Food Security and the Danish Ministry of Foreign Affairs for the support provided to producing this book. We also want to thank Jenny Glazebrook, the technical editor of this volume, for the exceptional effort which kept this difficult project together.

1

INTRODUCTION

Ian Christoplos and Adam Pain

Overview

This volume describes a range of different perspectives on food security, with an emphasis on the various meanings that are applied to 'food security'. It presents a number of thematic issues which are investigated and developed in a set of country case studies. Particular attention is paid to the range of trends that are seen to constitute food security 'crises', including those facing different populations today, which have been derived from historical processes and long-term trajectories. By understanding how food security crises are conceptualized, the book also looks at different scales: global, local and household responses to crises and risk in rural and urban contexts; arenas of national policy formation and global food regimes; and investment in land and productive technologies. Primary attention is given to the dynamic interplay between old and new challenges and how these different factors generate multiple risks and opportunities. The book explores how factors impinging on food production, access and consumption are embedded in powerful political and market forces and how these influence local actions.

This introductory chapter reviews how the concept of food security is changing in relation to the new challenges and frameworks for understanding vulnerability, livelihoods and crisis in the context of multiple long wave changes in demographics, markets and climate. It also reflects on the implications of the shifting strategies and roles of national governments to safeguard food availability, stabilize prices, respond to domestic and international shocks and preserve political stability. The implications of these efforts within changing international norms related to the right to food and land acquisitions are also considered. The interests of public and private actors in relation to food production and access, and also the new security agenda are reviewed. The chapter assesses the extent to which old food security issues are being reframed. Development and food security concepts are contrasted and guiding notions such as Sen's entitlement theory and neoclassical economic theories about market responses to scarcity are reconsidered in light of recent food crises. The chapter questions whether the language of resilience and adaptation is really new, and how these discourses may be used to disguise power relations and misguided tendencies to perceive food security as simply a managerial or technical issue. This chapter ends with an outline of the key themes that run through the book.

Why is food security back on the agenda?

Food security is high on the political agenda in many countries. There are fears about societal insecurity due to food price increases and hunger combined with grave scenarios regarding the effects of climate change. This fear is fed by general uncertainty about the impacts, on food prices and availability, of investments in biofuels and struggles to acquire rights to land and other resources. This has meant that food security is now recognized as a global and multifaceted challenge. It is also recognized that national food security and household food security for farming households are not equivalent (see India, Chapter 15) and there may even be trade-offs and difficult choices when decisions are made between, for example, enhancing national availability or helping smallholders meet their subsistence needs; between supplying growing urban populations with affordable food or addressing entrenched rural poverty; and between supporting smallholders versus encouraging investments in larger commercial enterprises. This contrasts to past and even present food insecurity interventions, which tended to focus on narrower goals and forms of intervention such as increasing the production levels of subsistence farmers through introducing new technologies, increasing national cereal self-sufficiency and food aid (see Nicaragua, Chapter 17).

The debate around food security has in part been driven by the emergence of new perspectives brought to the subject by an increasing number of researchers from a range of disciplines (politics, economics, nutrition) and practitioners from a range of sectors (agriculture, health, security, private sector development). Definitions and interpretation of the scope of the concept vary enormously. Claims are being made about different types of 'crisis' based on different interests, notions and beliefs on how risks of climate change, market volatility, globalization, conflict and other risks impact different aspects of food security. Indeed, the very term 'food security' can become a gloss that obscures the specific opportunities to meet nutritional needs in urban and rural areas, and the different roles and strategies of men and women as well as those reliant on farming versus those pursuing other livelihoods. It could be argued that three major narratives frame the meanings of food security – food as a global commodity, food as a product of farming, and environmental services and food as a basic right. This book brings together these different dimensions of food security to provide a basis for a more integrated understanding of the interplay of this range of risks. We do not claim to present a normative framework for convergence into a new food security paradigm, but we hope this volume can enhance understanding of the different entry points as a foundation for more holistic thinking, particularly in relation to the social, political and institutional aspects of food security.

A renewed concern for food security has come to prominence since the food price and financial crisis of 2008. Since then the ongoing crisis in the Horn of Africa, chronic conflicts in countries such as Afghanistan and urban crises such as the Haiti earthquake have demonstrated that food security has many dimensions. In addition, the 'war on terror' has added a further dimension by linking security and development as part of stabilization interventions in what have been labelled 'fragile states' (see Afghanistan, Chapter 14). Claims are being made regarding links between multiple crises based on scenarios about climate change and conflicts over scarce resources. In Latin America and India statements and commitments to 'end hunger' and protect the right to food have brought food entitlements into the political arena, at the same time as recognition is growing that remaining in farming is not necessarily an optimal solution for food security, even among the rural poor. The rising dominance of supermarkets as market players and food outlets and the changing concerns about consumption and safety (see Viet Nam, Chapter 13) have brought attention to the need for new perspectives on food security in relation to globalization, the effects of market liberalization on food markets, shifting consumption patterns, carbon footprints, new food production technologies and urbanization.

For national governments, keeping food prices low for urban populations often is (and has long been) more important than ensuring that farming is profitable or ensures subsistence in the rural hinterlands. A policy bias towards urban areas, combined with a demographic shift leading to increasing proportions of populations living in urban areas, has reinforced a preference toward ensuring price stability rather than profitable farming. Even in rural areas, shrinking farm sizes have meant that much of the food insecure population lacks sufficient land to produce a marketable surplus, and is thus more reliant for food security on price stability than on hypothetical benefits from producing for a market that is out of its reach. When governments have become nervous about sudden price spikes leading to unrest, stable national-level food availability has also taken centre stage politically. Old debates about choices between protectionism and free trade are being rekindled, with new twists, such as whether to invest in biotechnology versus organic production methods or whether to produce food or (bio)fuel. The implications of these choices for ensuring that citizens have enough to eat and can afford to pay for food, now and in the future, remain unclear. In the North and in urban areas of developing countries, there is also a growing recognition that waste and overconsumption of inappropriate foods have knock-on effects in global food systems.

Local governments are faced with helping their communities adapt to an increasing frequency of natural hazards related to climate change. Rising temperatures and environmental pressures, together with the disappearance of traditional markets for produce and labour, are bringing many local agricultural systems to the brink of 'tipping points', where past coping strategies are overwhelmed. These pressures are not just overwhelming farmers, but also the structures of agricultural extension, water management agencies and other institutions that are more accustomed to dealing with the 'old normal' than the 'new normal' of chronic, repeated, enduring and unpredictable crises.

Governments are not the only actors responding. One response to rising food prices and competition over resources is that private investors (sometimes in alliance with state authorities, see South Korea, Chapter 16) are seeking to gain control of agricultural land for producing food, fuel and profits through investments in large commercial farms that may either create new livelihoods and provide access to new technologies, or displace vast numbers of the rural poor (see Brüntrup, Chapter 5). In some countries that were, until recently, food insecure themselves, policy formation to encourage foreign investment to access food is driven by this historical experience.

In many parts of the world, food crises are now protracted, where threats to livelihoods and food insecurity and malnutrition have remained at unacceptably high levels for many years. This puts into question the assumptions that in the past justified responses based on short-term provision of food aid. Countries experiencing protracted food security crises are often also fragile states. Here, the challenges are how to provide long-term assistance in situations where humanitarian nutritional needs continue to exist (see Sudan, Chapter 8), and whether state-building modalities are appropriate and effective in terms of promoting societal and food security in a context of ongoing instability and political and economic upheaval.

There are increasing pressures from funders on humanitarian agencies to shift more quickly into development mode when conflicts subside, but in countries where agricultural policies emphasize market-driven models, the implications of rolling national policies out in war-torn areas are often overlooked. Assumptions that chronic conflict will merge seamlessly into development when displaced people return home raise serious questions about the extent to which these fragile areas can or should be subsumed into national policies focused on commercialization (see Uganda, Chapter 9). As agricultural services are rebuilt amid the uncertainties of post-conflict processes and as international agencies attempt to rethink their role, it is left to local level actors to deal with the ambiguities of addressing food security where food aid is no longer appropriate but where 'modernization' policies remain as pipe dreams.

Changing links between theory and narratives, old and new

As noted above, the new challenges to food security have led to new and contrasting definitions. A useful starting point for understanding where the different theories and narratives converge and diverge can be found in the relative emphases placed on pillars (or dimensions), agreed upon in the 2009 World Summit on Food Security (see Eide, Chapter 4) of availability, access, utilization and stability (FAO 2009)

Availability, and with that the drive for increased food production, has been perhaps the most traditional focus of food security efforts. Efforts to increase production and productivity dominated the discourse until Amartya Sen's concept of entitlements (Sen 1982) drew attention to the question of access, and the fact that few people starve because of a lack of availability of food. Growing urbanization and rural landlessness raised doubts about who was likely to benefit from increased production and productivity. Attention to entitlements led to moves away from faith that new technological advances in agriculture would eliminate hunger. Instead, increasing emphasis was given to ensuring that the poor could access food through securing a range of income and, where required, social transfers. This was evidenced first by a shift of attention away from food to livelihood security in the 1990s, and then by an emerging legal and ethical paradigm around the right to food. At the same time there was a decline in interest and public investment in agriculture, reflecting the rise of the liberal agenda and expectations that private sector development would generate new and more secure livelihoods. More attention was given to the factors that enabled people to buy food than to just produce it.

More recently, concerns regarding access have deepened with the recognition that those facing chronic food insecurity cannot secure food access solely, if at all, through market-based solutions, not least because markets are often the greatest source of risk to poor people. Risk has many dimensions related to gender, locality, resource tenure and other factors that must be addressed if nutritional levels are to be maintained. Social protection is being increasingly called for as a way to respond to risk, based on the understanding that there is a need to move beyond assumptions that a technical or market solution can be found for the poorest sectors of the population for either producing or buying food. There is also growing emphasis on nutrition security. The centrality of access to and utilization of food stuffs is epitomized by the 'South Asian enigma' of improved aggregate supply of foods and rapid economic growth, (see India, Chapter 15) combined with relatively poor progress in nutritional outcomes with both life course and intergenerational impacts. However, regardless of whether the solution has been envisaged as markets, social protection or nutrition, weak institutions and public service capacities have stood in the way of turning these approaches into changes on the ground (see Sierra Leone, Chapter 11).

Interventions for livelihood support are also being recast as ways to address concerns for youth employment and issues of inequality. This is partly in relation to the demographic factors behind the unrest such as that which contributed to the Arab Spring, and partly because of trends where the young in many countries are increasingly eager to leave agriculture.

Even those aspects of food security that remain associated with smallholder farming are being perceived differently, as is the assumption that reinforcing smallholder production should be the cornerstone of rural food security efforts. There is a growing recognition that smallholder farms are becoming too small to meet subsistence, but that smallholders are continuing to farm. This phenomenon has meant that the past assumptions of the poor being driven out of agriculture, leading to increasing farm sizes, have been disproven. Smallholders have hung on in agriculture but are becoming increasingly reliant on a range of other sources of income. Optimistic scenarios about these livelihood diversification efforts constituting a 'solution' for the rural poor are beginning to give way to recognition that these are more appropriately seen as deteriorating 'coping' efforts that are unlikely to meet nutritional needs in the long-term. This has implications for gender roles in ensuring food security, farming and livelihoods, as access to different forms of employment, markets, and responsibilities for generating cash and feeding the family is often regulated by gender (see Nepal, Chapter 12).

Though food access concerns are still strong, over the past decade there has been a shift back to recognizing that ensuring availability is, or will soon be, a serious problem at least at a regional level if not at a global level (see Willenbockel, Chapter 3). Narratives increasingly depict food security crises as being related to Malthusian scenarios regarding the effects of climate change, combined with demographic growth and scarcity of land and water. These narratives may have questionable empirical foundations, but they are powerful political messages that seem to be driving decision making about the assumed advantages of 'modern and efficient' mega-farms versus 'traditional and risk averse' smallholders. Fears have also grown regarding the shift in agriculture from food to fuel production and the search for ways to combine reductions in carbon emissions with increased production for a growing global population.

Acknowledgements that food security efforts in many countries have not led to sufficient or sustained reduction in hunger and malnutrition have resulted in a renewed attention to utilization and the consumption end of the equation. This may also be related to the new 'results agenda', as aid donors are asking for proof of whether or not they have received 'value for money' by measuring a given intervention's impact on saved lives and reduced hunger. The renewed attention to consumption and malnutrition is also because of the epidemic of obesity that is not just affecting high-income countries, but also middle-income countries on all continents.

This focus on food and nutrition security can in some instances be related to growing attention to rights-based approaches, where the 'bottom line' of whether people have enough to eat is becoming an indicator of if and how duty-bearers (states or their proxies where the state is extremely weak) are meeting this basic right. Even countries that do not formally recognize the right to food often recognize the political costs of failures to ensure that people have enough to eat.

Finally, and perhaps most importantly, the recognition of the need for stability, in availability, access and use, is coming to the fore. All of the factors above have generated anxiety about uncertain access and affordability of food for different groups of consumers. Stability was added to former international definitions of food security in the 2009 World Summit because of the uncertainties associated with price volatility, climate change, and state fragility. This emphasis on stability is related to the concept of resilience that has become an umbrella for newer development paradigms that acknowledge instability as the 'new normal'. Resilience may be an attractive way to draw attention to stability (or lack thereof), but it may also hide questions about 'whose resilience counts'. Calls for wiping out swidden agriculture and pastoralism through a shift towards industrial farming are justified by questionable claims that this is necessary to attain ecosystem resilience. Societal resilience based on food security is not necessarily congruent with strategies for more resilient ecosystems and the latter may exclude coping strategies that are under pressure, but may still provide more stable livelihoods than other alternatives.

Changing responses

How then should food insecurity be addressed? Will there be a return to the intervention modalities of the past, or will global, national and local actors find more innovative and appropriate ways to respond? The responsibilities of national and local institutions for preventing hunger and malnutrition are changing. The roles of the international community in humanitarian assistance, food aid, agricultural development and climate change adaptation are also in flux. Decades of food security assistance, such as agricultural support and food aid, have yielded mixed results (see Uganda, Chapter 9). Questions are being raised about the effectiveness and appropriateness of the technical approaches usually applied. At the same time, pressures to 'do something' in the face of crises have often led to a search for simple solutions, such as a temporary injection of food aid, distribution of a new seed variety, land concessions for large investors or closing borders to food exports. There are factors at international, national and local levels that are encouraging innovation in relation to past modalities, but also ample evidence that food security responses are often locked into repeating failures of the past or reviving past models when new approaches have not yielded intended results.

Responses by national governments include the following: food subsidies in Latin America and India, renewed investment in research and extension in Africa, combinations of protectionism and land acquisition in the face of price volatility and uncertain access to purchase (especially in net food deficit countries), capacity development in agricultural services, nutrition and social protection (see Sierra Leone, Chapter 11) and a range of technological and institutional fixes, from providing farmers with weather information by SMS (short message service) to agricultural extension methods such as farmer field schools.

New aid commitments are being made for food security. Some of this goes through traditional channels, such as the World Food Programme (WFP), while there is also an evident search underway to more broadly anchor food security by 'linking the dots' and reassessing the roles of different agencies and actors. New task forces have been created at global and national levels to ensure that efforts can be combined to find a more joined-up approach.

The depletion of food reserves and surpluses has completely changed the incentives that existed in the past for donor countries to see food aid (including disposal of their own surpluses) as a self-evident tool for addressing food insecurity (see Koning, Chapter 2). The use of food aid for development purposes, such as school feeding, has all but disappeared. Even in emergencies, efforts to move from food aid to provision of cash are gaining increasing attention.

This has in turn raised attention to the fact that cash for increasingly recurrent and protracted emergency responses is not necessarily so different from cash provided to the chronically poor to enable them to smooth income flows (and with that access to food). Interest has arisen in insurance as well as a way to strengthen people's entitlements in times of crisis. These similarities of response suggest a trend toward convergence around the need for social protection systems.

Food security as an entitlement is recognized by some political leaders as a central aspect of the social contract between states and citizens, and this may therefore be a driver for bringing together humanitarian, development, social protection and economic growth efforts (see Viet Nam and South Korea, Chapters 13 and 16). This is apparent at national levels, as politicians are worried about providing reasonably stable food prices and being seen as acquiescent to pressures to provide land to foreign investors. Globally, it is recognized that a food insecure population is a population that is also susceptible to radicalization. Food and political security are increasingly seen as being interlinked.

Issues, themes and cases

The structure and content of the chapters in this volume are not designed to catalogue food security concepts or experiences, but rather to stimulate critical reflection by presenting and contrasting evidence and different perspectives on key themes. The issue chapters (Part 2) are contrasted with country case study chapters (Part 3) that are used to explore empirically the convergence of diverse but specific risks in given contexts.

The country case studies describe the actions of different actors on different scales facing a convergence of diverse risks, and explore their implications for national and household food security (see Zambia, Chapter 10, for an account of household level water insecurity). Particular attention is given to looking at how countries are being affected by global structural changes, together with empirical evidence on how national and local actors are responding to these challenges. By exploring how different actors respond to emerging and converging risks to food security, the chapters unpack what the search for enhanced food security means today, and where research, development and humanitarian efforts may need to focus tomorrow. Six major themes run through the book.

Theme 1. Dynamics of change: From subsistence to livelihood security under conditions of increasing inequality

Access to labour, land and water are often stated to be the key determinants of the food security options of the rural poor. An increasing proportion of the rural poor and food insecure are functionally landless and rely on labour and other sources of income to meet food needs, evidenced by the increasing levels of rural income diversification from off- and non-farm activities. Farming careers are becoming less attractive to the daughters and sons of today's farmers. Thus the focus of food security efforts has, since the 1990s, shifted from shoring up subsistence agriculture to a recognition that food security in an increasingly urbanized, globalized and commercialized world is largely about livelihood security. This does not mean, however, that the importance of subsistence agriculture for rural food security is universally perceived to be minor. On the contrary, the exclusive nature of agri-food markets has meant that many small-scale producers either do not engage with risky food markets or are leaving or reducing commercial production to return to subsistence-based livelihoods. In addition, households migrating out of rural areas as a result of diminishing returns from agriculture or conflict have little choice but to engage in risky and uncertain activities in the informal urban economy. The rising numbers of urban food insecure provide a growing challenge (see Pain, Chapter 7).

The widespread rise in income inequality at national and global levels is a contributing factor to food and nutrition insecurity in both the South and the North. Gender inequalities are part of this equation of food insecurity and malnourishment, as in many cultures gender roles determine how much of food resources are devoted to consumption (see Nepal and India Chapters 12 and 15). When women's production and livelihood efforts are marginalized, this has immediate effects on food security. One response to the recognition of links between inequality and food security has been the emergence of efforts to position food security and freedom from hunger as a fundamental right (see Eide, Chapter 4). This has resulted in, for example, the National Food Security Act 2013 in India, which has taken steps to establish this basic right, although the bill is seen by many to be severely limited. The 'right to food' movement however has pushed the concept of food security from simply a condition of calorie deficiency to one in which rights to a broader set of social, economic and physical conditions are seen to be violated. Thus food security policy objectives reflect duty-bearers' responsibilities to address the conditions that give rise to the inequality that produces food insecurity in the first place.

Theme 2. Resource scarcity, competition for water and land acquisition

Poverty reduction in South Asia has long been closely connected to agro-ecological potential for growth in irrigation. Despite rapid growth in production and productivity, substantial proportions of South Asian populations remain absolutely poor and food insecure because of rising inequality and increasing demand for land, water and fuel intensive food and other agricultural products. In other regions, including much of Latin America (see Nicaragua, Chapter 17) and parts of Africa, production growth has long been related to expanding areas under production at the 'agricultural frontier' of shrinking forested areas. The potential to increase land areas to absorb growing populations is rapidly disappearing, which has led to a focus on productivity in existing agricultural areas. However, slow growth of agricultural productivity in many countries has limited growth in food entitlements for those whose incomes have grown slowly.

This has perpetuated and even intensified the prevalence of low food entitlements in a context where agricultural and water systems are experiencing increasingly severe stresses, chronically, seasonally or because of competition between different users (see Munk Ravnborg, Chapter 6). These pressures manifest themselves at both the local level (see Zambia, Chapter 10), and have implications for how countries experiencing resource scarcity design domestic policies and act at international levels (see South Korea, Chapter 16). Pressures on finite resources due to demographic and climate change have also meant that both local societal and armed conflicts are increasingly attributed to conflict over such resource access. The desire to ensure access to affordable food among their increasingly urban populations has led many governments and private investors to acquire their own rights to use resources beyond their national boundaries (see Brüntrup, Chapter 5 and South Korea, Chapter 16).

Water is crucial to nutrition security through its role in agricultural production and hence access to food and to the health environment in which food entitlements are transformed into nutritional and health attainments. Population growth and urbanization and changing food habits are producing increased stresses on water resources to meet increasing demands for agricultural products. A key issue is transjurisdictional water resource management in a context of increasing scarcity and increased variability of flow related to concerns with climate change impacts.

Theme 3. Globalization

The discourse on food security includes two seemingly diametrically opposed (but perhaps interrelated) perspectives on the forces that drive and perhaps respond to the factors that determine 'whose food security counts'.

On the one hand there is a view that the concept of national food security is outdated given the moves toward open markets in much of the world. Food prices are seen as being largely determined at a global level, and access to food through livelihoods is seen as also being regulated through markets. Subsistence, at both household and national levels, on the other hand, is thought of as a concern attached to an outdated perspective on how to ensure basic welfare in a growing global economy. However, assumptions about how spikes in food prices would translate into better prices (and incentives) for farmers in developing countries have been proven wrong, and there is a recognition that food insecure smallholders who do not produce surpluses and are net consumers, have no such opportunity to benefit from markets. In general the food price crisis of 2008 did much to debunk neoclassical economic assumptions about the forces of globalization (see Nepal Chapter 12). Although there is debate about how far global prices were transmitted into national economies, there is evidence that in landlocked countries such as Zambia this transmission was weak.

Globalization-driven models, often focusing on value chain development, have raised a major debate on the impact of market liberalization on food markets. This includes changing retail channels such as supermarkets, with their potential effects of marginalizing smallholders who cannot meet demands for quality, timeliness, certification and/or bulk production. These smallholders' livelihoods (and food security) may actually worsen because of the impacts of these innovation processes. However, the convenience and increased competition from new forms of retail may reduce food costs for consumers. Different forms of retail and longer market chains also create new incentives for production that endanger food safety and perhaps also weaken the monitoring of food quality.

There is also an ongoing debate about the lessons (learnt or perhaps not) from past agricultural transformations (e.g. the green revolution) and the policy measures that supported these for the new modernization agenda. At this point these trends in retail and consumer demands are beginning to suggest that there will inevitably be different roles for smallholders, as some can innovate to meet specific niche demands (see Viet Nam, Chapter 13) and others are likely to leave markets, relying more on a combination of subsistence production and wage labour (see Nicaragua, Chapter 17).

Theme 4. Uncertainty, volatility and resilience

The effects of climate change and volatility in food prices, together with other factors including state failure, are changing the extent and complexity of the landscapes of risk to household food security (see in particular Nepal, Chapter 12, where there was a collision of risks). Intensifying exposure to these risks and diminishing capacities to manage multiple risks may in turn result in behaviour wherein short-term coping becomes an obstacle to taking advantage of opportunities that could offer routes out of poverty.

For poor people, risks are likely to be multiple, covariant within communities and closely clustered because of the interlinked nature of activities through which livelihoods are pursued and the lack of diversity of options available to them. This clustering of activities increases the probability of risks being realized and exacerbates the intensity and significance of the impact of the shock, when it occurs.

In its narrowest usage, risk refers to an objective hazard that can be measured and assessed and probabilistic values given to its chance of occurrence. But risk also has to be understood in terms of its structural dimensions as caused by inequalities and exclusion, and needs to incorporate an institutional and relational account; this includes treating shocks much more as hazards that are chronic and to be negotiated, rather than as stochastic events. Thus a more expanded approach recognizes the objective dimensions of risks, but also takes account of the social and cultural processes that mediate exposure to the hazard.

The concept of resilience has rapidly entered the mainstream discourse. It is a term that is used to bring together the concerns about multiple risks and paths to manage these risks. A major focus is on climate change adaptation, but is also about uncertainty regarding financial systems, questions about macroeconomic growth models, as well as concerns about how to overcome the protracted nature of conflict. 'Resilience' remains a term with controversial connotations; it has drawn attention not only to the issues of calculable risks but also the capacities to deal with uncertainties that cannot be predicted or known about in advance. Some view it as a concept that points more to responsive change and include notions of flexibility, diversity and adaptive change as part of the recovery with potential transforming implications. Others see it simply in terms of ability to survive under difficult conditions. Yet others see it as a renewed call to better link humanitarian and development efforts. It is a term that needs to be used with care.

Theme 5. Innovation and agricultural modernization

As noted above, in the past there was a default assumption that agricultural modernization, and with that an increase in aggregate national production and productivity on individual farms, was the main route to enhanced food security. This assumption faded, to an extent, for two reasons. First with the rise of attention to Sen's entitlement theory there was a recognition that access to and utilization of food were as important as availability to food security.

The second reason can be attributed to the rise in the market liberalization agenda. It questioned the relevance of national food security as a goal based on the view that greater engagement with international markets would give national economies greater income and the ability to ensure food security through international trade. This has led to a resurgence of claims (see World Bank 2008) about the importance of market-driven agricultural modernization, but now couched more in terms of innovation, usually driven by markets but also implying close links to research and extension. Questions of whether this may have positive impacts on household food security have sometimes been overlooked because of a resurgent faith in production and productivity, as some policymakers and international institutions have seemingly forgotten Sen's observations. Sudan (Chapter 8) and the market-driven models for agricultural development in Afghanistan (Chapter 14) to a degree exemplify a techno-optimism by which some countries adopt policies that shift support to larger producers, i.e. those with the capacity to adopt new technologies and innovate along the lines that match policymakers' goals. This is in contrast to countries such as Viet Nam (Chapter 13), where the innovative capacity of smallholders has over time been increasingly recognized as a driver in achieving both national and household food security.

The resurgence of the agricultural modernization and innovation narrative may also be related to the multiple risk scenarios of the combined effects of climate change, demographic growth and the need for reducing greenhouse gas emissions. These seemingly intractable challenges may have turned an undefined notion of 'innovation' into a convenient black box for the grand, across-the-board solutions that are assumed necessary to avoid catastrophic consequences. Chapter 3 specifically reviews recent global food system scenario studies that explore these various risks and potential effects on food security outcomes.

Theme 6. Institutional landscapes and policy responses

The institutional landscape at a national and local level can be characterized in terms of the relative roles of the state, the market, civil society and household in ensuring well-being. In many states agricultural support policies, including subsidies and tariff protection, have historically been a key component for support to the agricultural sector. State-building in Europe and the United States was historically heavily dependent on protection measures for the domestic economy. Chapters 13 and 16 show how food security remains part of the policy repertoire of emergent strong states (e.g. Viet Nam, South Korea) that have adopted proactive agendas to invest in agriculture domestically and/or abroad to feed their citizens and promote economic development. India (Chapter 15) has a somewhat more equivocal position where elite capture of the political stage and enduring poverty have been countered to some degree by democratic protest and reasonably effective public food distribution systems, albeit with uneven social and geographical application. There are also countries in the South, both in Africa (Sudan, Chapter 8 and Uganda, Chapter 9) and Asia (e.g. Nepal, Chapter 12 and Afghanistan, Chapter 14), where states are weak and their authority is contested. Markets therefore experience little formal regulation and are subject to elite capture – the well-being of their citizens in turn being largely neglected. In such countries external actors such as the international financial institutions and bilateral donors, as well as international NGOs, are major players in the institutional landscape and policymaking related to food security. But this simple dichotomy must be understood in terms of the historical sweep of changes in policy models and in markets and other factors in relation to production as Koning discusses in Chapter 2.

Policies are (or at least should be) related to scenarios of food production and lessons can be drawn from Willenbockel (Chapter 3) on the extent to which policies are likely to change in relation to emerging and more nuanced understanding of food security and its scalar dimensions. This is beginning to displace simplistic and empirically unsupported Malthusian assumptions and technological optimism.

Setting the stage for understanding and responding to the changing institutional landscape and policies is another challenge. On the one hand there are the normative commitments related to the right to food. On the other there are the practices through which private sector actors, working within commercialization-oriented public policies, are increasingly gaining control over land, water and other natural resources. This questions the degree to which state action can drive change.

The book

The book starts with a historical overview of patterns and problems of hunger around the world. It explores the various interpretations and policy prescriptions that have been brought to bear on food crises and food insecurity over time and the implications for the future. This then leads into a chapter on scenario planning for food security, which is followed by a chapter on the emergence of the 'right to food' movement. Three chapters then follow that address particular challenges to food security linked to land acquisition, water security and crisis states. The next part of the book contains ten case studies drawn from contrasting countries that expand on the key themes highlighted in this introductory chapter. A final concluding chapter draws together the themes of the book.

References

FAO (2009) *Declaration of the World Summit on Food Security.* Online. Available <www.fao. org/fileadmin/templates/wsfs/Summit/Docs/Final_Declaration/WSFS09_Declaration. pdf> (accessed 21 March 2014).

Sen, A. (1982) *Poverty and Famines. An Essay on Entitlement and Deprivation,* Oxford: Clarendon Press.

World Bank (2008) *World Development Report 2008: Agriculture for Development,* Washington, DC: World Bank.

2

FOOD SECURITY AND FOOD CRISES

Evolving realities, evolving debates

Niek Koning

Introduction

Food crises have haunted humanity throughout the ages, yet acquiring an intellectual grasp of them is challenging. What constitutes a food crisis is already contentious; NGOs have sounded famine alarms where others only perceived moderate distress. Additionally, there are so many different types of crises: preindustrial 'old regime crises', modern famines resulting from war or oppression, hunger caused by poverty-amidst-plenty, the 'new food crisis' of obesity, the recent 'food price crisis' in international markets, and the impending food scarcity for which eco-pessimists are warning. Explanations, likewise, vary. Malthus' overpopulation view was opposed by Marx's class-based vision and Boserup's view that population growth caused innovation rather than crisis; experts disagree about whether today's famines in sub-Saharan Africa are to be blamed on economic crisis or political crime; and so on. Last but not least, there are vehement debates regarding how to remedy food insecurity. Some plead for state supported Green Revolution; others for macroeconomic liberalization with social protection; and others for participatory approaches and organic methods.

In this chapter, I attempt to make some sense of this bewildering reality. I begin with the preindustrial dynamic and the nineteenth-century breakthroughs that changed hunger into a problem of poverty in a world of plenty. I continue with the divergent pathways of different regions and the influence of changing metropolitan growth patterns. I conclude with a reconstruction of the hunger debate and the question of whether today's discourses and counter-discourses prepare us for future risks.

Preindustrial growth and 'old regime' crises

During the preindustrial era, expensive transport, slow innovation, and scarce fertilizer inhibited increases in food supply. Societies cycled between not-too-food-insecure population growth and acute food crisis. Thomas Malthus offered a biological explanation for this that resembled an ecological predator–prey model (humans as predators, natural resources as prey). However, the fluctuations of human societies involved more than just population-resource interactions. Our social nature added a political component. As long as a modicum of food security could be maintained, our social abilities assisted in solving cooperation problems. During subsistence crises, immediate survival – also at each other's expense – became of utmost importance. It undermined solidarity relationships and stimulated processes of exclusion that changed empathy into enmity – adding violent conflict to Malthus' arsenal of 'positive checks' that adjusted consumption and resources the hard way.

This political aspect may also be evident in other highly social animals. More specific for *Homo sapiens* were their intellectual and communicative abilities, which assisted them in finding transferable solutions for enhancing the carrying capacity of their natural resource base. It allowed humans to increase in numbers with a less than proportional increase in their ecological footprint. Until the industrial era, this process was mainly driven by population growth. This was Ester Boserup's message (Boserup 1965), and although it was formulated as a criticism toward Malthus, a synthesis of Boserup's and Malthus' viewpoints is possible (Lee 1988; Wood 1998). As long as the population pressure on resources did not become overwhelming, it stimulated technical–institutional innovation as a mitigating feedback effect. Once scarcity became acute, reinforcing feedback effects such as soil exhaustion, epidemic diseases, and conflicts emerged. This began a downward spiral that resulted in demographic crisis.

This Malthus–Boserup model may provide an acceptable description of simple societies regulated by kinship and 'big men' – the default form of human organization. However, agricultural intensification induced a shift from kin to class and a more structural differentiation between the powerful and the weak. Marx counter-contended that food crises were caused by exploitation rather than absolute shortage. Again, the opposition is not absolute. Peter Turchin's demographic-structural theory (Turchin and Nefedov 2009) combined Malthus' and Marx's perspectives. When adding Boserup's view, the following scenario arises. When population pressure began to increase, the elite could exact higher rents. This incited the elite to initiate innovations that facilitated intensification and were often imitated by common farmers so that additional mouths could be fed. Concurrently, the increased surplus extraction resulted in an expansion of the elite. When, in a next phase, the stress on resources further increased, the elite attempted to maintain their lifestyle even though production per worker decreased. This squeezed the margins for investment, reduced the demand for non-food products, and resulted in unemployment for artisans. Innovation yielded to complex forms of poverty-sharing in order to maintain an increasingly precarious equilibrium. Such equilibrium sooner or later collapsed (Tainter 1990), giving way to the horrors of Malthus' 'positive checks', which restored the balance between populations, elites and resources, making room for a new growth phase.

The nineteenth-century watershed

In the first half of the nineteenth century, a new Eurasian growth phase that had begun in the eighteenth century ran up against constraints. Although the potato famine (1845–50) demonstrated that Northwestern Europe was not unscathed, access to New World resources and a pattern of nuptiality that moderated birth rates through celibacy and late marriage preserved the region from the full impact of the predicament. This provided the window of opportunity for new breakthroughs which were based on fossil fuels and technical science and that revolutionized the potential food supply. Modern transport facilitated the expansion of food production to remote regions; artificial fertilizer enabled increases in yields; and fossil-based substitutes liberated for food production vast stretches of land that had previously been used for horse feed, fibres, or other non-food crops. The combined effect was to create room for 'modern economic growth' – growth not just in population but also in per capita incomes, which afforded an opportunity for ending hunger (Kuznets 1966). The increase in income was reinforced as modern growth included incentives for households to have fewer children. Therefore, an initial upswing in the population was followed by a spontaneous deceleration ('demographic transition') whereby incomes could increase even further.

The new breakthroughs had a global consequence for the nature of food insecurity. Until then, hunger had been a problem of poverty and scarcity. It now changed into a problem of poverty in a world of plenty. Whether this will remain so will be discussed at the end of this chapter.

Modern economic growth involved a transformation in the sectoral composition of the economy; the shares of industry and services in the GDP and employment increased relative to that of agriculture. This 'structural transformation' led an older generation of economists to believe that modern growth could be achieved simply by extracting resources from agriculture and transferring them to the industrial sector. This idea, famously expressed in Arthur Lewis' model of 'development with unlimited supplies of labour' (Lewis 1954), was revised by Johnston and Mellor who claimed a more active role for agriculture in boosting modern growth. Agriculture should develop in order to create opportunities for agro-industries, to feed industrial workers, and to generate savings for industrial investment (Johnston and Mellor 1961). Others added that agricultural development was also essential in providing market demand for emerging industries and engendering the human and social capital that these required (e.g. Timmer 1995). The question remained of how to start the agricultural booster itself. Classical economists had argued for leaving it to the market. Their neoclassical successors repeated this and only conceded some room for infrastructures as public goods. As many farm economists contended prior to the neoliberal era, this was too narrow a view. In a world of abundance, population growth failed to increase food prices; therefore, the incentive that had stimulated farm progress in the preindustrial era had disappeared. Rather, the opposite was true. Exploiting the new potential for farm production went with boom–bust cycles and recurrent overproduction in agricultural markets.[1]

Economies of scale in agriculture remained modest, so that a squeeze on farm profits was not compensated by a shake-out of small holdings. Wage increases reinforced the survivability of self-employed farmers. In the end, it was the large, rather than the small, farms that declined even though they were better equipped for technical progress. This evolution induced technical and institutional adjustments (downscaling of machines, marketing co-operatives, etcetera) that further strengthened the position of small farmers. Nevertheless, all of this failed to solve the problems of agricultural markets and did not change smallholders into pioneers of technical progress. The consequence was that agricultural development became vitally dependent on government intervention. The state had to provide for the hard and soft infrastructures that small farmers could not provide for themselves. By the same token, the state had to assume responsibility for stabilizing agricultural markets to overcome the risk-adversity of small farmers and provide them with sufficient margins for investment (Koning 1994).

Thus, while modern growth required agriculture to develop, agricultural development, in its turn, required government support. Consequently, modern economic growth – and the subsequent decrease in hunger – became vitally dependent on agricultural policy choices. In countries that supported smallholder modernization, this facilitated industrialization and normal-capitalist growth. In countries that failed to support their farmers, normal growth was hampered. The consequences of this latter development varied with local conditions. Where tenure relationships were highly unequal, the outcome tended to be disproportionate economic growth with considerable malnutrition persisting in the underclass. In circumstances where many people retained access to the land, socio-environmental crisis and widespread food insecurity prevailed. This occurred because farmers were unable to invest in sustainable intensification and not because options for sustainable intensification were exhausted. To distinguish this development from an old regime crisis, I call it an 'unsustainability spiral' (cf. Cleaver and Schreiber 1994).

The agricultural policy choices of different countries were shaped by various forces. First, there were international relationships. Countries subjected to foreign control had only minimal room for making their own choices. Conversely, hegemonic countries had opportunities for ensuring growth and food security that other countries had not. Second, there were metropolitan growth patterns. Metropolitan countries went through a social-imperialist, a Fordist, and a neoliberal growth pattern, respectively. These configurations not only influenced metropolitan farm policy choices, but they also affected the room that other countries had for pursuing active agricultural policies. Third, farm policy choices were influenced by sociopolitical characteristics of societies that followed from historical path-dependency and the ecological endowments of different regions (cf. Diamond 1998). Below I make a broad distinction between three ideal-types: farmer, landlord, and tribal societies. Farmer societies had a graded class stratification, kin-transcending interest articulation, and professional state bureaucracies imbued with *raison d'état*. Landlord societies had a polarized class structure and oligarchic states controlled by the landed elite. Tribal societies had kin solidarities and patrimonial political systems with state institutions owned, so to speak, by high-ranking members of the ruling clan. In the next sections, I will conduct a quick tour around the world to sketch how these conditions influenced the vicissitudes of various regions.

Independent farmer societies: The decrease of hunger in the West and Japan

Farmer societies existed in parts of Asia and in Western Europe where ecological endowments had long favoured population growth and agricultural intensification, and where local strongmen had been restrained by central rulers or had obtained attractive opportunities outside the realms of farming. These societies also existed in cooler North America and Oceania where European settlers had displaced indigenous populations, and a climate precluding tropical export crops had destined their economies to self-centred development. Of these societies, the West and Japan had been able to weather the crisis that hit other Eurasian societies in the earlier nineteenth century. The decline of international agricultural prices and other changes in the agro-economy that ensued in the later part of that century induced a class-based mobilization of farmers and accommodating political responses. This prompted the introduction of supportive farm policies. Countries with a comparative advantage in agriculture initially limited themselves to furthering farm progress and added price supports when prices collapsed again around 1930; other countries resorted to protection from the beginning. Britain was the only Western country that postponed protection although it no longer possessed a comparative advantage in farming. Agricultural free trade resulted in half a century of stagnation in farm output and productivity.[2] By World War I, approximately half of the British food supply was imported. Britain could afford to sacrifice its agriculture because it was the world's hegemonic power. Its industrial outlets did not rely on domestic farm incomes, and its empire and navy secured food imports. After 1900, Britain's hegemony began to erode. Only extensive American intervention rescued its population from starvation during the German submarine blockade in 1916–18. In the 1930s, the country belatedly began to support its farmers (Koning 1994; Offer 1989).

Free-market-minded historians and economists have explained agricultural protection as the result of the disproportional power of landed elites or agrarian pressure groups. In reality, farmer mobilization interacted closely with the ascent of parliamentary democracy and found a sympathetic ear with the public. In countries where large landowners subsisted, they lacked the power to shift the burden to rural workers. Therefore they opted for broad agrarian coalitions that demanded government support for the farming sector as a whole (Francks 2006; Koning 1994; Verdier 1994).

Supportive policies paved the way for smallholder modernization and normal-capitalist growth. The few famines that still occurred in the twentieth century resulted from wars and oppression, not from subsistence crisis (Ó Gráda 2009). Such events were no longer accepted as fate and prompted attempts to secure the right of food to civilian populations through international law such as the 4th Geneva

Convention. Apart from that, undernourishment diminished to minimal proportions, although it did not entirely disappear and was likely to re-emerge in certain situations. It became the object of public health institutions and nutritionists with their diagnoses of underweight, stunting and wasting and toolkit of supplementation, fortification and dietary diversification.

In time, this residual problem of undernutrition became overshadowed by a 'new food crisis' of obesity and other food related welfare diseases. As with undernutrition, this has a significant class aspect. In low-income countries, the affluent tend to be obese and the poor to be thin. In high-income countries, it is the other way around, presumably because the poor have more difficulty in maintaining a healthier lifestyle (McLaren 2007; Popkin and Gordon-Larsen 2004). The cross-cultural nature of the obesity epidemic suggests the influence of an underlying biological factor. Maybe the large brain relative to body size and erratic food supply of their Pleistocene ancestors have given humans special features such as a 'hungry gene' and a preference for high energy foods (Aiello and Wheeler 1995; Shell 2003). Or maybe humans just share with many other animals a fondness for the refined foods that modern techniques can provide (Wrangham 2009). Anyhow, in locations where supermarkets and chain restaurants have displaced more artisanal cooking, competition has pushed these enterprises toward the development of junk food that maximally satisfies this inborn predilection (Moss 2013).

Newly independent farmer societies: Post-colonial Asia and the Green Revolution

Outside Japan, many Asian farmer societies had succumbed to the nineteenth-century crisis. This hastened their subjection by foreign powers. Japan's colonial policies forcefully stimulated farm progress, thereby laying the foundations for the miracle growth of Korea and Taiwan after World War II (Francks et al. 1999). In the European colonies and in China, conversely, smallholders received scant support. As a consequence, the old regime crisis faded seamlessly into an unsustainability spiral (cf. Geertz 1963; Myrdal 1968). This is the reason why global hunger was long (and, partly, still is) concentrated in East and South Asia. A change would have to wait until after World War II when these countries regained independence. The political trajectories diverged between countries and involved brutal episodes such as the massacre of the rural underclass in Java in 1965 and the misguided Great Leap Forward effort at forced industrialization in China (1958–61) that caused a famine that killed between 15 and 30 million people (Ó Gráda 2009). Ultimately, however, similar pressures as those in the West and Japan led to a shift to more supportive farm policies. These facilitated a smallholder-based Green Revolution which subsequently led to modern growth and a reduction of hunger, and which turned China into a potential new superpower.[3]

The Green Revolution has provoked vehement debate. Some have hailed it as a triumph of modern breeding – though public investment and price policies were, in fact, just as important. Others have condemned it as an export product of American capitalism that aggravated, rather than alleviated, hunger by displacing small peasants and continuing exploitation. It is true that the Green Revolution was supported by the Rockefeller and Ford Foundations; that American officials hoped it would prevent a communist takeover; and that many poor peasants ceased farming. However, Green Revolutions had previously occurred in the West and in the Japanese Empire; a parallel Green Revolution occurred independently in communist China; and the Green Revolution encouraged non-farm growth which subsequently allowed many people to find more rewarding non-farm employment (e.g. Lipton and Longhurst 1989). The proportion of hungry people has strongly decreased in Green Revolution countries, which is in stark contrast to countries where a Green Revolution has not occurred. In India, which comprised landlord societies as well as farmer societies, economic development was geographically very uneven, but decreases in poverty were strongly related to the evolution of crop yields (Bhalla and Singh 2001; Ravallion and Datt 2002). Insofar as exploitative labour conditions persisted in the Green Revolution belts of this country, this was, to a large extent, because of the flood of extremely low cost, migrant labour from areas where a Green Revolution failed to occur.

Landlord societies: The consequences of inequality

Landlord societies existed in Latin America, parts of Eastern and Southern Europe, and in areas such as the Philippines, South Africa, and India's lower Gangetic plain. In some locations, they had evolved, as local strongmen possessed few alternatives outside agriculture or were not restrained by political centres. In other places, export demand had stimulated commercial estate farms that required constraints on labour mobility because of the abundance of land (cf. Domar 1970; Engerman and Sokoloff 2013).

In these societies, policies for smallholder modernization were not easily realized. The state was too intertwined with the elite, and the peasantry too weak to enforce effective reforms. Attempts at introducing such reforms – e.g. under Stolypin in Russia, Cárdenas in Mexico, or Frei and Allende in Chile – either failed or were soon eroded. Rather than support for smallholder modernization, the transformations in the global agro-economy induced adjustments that assisted large farms in surviving by shifting the problem of low agricultural prices onto the workers. Land reform often aided in this by severing traditional patronage relationships and creating a sector of mini-farms that became a reservoir of cheap labour for residual estate farms (De Janvry 1981). In an ensuing phase, these small croppers were often evicted to make room for cost-cutting mechanization that did not substantially raise land productivity. The long-term consequences were far-reaching. Extreme inequality resulted in an underinvestment in human capital. Rural poverty and labour displacement incited workers to move to the cities, but the same poverty constrained the market for domestic industries, so this flood was not absorbed by new employment opportunities. The outcome was the reproduction of extensive inequality both in the countryside and in the cities (Frankema 2009).

This evolution could have various sequels. At one extreme, the despair of the poor could explode into revolution. In the sociocultural make-up of landlord societies, this could lead to perverse outcomes. In Russia, Stalin effectively re-feudalized agriculture to extract a surplus from it for forced industrialization – a move that culminated in Europe's most deadly twentieth-century famine. Meanwhile, the disgruntled elite faction that had led the revolution changed into a new nomenklatura establishment. Stalin's successors relaxed the squeeze on farmers but still kept food prices low to legitimize their regime. Food consumption increased, but agricultural production lagged behind – a situation which hastened the final collapse of the Soviet Union (Cook 1992; Ellman 1984; Davies and Wheatcroft 2004).

At the other extreme, the elite could attempt to forestall radical reform by counter-revolution and repression as was the case with Apartheid in South Africa. In between these extremes, there was the more gradual reform that Latin American countries or the Indian state of West Bengal pursued for some time (Bértola and Ocampo 2012; Rawal and Swaminathan 1998). This was an uncertain middle ground where the ingrained oligarchic nature of the state and its continuing relationships with the elite could easily lead to regression.[4] The exclusionary, if not immiserizing, growth in many landlord societies has created a specific type of food situation: a double food crisis whereby a persistent undernutrition problem in the rural underclasses coincided with an obesity epidemic among the slightly better off (Monteiro et al. 2002).

Tribal societies: The increase of hunger in sub-Saharan Africa

Tribal societies had survived in certain locations where ecological conditions had hampered agricultural intensification – most importantly in sub-Saharan Africa. This was certainly not a homogeneous region. Northern Ethiopia possessed a feudal history, parts of the Sahel and the Great Lakes area had something between tribal and class structures, while the Kalahari Desert continued to be home to foraging groups. Nevertheless, much of the region had kin-oriented social relationships and patrimonial political structures (cf. Goody 1976). These features were reproduced because endemic diseases and fearsome warriors postponed the colonial scramble until the late nineteenth century when the trend change in international agricultural markets restricted the emergence of European-dominated landlord societies to a few areas that were settled by Caucasians. Once again, colonial governments failed to support indigenous smallholders. Where land confiscation for white settlers raised the population pressure on resources, an unsustainability spiral soon became apparent but, in many locations, land abundance still provided a safety valve.[5] For an extended period of time, therefore, hunger and malnutrition were less serious than in colonial Asia and pre-1949 China.

The great divergence ensued following national independence. While new Asian governments moved toward stimulating farm policies, the social-political fabric of sub-Saharan African societies generated fewer pressures for supporting smallholder modernization. Stimulating agriculture requires adequate government services, but the expansion of state apparatuses in the region was strongly influenced by the wish to create jobs for political clients, and farmers had to foot the bill (Bates 1981). While an accelerating population growth closed the safety valve of land abundance, unfavourable prices and deficient infrastructures prevented farmers from taking measures to make intensification sustainable. This caused a widening unsustainability spiral that swept the remainder of the economy along with it (Cleaver and Schreiber 1994; Koning and Smaling 2005; Van Donge et al. 2012). It entailed an increase in endemic hunger and had a degenerating influence on the political system, inducing violent conflicts that exacerbated food crises in which famine could even be employed as a weapon (cf. De Waal 1997; Edkins 2007).[6] As a consequence, sub-Saharan Africa is gradually replacing South Asia as the world's primary hunger belt.

Metropolitan growth patterns and international food regimes

While the nature of societies and international relations strongly affected the agricultural policy decisions of countries, the changing growth patterns of metropolitan capitalism also had a large influence. Prior to World War I, metropolitan powers sliced the world into exclusive spheres of influence. Multilateral regulation of international markets and support for smallholders in subordinate countries had very minimal opportunity to materialize in this setting. As previously indicated, this explains why many farmer societies in colonial Asia and China suffered from an unsustainability spiral during this period. Although many Western countries and Japan began to protect their own farmers, economic liberalism remained influential. The gold standard remained intact, and international trade expanded. In areas with abundant fertile land – the Punjab, the marshy plains of Southeast Asia, Latin America's southern cone – metropolitan investment and labour immigration led to burgeoning export-led growth.

The hegemonic shift from Britain to the United States and the upheavals of the 1930s and 40s signified the transition to a new social-Keynesian, or 'Fordist', pattern of metropolitan growth. Working class representatives were co-opted into ruling coalitions; states assumed a more active role in achieving new socio-economic aims; colonial blocks were dismantled; and multilateralism became more important. As part of these changes, metropolitan farm policies became more ambitious. Guided by an aim of 'parity' between farm and non-farm incomes, price supports were increased and the closing of 'unviable' farms was encouraged to allow mechanization of other smallholdings. To be sure, attempts at multilateral arrangements for stabilizing world markets were thwarted by agro-industrial lobbies that resisted controls on their own production volumes as well as supports for their imported tropical crops. The disciplines that the General Agreement on Tariffs and Trade imposed on national price supports were disregarded (Moyer and Josling 1990). The international commodity agreements that poor countries and the United Nations Conference on Trade and Development (UNCTAD) demanded were largely unrealized (Maizels 1992).

Nevertheless, the international dimension of Fordism gave poor countries increased space and some metropolitan support for shaping their own policies. This created a window of opportunity that Asian countries exploited for their Green Revolutions and that Latin America used for import-substitution industrialization but that sub-Saharan Africa failed to exploit.

During the first decades following World War II, Fordism developed without major disturbances. In the developed world, increasing wages ensured outlets for a growing supply of mass consumption goods. Given the evolved taste of humans, higher per capita incomes increased the demand for meat and dairy – even more as this was cheapened by intensive livestock systems. Meanwhile, supportive policies and high-yielding varieties generated a revolution in crop yields, first in metropolitan and then in Green Revolution countries. This period saw an accelerated world population growth, which peaked in the early 1960s, as well as a growing demand for livestock products that was caused by increasing incomes in the developed world. Thanks to the yields revolution, world agriculture could accommodate the ensuing rapid growth in the demand for food and feed at international prices that continued to decline.

The 1970s were a more troubled decade. Further wage increases began to squeeze profits. The United States had difficulty bearing the costs of hegemony, which resulted in growing budget and trade deficits. The reaction was a transition to a new, neoliberal pattern of metropolitan growth which was subsequently reinforced by the Soviet collapse. Beginning in the 1980s, the welfare state was rolled back and flexibilization and offshoring undermined trade union power. In farm policy, the parity aim was abandoned. Debt repayment problems were exploited to subject developing countries to the harsh therapy of structural adjustment. This especially affected sub-Saharan Africa where rapid public sector expansion in combination with sluggish growth created a growing fiscal crisis that undermined its recent autonomy. Structural adjustment entailed cuts in public investment and a wholesale dismantling of parastatal services. This exacerbated the unsustainability spiral in a manner that no 'poverty reduction' supplement to the policy could repair (Adejumobi 2006). The second victim was Latin America where reform attempts had foundered on rent-seeking, rural poverty and unsustainable external debt. Structural adjustment ended the pursuit of inward-oriented development and caused a slackening of economic growth and a new increase in inequality (Bértola and Ocampo 2012). Only Asian Green Revolution countries had developed enough to be less susceptible to the coercion by Western governments and Washington institutions. Their development had additionally reached a stage where they could afford some degree of liberalization without sacrificing farm progress (Dawe 2007).

Waving the free market flag did not end the reality of agricultural market failure. In the European Union and the United States, therefore, agricultural trade liberalization received a peculiar twist. While others had to decrease their customs defences, the WTO Agreement on Agriculture allowed countries that could afford it to whitewash protection through direct payments. The actual outcome was pseudo-liberal mercantilism and not agricultural free trade (Koning 2003).

Metropolitan growth patterns were one determinant of the configurations that have been called 'international food regimes' (McMichael 2009). The others were the global conditions of agricultural growth since the nineteenth-century breakthroughs and the divergent characteristics of regional societies and states that influenced the policy response to these conditions. We are currently witnessing new major changes including a global scramble for natural resources, the rise of China as a potential new superpower, and an increased role of geopolitics and state trading which might herald a new international food regime (McMichael 2012). Moreover, in the longer term, there is the possibility that the new potential for food production that was opened by the nineteenth-century breakthroughs might be depleted. This could change the situation completely, taking us back, in a way, to the realities of a preindustrial era. The question is whether old or new discourses on food security would prepare us for such a predicament. In the next sections, therefore, I consider how these discourses have evolved.

Framing the hunger debate

Around the year 1800, British classical economists contended that food security would be best served if market forces could do their harsh but necessary work. Competition should stimulate efficient farms; relief should be restricted to 'deserving poor'; and borders should be opened to allow international specialization according to comparative advantage. These prescriptions ruined domestic industries and reduced access for the poor to land in countries under British control, thereby contributing to the great famines in Ireland (1846–48) and India (1876–78). Yet, the classical precepts at least rightly reflected the nature of preindustrial agricultural growth, which was enabled by free market forces in which wealth-controlling rural classes were significant as early adopters. That neoclassical economists adhered to this vision after c. 1875, on the other hand, was denying the change in the dynamics of the global agri-food economy. The same applies for orthodox Marxists like Kautsky and Lenin who scorned any 'agrarian revisionism' within their ranks and maintained that class differentiation in agriculture would proceed similar to industry – a doctrine that Stalin subsequently exploited to justify his re-feudalizing collectivization in Russia.

Attempts to understand the new dynamics originated from the European Historical School, North American Institutional School, and Latin American Structuralist School of Economics, and from Russian farm economists such as Chayanov. Chayanov was executed by Stalin's henchmen, but economists of the Historical School contributed to the supportive farm policies in Europe around 1900, while structuralist economists encouraged the pursuit of agrarian reform and 'inward-looking development' in Latin America from the 1930s to the 1960s. In the United States, institutional economists assisted in shaping the agricultural legislation of the New Deal, which became the model for metropolitan farm policies in the Fordist era. The right-wing backlash against the New Deal after World War II forced the Institutional School underground. Nevertheless, leading American farm economists, for some time, retained the institutionalist focus on market failures and non-equilibrium dynamics (e.g. Cochrane 1959; Hathaway 1963).

These intellectual currents also inspired ideas about multilateral arrangements for stabilizing international markets. Institutional economists in the American New Deal Department of Agriculture and structuralist economists such as Raúl Prebisch called for international commodity agreements. Additionally, in Britain, John Keynes proposed an international buffer stock system in his blueprint for the Bretton Woods Conference. These ideas inspired the agricultural provisions in the General Agreements on Tariffs and Trade as well as UNCTAD's campaign for a coherent system of international commodity controls in the 1960s and 70s (Fantacci et al. 2012; Henningson 1981; Maizels 1992).

Institutional-evolutionary concepts also influenced the emerging discipline of development economics. Western scientists had long blamed the unsustainability spiral in European controlled Asia on the alleged inherent inertia of non-Western rural societies rather than on the neglect of smallholder needs (e.g. Boeke 1953; Geertz 1963). The Lewis model that treated agriculture as a reservoir of surplus labour that must be squeezed to the modern sector was in accordance with this (Lewis 1954). Not accidentally, the seminal paper that criticized these beliefs, claiming a more active role for agriculture in development (Johnston and Mellor 1961), was written by two American economists. The legacy of the Institutional School is clearly evidenced in their words. Additionally, structuralist-economic ideas regarding 'inward-looking development' gained significant influence through the UN Economic Commission for Latin America (e.g. Prebisch 1950). Views such as these inspired the World Bank and American governments to support land reform and Green Revolution policies in developing countries during these decades.

In the 1950s and 60s, this entire body of concepts received widespread support in metropolitan countries. In Europe, even the social democrats now fully endorsed the need for government support for smallholder modernization – as illustrated by Sicco Mansholt's acting as architect of the EU's common agricultural policy. Opposition remained largely restricted to the smallest farmers, who were supposed to relinquish space for more viable smallholdings. Beginning in the 1970s, however, agrarian Fordism provoked increasing criticism. One reason was the growing dumping of surpluses by Western countries. The interaction of Western support for the Green Revolution with the Cold War and the environmental problems caused by agro-chemicals, mechanization, and intensive livestock production also added to the censure.

Beginning c. 1980, the general crisis of Fordism entailed a U-turn in economic thinking. The synthesis of neoclassical and Keynesian thinking yielded to a mathematized form of pure and simple neoclassical economics. This 'microeconomic revolution' carried a new generation of development and agricultural economists with it. The former converted to the Washington consensus and structural adjustment. The latter reshaped their discipline to an applied form of neoclassical economics and professed that agriculture could best be left to the free market.[7] This conversion was coupled with no small amount of opportunism. By the 1980s, dumping of agricultural produce had resulted in mounting governmental cost as well as an agricultural 'trade war' between the European Union and the United States. The pressure to introduce effective production controls was increasing. In Europe, a milk quota system was introduced with cautious support from many dairy farmers. Nevertheless, agro-industrial lobbies and their allies continued to request protection with no strings attached. They exploited the free trade discourse to represent the shift to disguised dumping with direct payments through the WTO Agreement on Agriculture as a form of liberalization (Ingersent and Rayner 1999: Ch. 6; Paarlberg 1997). Rather than denouncing this mystification, agricultural economists supported it by producing an array of computable general equilibrium models that 'proved' WTO liberalization would entail significant welfare benefits for poor countries.[8]

Nevertheless, the ascent of neoliberal beliefs about agri-food issues was not solely due to economists and agro-industrial lobbies. Other social scientists and various shades of progressives, likewise, played a role. Prominent development NGOs demanded full liberalization of agricultural trade (e.g. Watkins and Fowler 2002). Many poverty experts endorsed liberalization if only it would be combined with safety nets and social insurance (e.g. Barrientos and Hulme 2009). Many sociologists and social ecologists contended that agricultural development in poor countries should be left to the invisible hand of local farmer initiatives and 'indigenous knowledge' (e.g. Altieri 1995; Barkin 2006; Chambers et al. 1989; Reijntjes et al. 1992; Scoones 1998; Tiffen et al. 1994). In regard to developed countries, they also championed alternative agri-food networks, organic agriculture, and locally differentiated rural development incentives against the 'industrialization' of agriculture by corporate interests and the state (e.g. Marsden 1995; Van der Ploeg et al. 2000; more generally, Pauli 2010). Notwithstanding the anti-globalist self-conception of some of these actors, they amended, rather than rejected, the Washington consensus and agricultural 'liberalization'.[9] In Europe, environmentalists could be co-opted for the official greenwashing of the mercantilism of direct payments – an episode culminating in the decisive support that the German Greens provided to the 'Fischler reform' in 2003 (Swinnen 2008; Syrrakos 2008). In a similar manner, some of these progressive 'criticisms' were integrated into neoliberal development policies, e.g. through the 'sustainable livelihoods framework' of the British Department for International Development and the new 'poverty reduction' edition of structural adjustment adopted by the World Bank in 1999.

This modified neoliberal thinking also became closely connected to the attempt to accelerate the reduction of hunger and poverty by procuring high-level commitments – an endeavour that culminated in the UN Millennium Development Goals of 2000. This campaign emulated the international law approach by which the West had attempted to preclude further man-made disasters following World War II. Whether this approach is feasible in poor countries in the new international context seems dubious. The goal of halving the proportion of hungry people by 2015 will not be realized. The claim that the goal will be 'within reach' depends entirely on a recent revision in the methodology by which undernourishment is measured. Neither the hard-neoliberal prescriptions of economists nor the soft-neoliberal solutions of their progressive 'critics' have reduced poverty and hunger as significantly as the Green Revolutions did in earlier decades. In sub-Saharan Africa – the main playground of these approaches – the percentage of under-nourished people has only slightly decreased while their absolute number has significantly increased (FAO 2012).

Why have so many progressives turned into soft-neoliberals on agri-food issues?

Neoliberalism is often associated with the manner in which economists who are trapped in linear equilibrium thinking and sensitive to commercial interests reacted to the crisis of Fordism. However, why have so many progressives turned into soft-neoliberals, especially on agri-food issues? As previously stated, some drawbacks of agrarian Fordism became salient during the 1970s – surplus dumping, pollution, and the growing emphasis on closing down small farms, while the parity objective was never fully realized. However, this does not explain why so many progressive academics came to oppose state-led smallholder modernization altogether. Why did so few of them support improvements in the existing model? Before examining the real-world adequacy of current discourses, let me first try to answer this question.

In my perception, two factors explain why many progressives have turned into soft-neoliberals regarding agri-food issues. One was the particular situation in Britain – the old imperial power that was home to many of these academics. Unlike its American counterpart, British progressivism was born in a country that was already the world's hegemonic power and whose agricultural frontier had long been closed. It harboured memories of the enclosures and the anti-Corn Law movement and lacked the American tradition of farmer-labour cooperation. In the early twentieth century, British Labourites joined forces with the City and older industries to defeat the protectionist 'Tariff Reform' coalition of farmers and steel manufacturers. Thus, they helped in sacrificing domestic agriculture to free trade (Koning 1994). Memories of this feat lingered in the collective consciousness of the British left and influenced its evolution following World War II. While their comrades in continental Europe accepted the need for supporting agricultural prices, British social democrats remained adamantly opposed to it. More than in any other location in the West, left-wing academics in Britain retained an orthodox-Marxist focus on intra-agrarian redistribution – ignoring the change in dynamic that had occurred after Marx's writings. Unaware of the special conditions that had allowed their country's exceptionalism in farm policies, they held their perspective valid for poor nations as well. Their conviction was confirmed by their interaction with academics from India's lower Gangetic plain – a landlord society that had been part of the British Empire. Critical intellectuals from this area naturally stressed the need for redistribution. Most influential was Amartya Sen who, as a schoolboy, was disturbed by the thousands dying in the streets during the Bengal Famine of 1943 while, in his own elite environment, no one suffered. Data suggesting (wrongly, as later became evident)[10] that there was no absolute food shortage led him to conclude that the famine was mainly caused by social inequality and speculation (Sen 1981). Sen was too cautious to extrapolate his insights into a wholesale rejection of government-led smallholder modernization, but others were less careful. The cross-pollination between British and Indian left-wing academia encouraged a criticism of the Green Revolution that threw out the baby with the bathwater (e.g. Patnaik 1999; Shiva 1991) and scorned farmer movements that pressed for supportive policies as a 'march of the kulaks' (e.g. Corbridge and Harriss 2000).[11]

A second and more general reason for the soft-neoliberal stance of many progressives is the desire for naturalness. Poor people tend not to worry very much about changes in the production systems, landscapes, and habitats that are connected to agricultural intensification. However, increasing incomes are correlated with a growing demand for food, lifestyles, and environments that are considered pure and not artificial (cf. Van Koppen 2000). The cross-cultural nature of this connection suggests the influence of an evolutionary psychological factor. A plausible hypothesis is that humans have an inborn predilection for the type of environment where their Pleistocene ancestors found food and shelter (Wilson 1984). With low income levels, this preference may be dominated by more basic necessities, but when increasing income allows people to climb Maslow's pyramid of needs, it becomes more prominent. This is especially evident in the formation of farm policies. During the New Deal, the inclusion of conservation in American farm legislation helped to broaden the support for the stabilization of agricultural markets. However, the current demand for naturalness has become so forceful in rich countries that this strategic balance is easily broken. Scores of people have become convinced that organic farming could feed the entire world; that Africa could flourish without fertilizer; and that sustainable chains could reverse the unsustainability spiral without government intervention.

It should be noted that both hard- and soft-neoliberal concepts regarding agri-food issues come disproportionally from metropolitan proponents. Apprenticeships at Western universities and the wish to tune in with donor priorities have led to the distribution of these ideas to poor countries. Nevertheless, intellectuals from poor countries – especially those with farmer societies or tribal societies – tend to adhere more strongly to the idea of government-led smallholder modernization than their metropolitan counterparts (e.g. Adesina 2010; Nwanze 2013).[12]

Longer-term risks

Now let me turn to the question of how current discourses on agri-food issues relate to recent and future developments in international food markets. Around the year 2000, international food prices reached an all-time low. Beginning in 2004, a cautious recovery set in and, from 2007, steep price spikes followed while price volatility strongly increased. The effects were not all negative. Just as in earlier decades with rising commodity prices, economic growth was revived in sub-Sahara Africa. However, the sudden nature of the price increases also had deleterious effects which were aggravated by the global financial crisis that began in 2008. Price peaks prompted food riots in many countries and allegedly contributed to the Arab revolutions. These events caused confusion in the international development community. Had the mantra not been that only the distribution of food, not food production, was the problem? Had economists not asserted that trade liberalization would stabilize markets as bumper harvests in some regions would compensate harvest failures in others? How, then, were these wild fluctuations and price spikes possible? A hunt for culprits soon started. Many blamed global warming, but there was little evidence for this. Others blamed speculation, but this was an amplifier rather than a primary cause of price instability. Again others blamed energy price rises and biofuel policies which, indeed, had an influence, no matter how much biofuel lobbies denied it.[13]

Other causes received less consideration. During the quarter century before 2007, extensive cuts in public sponsoring had undermined agricultural research efforts (Pardey et al. 2006). Falling prices – and in rich countries, the replacement of price supports with direct payments – had discouraged farmer investment. The growth in the agricultural capital stock and land productivity had diminished globally. In the developed world, the former had even decreased in absolute terms after 1990 (Von Cramon-Taubadel et al. 2009). Concurrently, the elimination of buffer stocks and other stabilizers made agricultural markets much more vulnerable to shocks.

Although these factors played a crucial role in the recent fluctuations in food prices, Western academics did not emphasize them. For decades, they had attacked government support for agriculture and extolled the virtues of either free markets or bottom-up initiatives. Conceding that low prices and rolling back the state had caused the 'food price crisis' amounted to admitting that they had been all wrong.

Meanwhile, the increased prices are now prompting a run on the margins that the world still possesses for inexpensive production growth. In tropical America, savanna land is being reclaimed. In the US and Argentina, farmers are increasing their fertilizer use. In sub-Saharan Africa, large stretches are being 'grabbed' by investors from land-poor Asian countries. In Community of Independent States countries, wheat land idled after 1991 is being brought back into production. One should not be surprised, therefore, if agricultural prices fall again sometime in the near future – especially if innovative extraction techniques and unconventional reserves (tar sands, shale gas, coal bed methane, underground coal gasification) would moderate fossil fuel prices for some time to come.

What will occur in the more distant future is less certain. In spite of the alarm about two billion more mouths to be fed and meat-eating Chinese, the completion of the demographic and nutrition transitions will slow the growth in demand for food and feed over this century (Alexandratos and Bruinsma 2012). However, the progressive depletion of easily exploitable fossil fuels may reverse the substitution of fossil- for biomass-based non-foods that began in the nineteenth century. By what amount the demand for bioenergy and biomaterials will increase is impossible to predict. It depends on the costs of novel energy sources, the dependence on liquid transport fuels, the scope for making liquid fuels from inorganic sources (e.g. by artificial photosynthesis), the development of new applications of plastics such as 3D-printing, and the prospects of recycling. Nevertheless, one might not expect the overall growth in phytomass demand for food, feed, and non-foods to be much slower than in past decades.

Will the supply response once again be enough to prevent a structural increase in food prices? Global land and water reserves have decreased. Breeders have ever more difficulty further increasing the yield potential of grains. The room for breeding higher-yielding varieties by improving the harvest index and growth patterns of plants seems to have grown thin, especially in wheat and rice (Fischer and Edmeades 2010). To be sure, considerable increases in global farm output are still possible by bringing actual yields closer to the potential yields of the best existing varieties. Technically speaking, world agriculture could still produce twice or triple the phytomass output that it is currently producing. Additionally, there are options for stretching the limits: improving conversion techniques (biorefinery, xth generation biofuels), increasing photosynthetic efficiency (C4 rice), producing algae (in ponds or photobioreactors), or a maricultural revolution (using fertile coastal waters, artificial upwellings, or some type of fertilization). However, many of these options are remote possibilities. Moreover, improving conversion techniques is a double edged sword. It does not just increase the supply of bio-based non-foods, but it also raises the demand for these products because it lowers their cost (Koning et al. 2008).

The problem is that a significant part of technical potential will not be economically feasible. Its full realization requires techniques, including near-perfect pest management, that do not yet exist. Moreover, diminishing returns push the economic optimum below the technical maximum. In locations where natural or social conditions lead to unfavourable price relationships and large risks, it may be rational for producers to adhere to traditional techniques that use less input but provide lower output. When rising energy prices make farm inputs more expensive, the total area for which this applies might even increase. Additionally, the human predilection for naturalness causes rising incomes to boost the demand for land for low-density housing, recreation, conservation, and land-intensive 'natural' foods. This competes with the employment of land for food quantity and has already caused a reduction in arable land in the developed world (Alexandratos and Bruinsma 2012: Table 4.8). The same will occur in other countries when incomes increase. In a country such as China, this is already evident.

The nineteenth-century breakthroughs have often been perceived as inaugurating an eternal cornucopia. What if these breakthroughs actually started a set of technological trajectories whose viability will be depleted in the course of this century? The issue is of significant consequence. If we continue to reside in an Amartya Sen-type of world where only the social distribution of food, but not food supply itself, is problematic, the groping toward an order where 'people, planet, and profit' are taken care of may prove resilient. Millennium Development Goal deadlines may be missed but, sooner or later, responsible policies will put an end to poverty and hunger. Conversely, a return to a world of scarcity could throw us back to something resembling the horrors of a preindustrial crisis. That would threaten to diminish us to the law of the jungle.

The biggest risk is not that the long-term decline in international food prices that began in the late nineteenth century would give way to a gradual long-term increase. It is that such a change would find humanity unprepared. In this regard, neoliberal approaches to agri-food issues provide a real hazard, because they could once more allow a cyclical downturn in prices to squeeze investment. If this were to be followed by a trend change, it could entail a longer and more severe surge in food prices than what we have experienced in recent years (Koning et al. 2008). This might prompt a chain of reactions that closes the window of opportunity for a soft landing. Think, for example, of an 'African Spring' that would lead to an armed conflict with China over the 'land grab' farms on which the consumption of the Chinese middle classes, and thereby the mandate of heaven of the Chinese communist party, has become dependent.

Conclusion

How is it possible to reduce hunger-by-poverty now and avoid global scarcity in the future? In a world with 10 billion people, a radical rejection of public coordination and modern techniques in the name of localism and naturalness is putting the cart before the horse. A set of reforms that can more realistically be expected to improve things is more beneficial. My program would comprise the following elements:

- Create a global social security system that grants people minimum access to the basic necessities in order to reduce hunger and poverty in the short term and moderate population growth in the longer term.
- Implement redistributive land reform in landlord societies coupled with effective support of smallholder modernization.
- Allow countries to pursue stimulating agricultural price policies including import tariffs.
- Create a system of international buffer stocks (already proposed by Keynes in 1943!) to stabilize international agricultural prices within desirable price bands. Defend the price floor by trade quotas or by converting surpluses into biofuel. Make biofuel production adjustable to defend the price ceiling.
- Require rich countries to co-finance rural infrastructure in poor countries, both to stimulate agricultural development and by way of employment projects that compensate poor consumers for the short-term effects of agricultural price supports.
- Mitigate the claim on phytomass for animal foods. Stimulate a shift from feedlot beef to pork, poultry, and fish. Graze ruminants on land that is only suitable for rough pasture. Develop meat substitutes that are actually attractive to consumers.
- Support research for raising humanity's capacity for biomass production. Develop farm techniques that reduce emissions but (unlike organic) provide

higher yields. Invest in innovations such as C4 rice, algae, and new marine systems.

- End fixed blending mandates and tax rebates for biofuel.
- Accelerate the development of novel energy technologies such as solar power and artificial photosynthesis and stabilize energy markets.

While free market liberalism or local-organic utopia have the advantage of simplicity, a program such as this involves sophisticated governance both on a national and an international level. One may question whether human societies will be able to manage this. Even if creating welfare for 10 billion people would require such a degree of complexity, isn't the attempt to achieve it simply doomed to entail another 'collapse of complex society' (cf. Tainter 1990)? Maybe. But we are not aliens looking at the human ecosystem from a different planet. We are part of it. It saddles us with the moral obligation to hand it down to the next generations unscathed. So we can only do our utmost. We may be aided by the deceleration of population and saturation of food demand in this century, and we may put our hope in human ingenuity. Here again, Britain may serve as an example – this time in a positive sense. While large parts of Europe suffered from subsistence crisis in the seventeenth century, this country innovated its production and institutions so as to allow a significant increase in its population and living standards (Allen 2009).

Looking at trends, humanity appears to have arrived at a critical juncture. If things go well, our grandchildren may live in a nature-friendly society where hunger has become a thing of the past. If they go wrong, civilizations may collapse by resource depletion and conflict. Nothing guarantees that the world's leading political classes will readily agree on action plans that will result in a positive outcome. Current debates regarding food security exhibit significant misunderstanding and confusion. Scientists should do their best to clarify the situation. This requires looking beyond academic hypes, mainstream convictions, and funder priorities to strive for a deeper, more encompassing, and transdisciplinary understanding of the issues at stake.

Notes

1 See Boussard et al. (2006) for a theoretical explanation and Harley (1980) for a historical account.
2 See Koning (1994: 76–7, 121–3) and literature referred to.
3 See e.g. Naughton (2007) for China, Thee (2002) and Timmer (2004) for Indonesia, and Bhalla and Singh (2001) and Chang (2009) for India.
4 That the reforms in West Bengal remained effective longer than those in Latin America may be because the former was part of a federative state that also included farmer societies and that pursued price and infrastructural policies to achieve a Green Revolution.
5 See Koning and Smaling (2005) and literature referred to.
6 However, I disagree with the idea of these authors that famines can be reduced to political crimes and have nothing to do with a broader social-ecological crisis.
7 See e.g. Gardner (1992) for a seminal paper reflecting this shift in perspective.

8 See e.g. FAO (2006) for a critical survey.
9 See Goodman (2004), Kay (2008), and Koning and Smaling (2005) for criticism of some of these ideas.
10 Sen was misled by a British report that underplayed the shortage in Bengal because the British government wanted to use shipping capacity for the war against Japan rather than for sending grain to hungry Bengal, and by propaganda of the Muslim League that wanted to mobilize the poor against the Hindu elite by accusing them of speculative hoarding (Ó Gráda 2009: Ch. 6).
11 Not accidentally, an American-born left-wing sociologist like Gail Omvedt judged the same movements much more positively (Omvedt 1988).
12 The demand of the international small farmers' organization Via Campesina for 'food sovereignty' likewise opposes the economic liberalism that so many self-nominated Western spokesmen for poor farmers have espoused.
13 See e.g. Baffes and Haniotis (2010) and HLPE (2011), for useful overviews of the debate.

References

Adejumobi, S. (2006) *Governance and poverty reduction in Africa: a critique of the Poverty Reduction Strategy Papers (PRSPs),* Paper presented to the 'Inter-Regional Conference on Social Policy and Welfare Regimes in Comparative Perspectives', Austin, USA, April 20–22.

Adesina, A.A. (2010) 'Conditioning trends shaping the agricultural and rural landscape in Africa', *Agricultural Economics,* 41(Supplement S1): 73–82.

Aiello, L.C. and Wheeler, P. (1995) 'The expensive-tissue hypothesis: the brain and the digestive system in human and primate evolution', *Current Anthropology* 36(2): 199–221.

Alexandratos, N. and Bruinsma, J. (2012) *World agriculture towards 2030/2050: the 2012 revision,* Rome: Food and Agriculture Organization.

Allen, R.C. (2009) *The British Industrial Revolution in global perspective,* Cambridge: Cambridge University Press.

Altieri, M.A. (1995) *Agroecology: the science of sustainable agriculture,* Boulder, CO: Westview Press.

Baffes, J. and Haniotis, T. (2010) 'Placing the recent commodity boom into perspective', in A. Aksoy and B. Hoekman (eds), *Food Prices and Rural Poverty,* Center of Economic and Policy Research and the World Bank, 41–70.

Barkin, D. (2006) *The new rurality: a framework for social struggle in the face of globalization,* Paper presented at the International Conference on Land, Poverty, Justice and Development, Institute of Social Studies, The Hague, 9–14 January.

Barrientos, A. and Hulme, D. (2009) 'Social protection for the poor and poorest in developing countries: reflections on a quiet revolution', *Oxford Development Studies,* 37(4): 439–56.

Bates, R.H. (1981) *Markets and states in tropical Africa: the political basis of agricultural policies,* Berkeley: University of California Press.

Bértola, L. and Ocampo, J.A. (2012) *The economic development of Latin America since independence,* Oxford: Oxford University Press.

Bhalla, G.S. and Singh, G. (2001) *Indian agriculture: four decades of development,* New Delhi: Sage Publications.

Boeke, J.H. (1953) *Economics and economic policy of dual societies as exemplified by Indonesia,* Haarlem: Tjeenk Willink & Zoon.

Boserup, E. (1965) *Population and technology,* Oxford: Blackwell.

Boussard, J.-M., Gérard, F., Piketty, M.G., Ayouz, M. and Voituriez, T. (2006) 'Endogenous risk and long run effects of liberalization in a global analysis framework', *Economic Modelling,* 23: 457–75.

Chambers, R., Pacey, A. and Thrupp, L.A. (1989) *Farmer first: farmer innovation and agricultural research,* London: Intermediate Technology Publications.

Chang, H.-J. (2009) *Rethinking public policy in agriculture: lessons from distant and recent history*, Rome: Food and Agriculture Organization.

Cleaver, K.M. and Schreiber, G.A. (1994) *Reversing the spiral: the population, agriculture, and environment nexus in Sub-Saharan Africa*, Washington, DC: World Bank.

Cochrane, W.W. (1959) *Farm prices – myth and reality*, Minneapolis: University of Minnesota Press.

Cook, E.C. (1992) 'Agriculture's role in the Soviet economic crisis', in M. Ellman and V. Kontorovich (eds), *The disintegration of the Soviet economic system*, London: Routledge, 193–216.

Corbridge, S. and Harriss, J. (2000) *Reinventing India: liberalization, Hindu nationalism and popular democracy*, Oxford: Polity Press.

Davies, R.W. and Wheatcroft, S. (2004) *The years of hunger: Soviet agriculture, 1931–1933*, The Industrialization of Soviet Russia, vol. 5, Houndsmills: Palgrave.

Dawe, D., (2007) 'The practical experience with agricultural trade liberalization in Asia', in N. Koning and P. Pinstrup-Andersen (eds), *Agricultural trade liberalization and the least developed countries*, Dordrecht: Springer, 175–95.

De Janvry, A. (1981) *The agrarian question and reformism in Latin America*, Baltimore: John Hopkins University Press.

De Waal, A. (1997) *Famine crimes: politics and the disaster relief industry*, Oxford: James Currey.

Diamond, J. (1998) *Guns, germs and steel*, London: Vintage.

Domar, E.D. (1970) 'The causes of slavery or serfdom: a hypothesis', *The Journal of Economic History*, 30: 18–32.

Edkins, J. (2007) 'The criminalization of mass starvations: from natural disaster to crime against humanity', in S. Devereux (ed.), *The new famines: why famines persist in an era of globalization*, Abingdon and New York: Routledge, 50–65.

Ellman, M.J. (1984) 'Fifty years of collectivised Soviet agriculture, 1929–1971', in M. Ellman, *Collectivization, convergence and capitalism; political economy in a divided world*, London: Academic Press, 58–71.

Engerman, S.L. and Sokoloff, K.L. (2013) 'Five hundred years of European colonization: inequality and paths of development', in C. Lloyd, J. Metzer and R. Sutch (eds), *Settler economies in world history*, Leiden and Boston: Brill, 65–104.

Fantacci, L., Marcuzzo, M.C. and Roselli, A. (2012) 'Speculation and buffer stocks: the legacy of Keynes and Kahn', *European Journal of the History of Economic Thought*, 19(3): 453–73.

FAO (2006) *Towards appropriate agricultural trade policy for low income developing countries*, FAO Trade Policy Technical Note no. 14, Rome: Food and Agriculture Organization.

FAO (2012) *The state of food insecurity in the world 2012*, Rome: Food and Agriculture Organization.

Fischer, R.A. and Edmeades, G.O. (2010) 'Breeding and cereal yield progress', *Crop Science*, 50(Supplement 1): S86–S98.

Francks, P. (2006) *Rural economic development in Japan: from the nineteenth century to the Pacific War*, London: Routledge.

Francks, P., Boestel, J. and Kim, C.H. (1999) *Agriculture and economic development in East Asia: from growth to protectionism in Japan, Korea and Taiwan*, London and New York: Routledge.

Frankema, E. (2009) *Has Latin America always been unequal? A comparative study of asset and income inequality in the long twentieth century*, Leiden and Boston: Brill.

Gardner, B.L. (1992) 'Changing economic perspectives on the farm problem', *Journal of Economic Literature*, 30(1): 62–101.

Geertz, C. (1963) *Agricultural involution: the processes of ecological change in Indonesia*, Berkeley and Los Angeles: University of California Press.

Goodman, D. (2004) 'Rural Europe redux? Reflections on alternative agro-food networks and paradigm change', *Sociologia Ruralis*, 44(1): 3–16.

Goody, J. (1976) *Production and reproduction: a comparative study of the domestic domain,* Cambridge: Cambridge University Press.

Harley, C.K. (1980) 'Transportation, the world wheat trade, and the Kuznets cycle, 1850–1913', *Explorations in Economic History,* 17(3): 218–50.

Hathaway, D.E. (1963) *Government and agriculture: economic policy in a democratic society,* New York: Macmillan.

Henningson, B. (1981) *United States agricultural trade and development policy during World War II: the role of the Office of Foreign Agricultural Relations,* Unpublished Ph.D. dissertation, University of Arkansas.

HLPE (2011) *Price volatility and food security; a report by the High Level Panel of Experts on Food Security and Nutrition of the Committee on World Food Security,* Rome: Food and Agriculture Organization.

Ingersent, K.A. and Rayner, A.J. (1999) *Agricultural policy in Western Europe and the United States,* Cheltenham and Northampton: Edward Elgar.

Johnston, B.F. and Mellor, J.W. (1961) 'The role of agriculture in economic development', *American Economic Review,* 51: 566–93.

Kay, C. (2008) 'Reflections on Latin American rural studies in the neoliberal globalization period: a new rurality?', *Development and Change,* 39(6): 915–43.

Koning, N. (1994) *The failure of agrarian capitalism: agrarian politics in the United Kingdom, Germany, the Netherlands and the USA, 1846–1919,* London and New York: Routledge.

Koning, N. (2003) 'Agriculture and the WTO: time to reconsider the basics?', *EuroChoices,* 2(3): 26–31.

Koning, N. and Smaling, E. (2005) 'Environmental crisis or "lie of the land"? The debate on soil degradation in Africa', *Land Use Policy,* 22(1): 3–22.

Koning, N.K., van Ittersum M.K., et al. (2008) 'Long-term global availability of food: continued abundance or new scarcity', *NJAS-Wageningen Journal of Life Sciences,* 55: 229–92.

Kuznets, S. (1966) *Modern economic growth: rate, structure and spread,* New Haven and London: Yale University Press.

Lee, R.D. (1988) 'Malthus and Boserup: a dynamic synthesis', in D. Coleman and R.S. Schofield, *The state of population theory: forward from Malthus,* Oxford: Basil Blackwell, 96–103.

Lewis, W.A. (1954) 'Economic development with unlimited supplies of labor', *Manchester School* 22: 139–91.

Lipton, M. and Longhurst, R. (1989) *New seeds and poor people,* London: Unwin Hyman.

Maizels, A. (1992) *Commodities in crisis: the commodity crisis of the 1980s and the political economy of international commodity policies,* Oxford: Clarendon Press.

Marsden, T. (1995) 'Beyond agriculture? Regulating the new rural spaces', *Journal of Rural Studies,* 11(3): 285–96.

McLaren, L. (2007) 'Socioeconomic status and obesity', *Epidemiology Reviews,* 29: 29–48.

McMichael, P. (2009) 'A food regime genealogy', *Journal of Peasant Studies,* 36: 139–69.

McMichael, P. (2012) 'The land grab and food regime restructuring', *Journal of Peasant Studies,* 39(3/4): 681–701.

Monteiro, C.A., Conde, W.L. and Popkin, B.M. (2002) 'Is obesity replacing or adding to undernutrition? Evidence from different social classes in Brazil', *Public Health Nutrition,* 5(1a): 105–12.

Moss, M. (2013) 'The extraordinary science of addictive junk food', *New York Times,* 20 February.

Moyer, H.W. and Josling, T.E. (1990) *Agricultural policy reform: politics and process in the European Community and USA,* Brighton: Harvester Wheatsheaf.

Myrdal, G. (1968) *Asian drama: an inquiry into the poverty of nations,* New York: Pantheon.

Naughton, B. (2007) *The Chinese economy: transitions and growth,* Cambridge, MA: The MIT Press.

Nwanze, K. (2013) *African agricultural development: opportunities and challenges,* Statement by IFAD president at the 6th Africa Agriculture Science Week and FARA general assembly, Accra, July.

Offer, A. (1989) *The First World War: an agrarian interpretation,* Oxford: Oxford University Press.

Ó Gráda, C. (2009) *Famine: a short history,* Princeton: Princeton University Press.

Omvedt, G. (1988) 'The "new peasant movement" in India', *Bulletin of Concerned Asian Scholars,* 20(2): 14–23.

Paarlberg, R. (1997) 'Agricultural policy reform and the Uruguay Round: synergistic linkage in a two-level game?', *International Organization,* 51(3): 413–44.

Pardey, P.G., Beintema, N., Dehmer, S. and Wood, S. (2006) *Agricultural research: a growing global divide?* Washington, DC: International Food Policy Research Institute (IFPRI).

Patnaik, U. (1999) *The long transition – essays on political economy,* New Delhi: Tulika.

Pauli, G. (2010) *10 years, 100 innovations, 100 million jobs: report to the Club of Rome,* Taos: Paradigm Publications.

Popkin, B.M. and Gordon-Larsen, P. (2004) 'The nutrition transition: worldwide obesity dynamics and their determinants', *International Journal of Obesity,* 28: 2–9.

Prebisch, R. (1950) *The economic development of Latin America and its principal problems,* New York: Economic Commission for Latin America.

Ravallion, M. and Datt, G. (2002) 'Why has economic growth been more pro-poor in some states of India than others?', *Journal of Development Economics,* 68(2): 381–400.

Rawal, V. and Swaminathan, M. (1998) 'Changing trajectories: agricultural growth in West Bengal, 1950 to 1996', *Economic and Political Weekly,* 33(40): 2595–602.

Reijntjes, C., Haverkort, B. and Waters-Bayer, B. (1992) *Farming for the future: an introduction to low-external-input and sustainable agriculture,* London and Basingstoke: Macmillan.

Scoones, I. (1998) *Sustainable rural livelihoods: a framework for analysis,* Working Paper No. 72, Institute of Development Studies, University of Sussex.

Sen, A. (1981) *Poverty and famines: an essay on entitlement and deprivation,* Oxford: Clarendon Press.

Shell, E.R. (2003) *The hungry gene: the inside story of the obesity industry,* New York: Grove Press.

Shiva, V. (1991) *The violence of the Green Revolution: third world agriculture, ecology and politics,* London: Zed Books.

Swinnen, J.F.M. (2008) *The political economy of the 2003 reform of the common agricultural policy,* LICOS Discussion Paper 215, Katholieke Universiteit Leuven.

Syrrakos, B. (2008) 'An uncommon policy: theoretical and empirical notes on elite decision-making during the 2003 CAP reforms', in J.F.M. Swinnen (ed.), *The perfect storm: the political economy of the Fischler reforms of the Common Agricultural Policy,* Brussels: Centre for European Policy Studies, 115–34.

Tainter, J.A. (1990) *The collapse of complex societies,* Cambridge and New York: Cambridge University Press.

Thee, K.W. (2002) 'The Soeharto era and after: stability, development and crisis, 1966–2000', in H.W. Dick, V.J.H. Houben, J. Lindblad and K.W. Thee (eds), *The emergence of a national economy: an economic history of Indonesia, 1800–2000,* Crown West, Australia: Allen & Unwin, 194–243.

Tiffen, M., Mortimore, M. and Gichuki, F. (1994) *More people, less erosion: environmental recovery in Kenya,* London: Wiley.

Timmer, C.P. (1995) 'Getting agriculture moving: do markets provide the right signals?', *Food Policy,* 20: 455–72.

Timmer, C.P. (2004) 'The road to pro-poor growth: the Indonesian experience in a regional perspective', *Bulletin of Indonesian Economic Studies,* 40(2): 177–207.

Turchin, P. and Nefedov, S.A. (2009) *Secular cycles,* Princeton: Princeton University Press.

Van der Ploeg, J.D., Renting, H., Brunori, G., Knickel, K., Mannion, J., Marsden, T., de Roest, K., Sevilla-Guzman, E. and Ventura, F. (2000) 'Rural development: from practices and policies towards theory', *Sociologia Ruralis,* 40(4): 391–408.

Van Donge, J.K., Henley, D. and Lewis, P. (2012) 'Tracking development in Southeast Asia and Sub-Saharan Africa: the primacy of policy', *Development Policy Review,* 30(Supplement 1): S5–S24.

Van Koppen, C.S.A. (2000) 'Resource, arcadia, lifeworld: nature concepts in environmental sociology' *Sociologia Ruralis,* 40(3): 300–18.

Verdier, D. (1994) *Democracy and international trade: Britain, France, and the United States, 1860–1990,* Princeton: Princeton University Press.

Von Cramon-Taubadel, S., Anriquez, G., de Haen, H. and Nivyevskiy, O. (2009) 'Investment in developing countries' food and agriculture: assessing agricultural capital stocks and their impact on productivity', in *How to feed the world in 2050,* Proceedings of a technical meeting of experts, 24–26 June, Rome, Italy: FAO, 1–29.

Watkins, K. and Fowler, P. (2002) *Rigged rules and double standards,* Oxford: Oxfam-International.

Wilson, E.O. (1984) *The biophilia hypothesis,* Cambridge, MA: Harvard University Press.

Wood, J.W. (1998) 'A theory of preindustrial population dynamics', *Current Anthropology,* 39: 99–135.

Wrangham, R. (2009) *Cooking: how cooking made us human,* Boston: Basic Books.

PART I
New trends and challenges

3

SCENARIOS FOR GLOBAL AGRICULTURE AND FOOD SECURITY TOWARDS 2050

A review of recent studies

Dirk Willenbockel

1. Introduction

The global food system will face an unprecedented concurrence of pressures over the next decades. The combination of population growth and rising per-capita incomes that will be accompanied by a shift towards more livestock-intense diets in parts of the world will translate into a substantial increase in the demand for agricultural output between now and the middle of the century. These demand-side driving forces are bound to intensify the competition for land and water, particularly in low-income regions with high population growth and a high present incidence of undernutrition.

A further set of challenges to future food security emerge from climate change. Long-run agricultural productivity trends as well as short-run yield variability are directly affected by climate change and the associated expected increases in extreme weather events. A growing number of studies suggest that climate change may well reduce the productivity of farming in precisely those regions of the world where food insecurity is most prevalent.

At the same time the global food system and its impact on land use contribute significantly to global greenhouse gas emissions – a fact which severely complicates the challenge of securing access to affordable nutritious food for a growing world population in a sustainable manner. Moreover, climate change mitigation policies aimed at the energy sector that raise fossil fuel prices affect bioenergy demand and further intensify the competition for land.

Deciding how to balance the multiple pressures and competing demands on the global food system is a major task facing policymakers today. Rational forward-looking decision making requires information about the likely consequences of alternative policy strategies to address the challenges to food security over the coming decades.

Given the complex dynamic interactions among the socio-economic drivers and bio-physical processes that co-determine food security outcomes and the numerous uncertainties surrounding these drivers, the creation of long-run projections for the global food and farming system is a difficult task. Contemporary studies that take on this task have used scenario analysis in conjunction with global simulation models to explore alternative plausible futures.

This chapter provides a concise selective review of recent global food system scenario studies from a food security perspective. The review covers six major studies published since 2005 that provide a quantified outlook up to 2050. It includes the long-run scenarios developed as part of four international assessments, namely

- the Millennium Ecosystem Assessment (Carpenter et al. 2005 – henceforth MA);
- the International Assessment of Agricultural Science and Technology for Development (Rosegrant et al. 2009 – IAASTD);
- the Comprehensive Assessment of Water Management in Agriculture study (de Freiture et al. 2007 – CAWMA); and
- the Global Environmental Outlook 4 (Rothman et al. 2007 – GEO4), as well as the scenarios reported in
- the International Food Policy Research Institute's report 'Food Security, Farming, and Climate Change to 2050' (Nelson et al. 2010 – IFPRI) and
- the 2012 revision of the Food and Agriculture Organization's global long-run projections (Alexandratos and Bruinsma 2012 – FAO).

The following section outlines the scope and objectives of the various studies and compares and contrasts the approaches to scenario development they adopt. Section 3 provides a selective synopsis of the assumptions for key drivers of food demand and supply at a global scale including population growth, per-capita income growth and agricultural productivity growth. Section 4 gives a brief sketch of the scenario projections for food security, international trade in food commodities as well as for envisaged qualitative changes in food production and distribution systems towards 2050 and Section 5 draws conclusions.

2. A perspective on scope, objectives and scenario design

Scope and objectives of the scenario studies

The six studies differ in terms of their focus and objectives, and these differences influence the choice of scenario design. The broad remit of the MA Scenarios Working Group was to assess future changes in world ecosystems and the consequences of these changes for human well-being, and to inform decision-makers at various scales about possible response strategies. The MA is primarily geared towards the information requirements of the various United Nations (UN) conventions on biodiversity, desertification, wetlands and migratory species. The United Nations Environment Programme's (UNEP) GEO4 assessment is likewise broadly concerned with environmentally sustainable development, and in both studies food provision is conceptualized as one among other ecosystem services.

In contrast, in the other four studies under review agriculture and food security take the center stage. The objective of IAASTD is to assess the impacts of past, present and future agricultural knowledge, science and technology (AKST) on the reduction of hunger and poverty; improvement of rural livelihoods and human health; and equitable, socially, environmentally and economically sustainable development. IAASTD was initiated in 2002 by the World Bank and the FAO as an intergovernmental consultative process co-sponsored by other United Nations bodies including the Global Environment Facility (GEF); UN Development Programme; UNEP; and the UN Educational, Scientific and Cultural Organization; and the World Health Organization.

CAWMA aims to assess the current state of knowledge on how to manage water resources to meet the growing needs for agricultural products, to help reduce poverty and food insecurity, and to contribute to environmental sustainability in order to enable better investment and management decisions in water and agriculture in the future. CAWMA was co-ordinated by the International Water Management Institute and co-sponsored by the Consultative Group on International Agricultural Research (CGIAR), the FAO and the UN biodiversity and wetlands conventions.

The objective of the IFPRI study is to provide an end-of-decade assessment of the challenges to global food security up to 2050 with a particular emphasis on the nexus between climate change, climate change adaptation and agricultural yields. The study has been co-sponsored via the UK Government Office for Science Foresight Programme and early results from the IFPRI study served as an input to the influential Foresight Global Food and Farming Futures Report (Government Office for Science 2011; Godfray et al. 2010).

The FAO 2012 projections provide an update of earlier 2006 FAO long-run projections based on revised and more recent data, 'aiming at describing the future as it is likely to be to the best of our knowledge at the time of carrying out the study, and not as it ought to be from a normative point of view' (Alexandratos and Bruinsma 2012: 137).

Scenario design

With respect to the approach to scenario design adopted in the various studies, it is helpful to distinguish between 'business-as-usual' trend projections without major shifts in policy orientations or step changes in human behaviour, and exploratory scenario studies that consider a broader range of alternative conceivable futures (Reilly and Willenbockel 2010).

Examples of the former are the FAO projections, the IAASTD reference world scenario, and the IFPRI and CAWMA baseline scenarios. Exploratory scenarios contrast different possible trajectories for the main drivers of change in the global agri-food system including population growth, aggregate real income growth, policies towards agricultural productivity growth, trade and international development policy orientations, attitudes towards ecological sustainability, and climate change impacts on yields. Among the studies in this category, the MA and GEO4 scenarios try to underpin the different assumptions for the driver paths with more (MA) or less elaborate (GEO4) qualitative narratives that outline envisaged major shifts in institutional and socio-political frameworks as well as in value systems. The four MA scenarios are framed in terms of contrasting evolutions of governance patterns for international cooperation and trade (globalized versus regionalized) and opposite approaches towards ecosystem management (pro-active versus reactive). Similarly, in GEO4 the scenarios are defined by different policy approaches and societal choices, 'with their nature and names characterized by the theme that dominates the particular future envisioned, such as what comes *first*'[1] (Table 3.1).

Other scenarios in this category – including the IAASTD AKST scenarios, the IFPRI productivity improvement simulation runs and the CAWMA water management investment scenarios – focus on an exploration of alternate investment strategy options for the agricultural sector while adopting *ceteris paribus* assumptions for other drivers of system change. Table 3.1 provides a brief overview of the distinguishing key features of the scenarios reviewed here in terms of their assumptions about policy shifts affecting the food system.

TABLE 3.1 Policy orientations in the scenarios

Study	Scenario	Key Characteristics
FAO		Projection without major shifts in policy orientations
MA	Global Orchestration – GO	Global cooperation and a reactive approach towards ecosystem management
	Techno Garden – TG	Proactive technology- and market-based approach to ecosystems and high levels of international cooperation
	Adapting Mosaic – AM	Emphasis on local approaches and local learning to the improvement of ecosystem services with low levels of international cooperation
	Order from Strength – OS	Reactive approach to ecosystem stresses, high trade barriers and low levels of global cooperation
GEO4	Markets First	Emphasis on market-based solutions, increased role of the private sector in previously government-led areas, liberalised trade, and 'commoditization of nature'
	Policy First	Centralized policy-led approach to balancing strong economic growth with 'a lessening of the potential environmental and social impacts'. Similar to MA GO

(Continued)

TABLE 3.1 (Continued)

Study	Scenario	Key Characteristics
	Security First	Government and private sector compete for control in efforts to improve human well-being for 'mainly the rich and powerful in society'. Similar to MA OS
	Sustainability First	Actors at all levels follow through on pledges made to date to address environmental and social concerns
CAWMA	Baseline Scenario	Projection without major shifts in policy orientations
	Rainfed High Yield	Investments in rainfed areas: water harvesting, supplemental irrigation
	Rainfed Low Yield	Pessimistic case: upgrading rainfed agriculture is not successful
	Irrigation Area Expansion	Expansion of irrigated areas
	Irrigation Yield Improv.	Improvements in the performance of existing irrigated areas
	Trade Scenario	Increased agricultural trade from water-abundant to water-scarce countries
	Comprehensive Scenario	Optimal region-specific combination of rainfed, irrigation and trade strategies
IFPRI	Baseline	Projection without major shifts in policy orientations
	Optimistic	Lower population growth and higher per-capita GDP growth than baseline
	Pessimistic	Higher population growth and lower per-capita GDP growth than baseline
	Productivity Improvement	Various measures to raise agricultural productivity
	Perfect Mitigation	Baseline without climate change impacts on crop yields
IAASTD	Reference Run	Current policy pathways are expected to continue out to 2050
	AKST_high_pos	High levels of agricultural R&D investment with high growth of complementary investments
	AKST_low_neg	Low levels of agricultural R&D investment with decelerating growth of complementary investments

The use of simulation models in the scenario studies

The MA, GEO4 and IAASTD employ various ensembles of 'soft-linked'[2] simulation models with a common core that consists of the global integrated assessment model IMAGE (Integrated Model to Assess the Global Environment) developed at the Dutch National Institute for Public Health and the Environment and the global multi-market partial-equilibrium model IMPACT (International Model for Policy Analysis of Agricultural Commodities and Trade) maintained at IFPRI. IMAGE is designed to capture interactions between economic activity, land use, greenhouse gas (GHG) emissions, climate, crop yields and other environmental variables. It includes a stylized multi-region computable general equilibrium model of global trade and production, a carbon-cycle module to calculate GHG emissions resulting from economic activity including energy and land use, a land-use module and an atmosphere–ocean climate module that translates GHG emissions into climate outcomes. The model-determined temperature and precipitation outcomes in turn feed back into the performance of the economic system via agricultural productivity impacts. The global agricultural market model IMPACT is designed to provide disaggregated projections for production, demand for food, feed and other uses, prices, trade, crop area and yields across over 30 food commodity groups and also derives impacts on childhood malnutrition from these projections.

In the three scenario studies this common core of models is further soft-linked to a varying range of further simulation models to arrive at projections for other variables of interest and to downscale[3] results for particular regions. Thus, for instance both MA and GEO4 use WaterGAP hydrology and water-use model simulations to assess water stress and the EcoPath/EcoSim modelling suite to project marine ecosystems. Both MA and IAASTD employ the GLOBIO model to simulate biodiversity impacts.

In contrast to these three studies, the projections of the IFPRI and CAWMA scenario are each based on single partial-equilibrium models, namely the aforementioned IMPACT model and IWMI's WATERSIM model respectively. WATERSIM consists of a food production and demand module and a water supply and demand module based on a water balance and water accounting framework.

The quantitative FAO projections are not based on the simulation of a formal documented behavioural model but are derived by combining simple demand projections from Engel demand functions[4] and exogenous assumptions on population and GDP growth with an iterative process of adjustments to yield and area change projections by commodity and region involving expert judgements. The FAO supply utilization account framework is used to establish consistency between source and use projections (Alexandratos and Bruinsma 2012: 137–139). This approach does not generate projections for the evolution of food prices over time.

The simulation results from any dynamic global simulation analysis for a long-term horizon of several decades are surrounded by numerous uncertainties – about the adequacy of the model structure to capture the key factors at work, about model parameters, about the evolution of the main drivers of change in agricultural systems that enter the simulation analysis exogenously,[5] as well as about the presence of potential nonlinearities due to the occurrence of tipping points beyond which fundamental change in systems behaviour might occur. The results are necessarily contingent on the current state of scientific knowledge used in the course of the development and numerical calibration of the model components. As a case in point, the skill of the models in representing climate change impacts on agricultural productivity is necessarily constrained by the state of the art in climate science and crop science, where gaping holes in knowledge – e.g. with respect to future regional precipitation patterns and extreme weather event frequencies or the strength of CO_2 fertilization effects in a changing climate – persist.

3. Assumptions for key drivers of change in the global food system

Population and income growth

Table 3.2 reports the assumptions about global population and per-capita real income growth – the two main drivers of food demand – underlying the various scenarios. In the MA, the predicted world population in 2050 ranges from 8.1 billion in the GO to 9.6 billion in the OS scenario. The main reason for the divergence is that GO assumes higher economic growth and higher human capital investments in education and health than OS does and hence a faster transition towards lower fertility and mortality rates in developing regions. Trade liberalization, international economic cooperation, and technology exchange foster economic performance in the two MA scenarios with globalized governance (GO and TG), while trade barriers and inward-oriented policies are assumed to contribute to lower growth rates in the OS and AM scenarios. Growth rates are higher in GO compared to TG, because in the latter investments in environmental technologies are favoured at the expense of human capital investments.

The differences in population growth across the four GEO4 scenarios are based on a similar reasoning. In the IFPRI study, the assumed differences in population and income per capita are simply the defining characteristic that differentiates the three scenarios, but in contrast to the MA and GEO studies these differences are not underpinned by an elaboration of the socio-political developments that co-determine demographic trends and economic performance.

From a food security perspective, it is worth emphasizing that the major portion of the projected global population growth will be located in the least-developed countries of sub-Saharan Africa and South Asia.

Agricultural productivity growth

On the supply side, a key direct driver of change in food system performance is agricultural productivity growth. The last column of Table 3.3 displays the average annualized growth rates of cereal yields between the study-specific base year (third

TABLE 3.2 Population 2050, per-capita income growth and calorie intake

Study	Scenario	Population 2050 million	Average Per-cap. GDP Growth Base Year to 2050 % p.a. (Base Year)	Calorie Intake 2050 World kcal/capita/ day	Calorie Intake 2050 SSA kcal/capita/ day
		2005: 6,900		2005/07: 2,770	2005/07: ≈2,300
FAO		9,150	1.3 (2005)	3,070	≈2,700
MA	Global Orchestration	8,100	2.7 (1997)	3,521	>3,000
	Techno Garden	8,800	2.2 (1997)	3,210	<3,000
	Adapting Mosaic	9,500	1.7 (1997)	2,920	<2,500
	Order from Strength	9,600	1.28 (1997)	2,953	<2,500
GEO4	Markets First	9,200	≈2.6 (2005)	≈4,000	≈3,500
	Policy First	8,600	≈2.8 (2005)	≈4,200	≈3,700
	Security First	9,700	≈1.4 (2005)	≈3,200	≈2,400
	Sustainability First	8,000	≈2.5 (2005)	≈4,400	≈4,000
CAWMA	All Scenarios	8,900	2.2 (2000)	2,970	na
IFPRI	Baseline	9,096	2.5 (2010)	↑	↑
	Optimistic	7,913	3.2 (2010)	↑	↑
	Pessimistic	10,399	0.7 (2010)	↓	↓
IAASTD	Reference Run	8,200	≈2.0 (2000)	≈3,000	2,738
	AKST_high_pos	8,200	na	>3,000	4,700
	AKST_low_neg	8,200	na	<3,000	1,600

Notes: rf: rainfed cereal yields; irr: irrigated cereal yields; na: not available; p.a.: per annum. GEO4 reports neither of the indicators tabulated here.

column) and 2050 as a rough aggregate indicator of differences in agricultural productivity growth across the various scenarios. As the base years range from 1997 to 2010 across the scenario studies, comparisons across different scenarios within the same study are more straightforward than cross-study comparisons.

Climate change

With the exception of FAO and presumably CAWMA,[6] all scenario projections take account of climate change impacts on agricultural productivity. In the MA scenarios, global mean temperature increases for 2050 range from +1.6°C (TG) to +2.0°C (GO) over pre-industrial levels. Yield impacts are based on IMAGE model

TABLE 3.3 Assumed cereal yield growth and projections for staple crop prices

Study	Scenario	Base Year	World Market Price Change Base to 2050 %			Cereal Yield Growth Base to 2050 % p.a.
			Rice	Wheat	Maize	
FAO		2005/07	na	na	na	0.66
MA	Global Orchestration	1997	−31.6	6.3	38.8	≈1.00
	Techno Garden	1997	−5.6	−12.7	−11.7	≈0.82
	Adapting Mosaic	1997	56.1	41.3	53.4	≈0.57
	Order from Strength	1997	46.0	14.7	19.4	≈0.41
CAWMA	Rainfed High Yield	2000	na	na	na	rf 1.09 irr 0.61
	Rainfed Low Yield	2000	na	na	na	rf 0.37 irr 0.58
	Irrigation Area Expansion	2000	na	na	na	rf 0.36 irr 1.00
	Irrigation Yield Improvement	2000	na	na	na	rf 0.38 irr 1.15
	Trade Scenario	2000	na	na	na	rf 0.93 irr 0.58
	CAWMA	2000	na	na	na	rf 0.92 irr 0.88
IFPRI	Baseline	2010	54.8	54.2	100.7	na
	Optimistic	2010	31.2	43.5	87.3	na
	Pessimistic	2010	78.1	58.8	106.3	na
	Productivity Improvement	2010	31.2	20.0	59.8	na
	Perfect Mitigation	2010	19.8	23.2	32.2	na
IAASTD	Reference Run	2000	21.5	61.6	41.7	1.02
	AKST_high_pos	2000	−53.8	−48.2	−73.1	1.63
	AKST_low_neg	2000	303.3	780.8	1291.5	0.41

projections that suggest positive effects for the USA and the former Soviet Union but negative for other regions, including in particular South Asia, with strong adverse impacts on rice and temperate cereal yields. In GEO4, global mean temperature increases range from +1.7°C (Sustainability First) to +2.2°C (Markets First) by 2050 relative to the pre-industrial mean. Yield impacts relative to a reference case without climate change are not reported for this study. IAASTD assumes a mean global rise of +1.7°C over pre-industrial levels by 2050. Agricultural impacts are projected to be negative for dryland areas in Africa, Asia and the Mediterranean area. Yield impacts of climate change by 2050 are characterized as 'still relatively small, apart from some crucial regions like South Asia'.[7] In the IFPRI scenarios, which consider climate projections arising from the combination of two Intergovernmental Panel on Climate Change (IPCC) emission scenarios (A1B, B1) with two global circula-tion models (CSIRO, MIROC), mean global surface temperature changes by 2050 range from +1.0 to +3.0°C relative to the late twentieth-century mean (i.e. about +1.6 to +3.6°C compared to pre-industrial levels). The global average yield changes compared to the counterfactual Perfect Mitigation (i.e. zero climate change) sce-nario for 2050 range from −2.0 to −12.0 per cent for maize, from 0 to −12.1 per cent for rice and from −4.1 to −13.2 per cent for wheat.

All studies under review emphasize the high degree of uncertainty surrounding projections of climate change impacts on agricultural productivity, which arises *inter alia* from the high variance in projections of future precipitation patterns across climate models and from the current lack of scientific consensus about the potential magnitude of benign carbon fertilization effects on crop yields.

The IPPC Fourth Assessment Report anticipates with high confidence that 'pro-jected changes in the frequency and severity of extreme climate events will have more serious consequences for food and forestry production, and food insecurity, than will changes in projected means of temperature and precipitation' (Easterling et al. 2007); and the more recent IPCC Special Report on Managing the Risks from Extreme Events (IPCC 2012) lends further support to this conclusion. While most of the scenario studies under review mention that an increased frequency of droughts and floods in a changing climate will affect food security outcomes, the quantitative projections do not take account of this additional climate change impact channel.[8]

In view of the current deadlock in global mitigation negotiations, it is worth pointing out that the global temperature projections for 2050 at the lower end of the spectrum in the reviewed scenarios already appear highly over-optimistic, if not obsolete. Climate science is adamant that annual global CO_2 emissions would have to peak by 2020 at the latest and need to drop steadily in subsequent decades in order to maintain a reasonable chance to achieve the aim of limiting the aver-age global temperature rise to +2°C above pre-industrial levels. It is now clear that even under optimistic assumptions about further progress in post-Durban climate diplomacy, no binding global deal covering the major greenhouse gas emitters will be in force prior to 2020 – hence the +2°C goal is now illusory and scientists are getting serious about contemplating human development prospects in a +4°C+ world.[9]

Biofuel demand

The links between climate change and agriculture go beyond productivity effects on the supply side. Agricultural production and the associated change in land use contribute significantly to greenhouse gas emissions – a fact which severely com-plicates the challenge of securing access to affordable nutritious food for a growing world population in a sustainable manner. Climate change mitigation policies aimed at a low-carbon transition in the energy sector affect food security out-comes not only through their impact on prices for energy-intensive inputs such as fertilizers, but – potentially far more importantly – through their impact on future bioenergy demand.

In the MA GO scenario, global bioenergy production rises by a factor of six towards 2050 relative to the 1997 level, driven by price increases for fossil fuels as a result of high energy demand growth that follows from the assumption of high per-capita income growth. The regions making the biggest contribution to the increase in biofuel feedstock production are Asia, Middle East and North Africa, and sub-Saharan Africa, and as a consequence a high rate of deforestation is projected for these regions. In contrast, in the MA TG scenario with its lower GDP growth, biofuel production expands by a factor of four and is primarily driven by climate mitigation policy. In the GEO4 Policy First and Sustainability First scenarios, climate change mitigation policies likewise induce significant increases in biofuel demand with adverse effects on forest land cover. Indeed, in the Policy First scenario, 'nearly all of Africa's forests are lost' (Rothman et al. 2007: 418) as a result of the intensified competition for land between food and biofuel crops. The IFPRI scenarios assume that beyond 2025 traditional biofuels are entirely replaced by second generation biofuels, but the interactions between economic growth, energy demand, climate mitigation policy, biofuels and land use change are not explored in this study. The CAWMA study notes the role of biofuels in the future competition for land and water, but the respective assumptions underlying the CAWMA scenario simulations are not spelled out. Finally, the FAO 2050 projections assume no further growth in bioenergy use beyond FAO medium-run projections for 2019, which are based on biofuel mandates already in place or announced.

4. A brief synopsis of projections towards 2050

This section provides a concise selective synopsis of scenario projections for the middle of the twenty-first century with a focus on food security outcomes, international trade in food commodities and envisaged shifts in the modes of food production and distribution systems.

Food security

Table 3.2 displays the simulated results for average calorie intake per capita per day in 2050 at a global scale. None of the studies envisages a plain world-wide Malthusian doom scenario. However, the global averages mask considerable variations in food security outcomes towards the middle of the century at sub-global scales across the simulations, as illustrated in the last column of Table 3.2 for the case of sub-Saharan Africa (SSA).[10]

While most of the scenarios suggest improvements in food security along with reductions in malnutrition for all developing sub-regions relative to the base year, the MA Order from Strength scenario as well as the IFPRI pessimistic case, the GEO4 Security First and the IAASTD AKST_low_neg scenario suggests that a combination of continued high population growth, adverse climate change impacts and low investments in yield growth in the absence of effective international development cooperation may lead to opposite outcomes for SSA – and similar conclusions apply to low-income South Asia, the other hotspot of persisting or worsening food security problems in these pessimistic scenarios.

In stark contrast, the reported calorie consumption figures for the IAASTD AKST_high_pos and GEO4 Sustainability First scenarios, taken at face value, would seem to imply that by 2050 nutrition problems in SSA will take the form of obesity problems.

Three of the MA scenarios project absolute decreases in the absolute number of malnourished children in 2050 relative to 1997 baseline numbers, while in the MA OS this number increases by 18 million in SSA and by 6 million in South Asia as a result of depressed food supplies, high food prices and low investments in maternal and child care as well as health and sanitation services. Under the AM scenario, the number of malnourished children increases by 6 million children in sub-Saharan Africa, but declines by 14 million in South Asia.

In the FAO projection the incidence of undernourishment in today's developing countries as a group drops from 15.9 per cent of the total developing country population in the 2005/07 baseline to 7.1 per cent in 2050 (that is an absolute decline by 509 million people). In SSA, undernourishment drops from 27.6 to 7.1 per cent (–82 million), in South Asia from 21.8 to 4.2 per cent (–238 million), in East Asia from 11.0 to 2.8 per cent (–217 million), in Latin America and the Caribbean from 8.5 to 2.5 per cent (–29 million) and in the Middle East and North Africa (MENA) region from 7.4 to 3.4 per cent (–7 million).

With respect to the projected remaining levels of undernourishment in 2050 (318 million people –119 million in SSA, 93 million in South Asia and 62 million in East Asia), the FAO study points out that despite the assumptions about positive per-capita income growth across all regions, 15 developing countries will still show a per-capita income below $1,000 in 2050 according to the projections[11] and the persistence of undernourishment in 2050 is a reflection of this prospect. While the FAO study robustly dismisses the notion of an emerging *global* food availability problem as a result of population growth, the persistence of undernourishment is *not* reduced to a mere problem of access that is unrelated to supply side constraints:

> production constraints are and will continue to be important determinants of food security; however, they operate and can cause Malthusian situations to prevail, at the local level and often because in many such situations production constraints affect negatively not only the possibility of increasing food supplies but can be veritable constraints to overall development and prime causes of the emergence of poverty traps.
>
> *(Alexandratos and Bruinsma 2012: 10)*

In the IAASTD reference run, the global childhood malnutrition headcount is projected to decline by 50 million children relative to the 2000 baseline to 99 million children by 2050. However, in sub-Saharan Africa child malnutrition rises by

11 per cent to 33 million in this scenario. Average per-capita calorie availability for South Asia in 2050 is with 2746 kcal/day, very close to the projection for SSA in the reference scenario. Under the AKST_high_pos scenario, food security outcomes improve significantly.

In the IFPRI study, daily calorie availability in low-income developing countries as a group rises by 6.8 per cent in the baseline, by 9.7 per cent in the optimistic scenario and by up to 26.9 per cent in the productivity improvement scenarios while dropping by 6.2 per cent in the pessimistic scenario between 2010 and 2050. The corresponding figures for the projected percentage change in the number of malnourished children are −8.6 (baseline), −36.6 (optimistic), −22.6 (productivity improvement) and +18.1 (pessimistic) respectively.

Staple crop prices

Table 3.3 reports world market price projections for the main traded staple crops between base year and 2050 as a key indicator of the extent to which global supply keeps up with rising demand, as far as price results are presented in the scenario reports. Because of the differences in base years across the scenarios, comparisons of the price results across the different scenarios are far from straightforward. It is not obvious that rebasing to a common base year using observed prices for a particular recent year would resolve this problem, given that prices in the long-run simulation models should appropriately be interpreted as long-run trend prices that do not reflect short-run deviations from the underlying long-run trend due to temporary shocks. However, it is clear from the figures in Table 3.3 that long-run price projections vary widely across scenarios, and that – unsurprisingly – differences in the assumptions about long-run agricultural productivity growth are one of the main explanatory factors for these variations.

How sensitive are projections to the choice of model?

This cursory synoptic glimpse at recent long-run food system scenarios highlights the wide range in assumptions about the main drivers of change within and across the studies. For the exploratory scenarios this is as it should be – it is after all precisely the high degree of uncertainty about the future pathways for the main drivers of change and the need to explore alternative policy options that motivates the adoption of a multi-scenario approach in the first place (Reilly and Willenbockel 2010). But even in the various baseline scenarios that aim to employ 'middle-of-the-road' trend projections and envisage no major policy shifts, the exogenous driver assumptions about population, GDP and agricultural productivity growth differ substantially.

Therefore, this synopsis does not show to which extent the large variation in existing long-run projections is attributable to methodological differences across model ensembles used in the various studies as opposed to differences in these exogenous driver assumptions. To address this crucial question for further progress in model-based food system scenario analysis, the Agricultural Model Intercomparison and Improvement Project (AgMIP), hosted by Columbia University, includes an ongoing study in which ten different global models with a detailed representation of the food sector are subjected to a harmonized set of driver pathways up to 2050 and report simulation results for a standardized set of geographical regions and variables.

Initial results from this ongoing project reported by von Lampe et al. (2013; 2014) and Nelson et al. (2014) show that substantial cross-model differences in long-run projections persist even under harmonized assumptions about these main drivers of change. From a conceptual perspective, the differences in projections across models under these harmonized assumptions can be traced back to differences in the explicit or implicit assumptions about the effective size orders of four sets of elasticities that determine the responsiveness of the food system to external pressures: the effective price and income elasticities of demand for agricultural commodities, the effective land supply elasticities, and the effective price elasticities of agricultural productivity growth. Thus, for the simulation behaviour of the models at broad regional scales, the choice of numerical values for the parameters that co-determine these elasticities are ultimately more important than the prior choice of methodological framework (Hertel 2011; Robinson et al. 2014). Thus, a priority area for future research must be to extend and consolidate the empirical knowledge base for informed judgements about plausible ranges for these elasticities.

To a large extent, future model improvements are necessarily contingent on parallel progress in empirical research across a range of disciplines including, *inter alia,* climate science, crop and livestock science and economics to advance the state of scientific knowledge, on which modellers must rely in the course of the numerical parameterization of the model components. More dialogue across modelling teams – in particular about best practice in incorporating the existing empirical evidence regarding the key parameters that determine simulation results – is essential to narrow down the variance in projections. The satisfactory integration of water availability and use in global long-run simulation models and model validation remain major challenges for future research.

Projections for global trade of food commodities

All scenarios under consideration – including scenarios that envisage the imposition of additional barriers to trade such as the MA Order from Strength (OS) and the GEO4 Security First futures – project increases in the global volume of international trade in food commodities. Given the underlying assumptions about economic growth in all scenarios, this commonality is not particularly surprising. However, scenarios differ considerably with respect to the evolution of the shares of trade to global food production over time and with respect to changes in regional trade patterns.

In the scenarios that envisage further progress in multilateral trade liberalization, such as the MA Global Orchestration (GO) and the GEO4 Market First scenario, agricultural trade expansion is especially pronounced – e.g. under MA GO global grain trade rises by over 200 per cent and trade in meat products by around 670 per cent, and cereal and global livestock product trade are projected to rise by well over 200 per cent between the 1997 baseline and 2050. It is noteworthy that in this scenario sub-Saharan Africa turns from a net grain importing to a net exporting region by 2050. Net cereal exports from the OECD region are projected to rise in response to rising import demand from Asia and the MENA region. Net exports of meat products increase particularly in Latin America, primarily driven by rising import demand from Asia.

Despite the assumption of high trade barriers in the MA OS scenario, total trade in food commodities still more than doubles relative to 1997, as high population growth coupled with low agricultural productivity in Asia entails a rising net import demand for this region. Trade growth in the MA Techno Garden (TG) scenario is also less strong than under MA GO because the TG scenario assumptions include preference shifts towards less meat-intensive diets in high-income regions and somewhat lower average per-capita income growth, although the effect of the latter on food demand and trade volumes is partially offset by higher population growth than under OS (Table 3.2). The MA Adapting Mosaic scenario assumes more localized food production and distribution and lower income growth than under GO, and hence food commodity trade growth is significantly lower despite higher population growth. The adoption of appropriate technologies and conservation strategies is seen to lead to a small net cereal export surplus for sub-Saharan Africa by 2050 under this scenario.

In the IAASTD reference scenario, global cereal trade is projected to increase by 155 per cent and trade in meat products by over 310 per cent between this study's 2000 baseline and 2050. Similar to the MA scenarios, the trade expansion is primarily driven by rising import demand in the 'Central and West Asia and North Africa' and in the 'East and South Asia and Pacific' regions, while this import demand is primarily matched by rising exports from the 'North America and Europe' and the 'Latin America and Caribbean' regions. In contrast to the MA GO and TG scenarios, large increases in net food imports are projected for sub-Saharan Africa. The narrative for this scenario emphasizes the 'increasingly critical role' of international trade in meeting the food consumption needs of countries in which domestic food production does not grow rapidly enough to match the growth in domestic demand. The exports from USA, Brazil and Argentina are seen as a 'critical safety valve' in providing relatively affordable food to the net importing countries.

The CAWMA study likewise emphasizes the role of international trade for food security. The scenario chapter refers to the notion of virtual water which denotes the water used to produce imported food and cites estimates suggesting that without international trade in cereal crops, irrigation water consumption in 1995 would have been higher by 11 per cent. In the CAWMA pessimistic yield scenario, global food trade increases from 14 per cent of total agricultural production in the base year to 22 per cent in 2050, as regions lacking potential to expand rainfed areas increase their food imports. The MENA region is projected to import over two-thirds of its demand for agricultural commodities and land-constrained South and East Asia will import 30–50 per cent of its domestic demand, while net exports from Latin America, Central Asia and Eastern Europe rise correspondingly.

In contrast, in the CAWMA irrigation expansion scenario the ratio of trade to global production does not change significantly from the current level. The CAWMA trade scenario that assumes a further expansion in food export volumes from water-abundant to water-scarce regions suggests that, in principle, future world food demand could be satisfied through international trade without worsening water scarcity and without requiring additional irrigation infrastructure. However, the partial-equilibrium simulation analysis upon which this conclusion is based ignores balance-of-payments constraints and thus does not address the question of how water-scarce countries will pay for the additional imports. With a view to this limitation of the analysis, the authors of the CAWMA study conclude that it is unlikely that food trade alone will solve problems of water scarcity in the near term.

The IFPRI study emphasizes the essential role international trade can play in compensating for climate change effects on agricultural production, as trade flows can partially offset local climate change productivity effects, allowing regions of the world with positive or less negative effects to supply those with more negative effects.

The FAO projections for international food trade envisage a continuation of current trends. Today's developing countries as a group remain net importers of wheat and coarse grains and net exporters of rice. Traditional net exporters including North America, the EU and Australia remain exporters. The Russian Federation and Ukraine are projected to supply a rising share of world wheat and grain exports. Increases in meat import demand by today's developing countries will be matched by exports from the same country group.

Envisaged shifts in agricultural production and distribution systems

As far as the qualitative evolution of food production and distribution modes towards 2050 underlying the quantitative projections is addressed at all, the scenario narratives devote generally only little space to an articulation of such envisaged changes. In this respect, the MA is most explicit among the scenario exercises reviewed here. In view of the contentious contemporary debates about the future of smallholder agriculture in sub-Saharan Africa, it is interesting to have a closer look at the alternative futures for the region in terms of agricultural production and distribution systems sketched in the MA scenario report.

In the MA GO scenario with its emphasis on multilateral trade liberalization, global cooperation and a reactive approach towards environmental management, by 2050 agricultural output in both developed and developing regions is mostly produced on large, highly mechanized farms. Low-intensity farming continues only as a lifestyle choice in developed countries and on marginal lands in least-developed areas.

Under the MA TG scenario, the projected removal of subsidies and other agricultural trade barriers, in combination with an increasing spread and development of locally adapted genetically modified crops, is likewise envisaged to trigger a global transformation of agriculture involving an intensification of farm production in Asia, Africa and Latin America. More specifically, the elimination of agricultural trade barriers attracts investments from agri-business and supermarket chains into African as well as into Latin American and Eastern European agriculture, and drives the aforementioned agricultural intensification and adoption of locally adapted genetically modified crops in these regions. Remarkably, with the help of these foreign direct investment inflows and successful regional economic integration efforts, sub-Saharan Africa is seen to turn into 'one of the globe's "breadbaskets" with some of the cleanest cities and most rational land use in the world' (Carpenter et al. 2005: 259).[12] In developed regions, the assignment of property rights generates incentives for farmers to dedicate land increasingly to the provision of multiple ecosystem services. Crop yield growth over time is assumed to be lower in today's high-income regions because of a greater focus on organic farming. However, investments in biotechnology and other crop innovations resulting from investments in agricultural research are sufficient enough to bring about significant crop yield improvements.

Under the MA AM scenario, food production is geared primarily towards local markets. Investment in food production technologies is low, there are no breakthroughs in yield-enhancing technologies and little growth occurs in irrigated land area at a global scale. However, for sub-Saharan Africa this scenario envisages the highest yield growth among all regions as a result of the adoption of 'successful local adaptation mechanisms'.[13]

Under the MA OS scenario, agricultural production growth in sub-Saharan Africa is primarily achieved through area expansion, as investments in yield improvements are by assumption very low in this region as in other world regions. The contrast to the MA TG scenario is stark. In the OS narrative, the insufficient productivity growth exacerbated by adverse climate change impacts is envisaged to raise food insecurity, particularly in southern Africa. This triggers a mass migration from southern to West and East Africa, leading to social unrest and civil war in the latter regions.

In the CAWMA Comprehensive Assessment scenario – which considers agricultural investment strategies with location-specific optimal combinations of productivity improvements in rainfed settings, expansion of irrigated areas, yield improvements in existing irrigated areas and trade expansion – investments in sub-Saharan Africa are targeted toward improving rainfed smallholder agriculture with an emphasis on poverty alleviation. In this scenario the area under irrigation is envisaged to increase by 80 per cent from its present small base, mainly through small-scale informal irrigation geared to producing high-value cash crops such as sugar, cotton and fruits. At the same time, smallholders are seen to produce labour-intensive crops for local markets. This approach requires supporting investments in physical and institutional infrastructure that is loosely described to include 'favorable policies, credit, subsidies, education, and healthcare, capable government institutions, and water user associations' (de Freiture and Wichelns 2007: 135) to ensure economic feasibility. Under this strategy, investments in smallholder agriculture serve as a necessary first step to promote rural growth and poverty alleviation. In the longer run, with rising incomes and the associated increased diversification of economic activity, the number of people engaged in farming is envisaged to decline while farm sizes increase.

As van Vuuren, Ochola and Riha (2009) note in their review of pre-2009 long-run food system scenario exercises for the IAASTD, the focus of these studies is primarily on the evolution of production and consumption and not on changes in the distribution component of food value chains.[14] However, the authors point out that in this respect some trends not explicitly spelled out in the scenario narratives can be hypothesized. In particular, obvious implications of the scenario projections suggesting a significantly increased role for international trade in food commodities – including growth in exports from low-income countries as outlined above – would include an increasing importance of multinational companies, an increasing focus on quality and safety standards and with it an increasing commercialization of upstream production processes in these countries. The continuation of urbanization trends in developing regions implied by the driver assumptions about income growth, population growth and agricultural productivity growth entails a continuation of observable trends with respect to the role of supermarkets and other retailers in food distribution.

5. Conclusions

From a food security perspective, a number of general messages emerge from the preceding synopsis of global long-run scenario studies. To begin with, apocalyptic visions in which population growth meets with inescapable *global* bio-physical resource constraints to cause mass starvation are notable by their absence. Most of the scenario projections under review project significant increases in global average calorie availability per head towards 2050.

However, as Alexandratos and Bruinsma (2012: 10) put it, food security is only weakly linked to the capacity of the world as a whole to produce food. The scenarios draw attention to persisting and emerging serious challenges at sub-global levels, particularly in low-income regions with continued population growth, that are predominantly located in sub-Saharan Africa and South Asia.

In cases where scenario projections imply catastrophic deteriorations in food security outcomes at regional scales – as in the IAASTD AKST_low_neg scenario where average daily calorie availability in sub-Saharan Africa drops by 1,100 kcal per capita to levels well below minimum daily requirements – it is an assumed deceleration of growth in agricultural research and development and lack of investment in complementary infrastructure and social services, rather than the presence of insurmountable regional bio-physical resource constraints as such, that are framed to lead to these dismal outcomes.

Most of the scenario reports recognize explicitly that the principal underlying causes of food insecurity today are economic and social rather than natural resource-related constraints on food availability, e.g. the MA report cites poverty, inequity and deprivation of the opportunity to earn income or to obtain land rather than environmental factors as the principal causes of persistent hunger (Carpenter et al. 2005: 499). All scenarios assume substantial per-capita income growth in all regions up to 2050, and in all but the least optimistic scenarios the incidence of chronic hunger and undernourishment declines.

In line with their respective remit and core target audience, both the MA and GEO4 scenario studies emphasize the challenge of expanding food production in an ecologically sustainable manner. They also highlight the linkages between the degradation of ecosystems (deforestation, soil erosion, water pollution, desertification, biodiversity loss, marine fishery depletion) as a result of unsustainable practices, and poverty for households whose livelihoods depend directly on ecosystem services. One of the key messages of the IAASTD scenario chapter is that research and development efforts towards improved agricultural knowledge, science and technology are required to reduce the tradeoffs between agricultural growth and environmental sustainability. The CAWMA study, the results of which have informed the IAASTD scenarios, focuses specifically on the need to address water scarcity problems through regionally differentiated water management investment strategies with the potential to raise yields, along with the efficiency of water use in agriculture, significantly. The IFPRI scenarios serve to draw particular attention to the challenges posed by adverse climate change impacts on agriculture in the context of population growth. While the challenges ahead are characterized as daunting, the main conclusion of the IFPRI study is that a combination of policies that encourage broad-based economic growth and foster agricultural productivity growth could lead to a substantial decline in the number of malnourished children – the main food insecurity metric reported in this study – between today and the middle of the century.

Thus, the common broad generic message emerging from the scenario exercises under review is that the challenges to food security over the coming decades are serious and call for sustained attention, but policy options exist to address these challenges and deserve a high priority on the agenda for international development cooperation.

The fact that all the scenario projections are surrounded by numerous uncertainties does not constitute a reason for delaying action, as efforts to speed up agricultural productivity growth in food-insecure low-income regions through some form of sustainable intensification is a reasonable policy strategy, both from a food security perspective and from a wider poverty reduction perspective, regardless of how these uncertainties are resolved with the passage of time.

However, to avoid potentially severe deteriorations in food security outcomes in the decades beyond 2050 for low-income populations in arid and semi-arid zones particularly exposed to harmful climate change impacts, parallel decisive progress in climate change mitigation action is urgently required.

Notes

1 Rothman, Agard and Alcamo (2007: 451), italics in original.
2 In soft-linked model ensembles, output variables from one model are used to inform the selection of values for the input variables or parameters of another model, but the

different models are not formally merged – or hard-wired – into a single consistent simultaneous-equation system. Because of the heterogeneity of scales, accounting methods and conceptual frameworks across different models, the soft-linking approach is associated with substantial problems in achieving consistency and is susceptible to error propagation. The scientific basis for linking models across disciplines and scales is still weak and requires specific attention in future research.

3 Downscaling refers to the process of disaggregating variables towards a more detailed spatial or commodity classification scale.

4 An Engel curve describes the relation between demand for a commodity and income for given prices.

5 The time paths for exogenous drivers are not determined within the modelling framework but are taken from outside sources (e.g UN population projections or World Bank long-run GDP trend projections) and enter the simulation model in the form of given assumptions.

6 In an appendix note, Alexandratos and Bruinsma (2012: 92–93) concede that the FAO projections make no attempt to incorporate climate change impacts and end with the remarkable admission that '(i)n principle, a scenario that assumes no climate change has no place in the array of scenarios to be examined'. The CAWMA study points out that climate change adds to the complexity of water resources planning and management and includes a selective review of the literature on climate change and agricultural performance, but there is no indication that climate change impacts have been incorporated in the quantitative projections.

7 Rosegrant, Fernandez and Sinha (2009: 327). The unsharp nature of the cited statement is typical for the descriptions of the assumed influence of climate change in most of the scenario studies under review. With the exception of the IFPRI study, none of the scenario reports provides a systematic quantitative comparison relative to a hypothetical reference case without climate change.

8 A partial exception is the IFPRI study, which explores the price implications of a prolonged drought in South Asia in a separate scenario. For a brief selective review of the current state of science concerning projections of changes in the frequency of extreme events due to anthropogenic climate change and a set of explorative simulation scenarios examining the potential food price impacts of extreme weather events using a global general equilibrium model, see Willenbockel (2012). This study also examines the contribution of weather extremes to the recent observed food price hikes. For an example of recent progress in model-based economic climate change impact analysis at country and sub-regional levels incorporating projections of extreme weather shocks, see Robinson, Willenbockel and Strzepek (2012).

9 See e.g. New et al. (2011) and Thornton et al. (2011). The latter study envisages a bleak future for agriculture in sub-Saharan Africa. As emphasized in van der Mensbrugghe et al. (2011), the emission scenarios used in the IPCC Fourth Assessment Report were generated around 2000 and have significantly underestimated both actual output and emission growth over the last decade, notwithstanding the recent financial crisis. If this pattern continues, it puts the world on a trajectory of much higher temperature changes than the IPCC median of about 3°C by the end of the century, and a 2.5°C increase is more likely to be reached in 2050 rather than in 2080.

10 Since the broad composite sub-regions for which scenario results are reported differ across the various studies, it is not possible to provide a systematic and concise tabular synopsis of food security outcomes or other scenario results at sub-global scales. However, the following overview highlights scenario projections for regional food insecurity hotspots – which are predominantly located in sub-Saharan Africa – as far as these are reported in the scenario narratives.

11 Down from 45 countries in the 2005/07 baseline.

12 See Willenbockel (2009) for further reflection on the cogency of the pathway leading to this outcome offered by the MA narrative for this scenario.

13 Carpenter et al. (eds) (2005: 366). The nature of these local adaptation mechanisms is not further specified in the MA scenario report.

14 Wood et al. (2010) argue that the focus on production and neglect of food distribution channels in these international assessments entails the danger that potentially significant and cost-effective interventions for improving food security outcomes – such as the design of technologies to reduce post-harvest storage and processing losses and the commercialization of food processing and preparation methods that reduce waste and the loss of key nutrients – receive inadequate attention, which may lead to a relative overinvestment in the expansion of production potential *per se,* since that is the factor most readily observed and modelled.

References

Alexandratos, N. and Bruinsma, J. (2012) *World Agriculture: Towards 2030/2050 – The 2012 Revision,* ESA Working Paper No. 12-03, Rome: Food and Agriculture Organization.

Carpenter, S.R., Pingali, P.L., Bennett, E.M. and Zurek, M.B. (eds) (2005) *Ecosystems and Human Well-being: Findings of the Scenarios Working Group of the Millennium Ecosystem Assessment,* Millennium Ecosystem Assessment Series Vol. 2, Washington, Covelo and London: Island Press.

de Freiture, C. and Wichelns, D. (2007) 'Looking ahead to 2050: scenarios of alternative investment approaches', in D. Molden (ed.), *Water for Food, Water for Life: A Comprehensive Assessment of Water Management in Agriculture,* London: Earthscan.

Easterling, W.E., Aggarwal, P.K., Batima, P., Brander, K.M., Erda, L., Howden, S.M., Kirilenko, A., Morton, J., Soussana, J.-F., Schmidhuber, J. and Tubiello, F.N. (2007) 'Food, fibre and forest products', in M.L. Parry, O.F. Canziani, J.P. Palutikof, P.J. van der Linden and C.E. Hanson (eds), *Climate Change 2007: Impacts, Adaptation and Vulnerability. Contribution of Working Group II to the Fourth Assessment Report of the Intergovernmental Panel on Climate Change,* Cambridge, UK: Cambridge University Press.

Godfray, C., Beddington, J.R., Crute, I.R., Haddad, L., Lawrence, D., Muir, J.F., Pretty, J., Robinson, S., Thomas, S.M. and Toulmin, C. (2010) 'Food security: the challenge of feeding 9 billion people', *Science* 327: 812–18.

Government Office for Science (2011) *Foresight. The Future of Food and Farming: Challenges and Choices for Global Sustainability,* Final Project Report, London: Government Office for Science.

Hertel, T.W. (2011) 'The global supply and demand for agricultural land in 2050: a perfect storm in the making?' *American Journal of Agricultural Economics* 93(2): 259–75.

IPCC (2012) *Managing the Risks of Extreme Events and Disasters to Advance Climate Change Adaptation,* A Special Report of Working Groups I and II of the Intergovernmental Panel on Climate Change, Cambridge and New York: Cambridge University Press.

Nelson, G.C., Ahammad, H., Deryng, D., Elliott, J., Fujimori, S., Havlik, P., Heyhoe, E., Kyle, P., von Lampe, M., Lotze-Campen, H., Mason d'Croz, D., van Meijl, H., van der Mensbrugghe, D., Müller, C., Robertson, R., Sands, R.D., Schmitz, C., Tabeau, A., Valin, H. and Willenbockel, D. (2014) 'Assessing uncertainty along the climate-crop-economy modeling chain', *Proceedings of the National Academy of Sciences of the United States of America* 111(9): 3274-90.

Nelson, G.C., Rosegrant, M.W., Palazzo, A., Gray, I., Ingersoll, C., Robertson, R., Tokgoz, S., Zhu, T., Sulser, T.B., Ringler, C., Msangi, S. and You, L. (2010) *Food Security, Farming, and Climate Change to 2050: Scenarios, Results, Policy Options,* Washington, DC: International Food Policy Research Institute.

New, M., Liverman, D., Schroeder, H. and Anderson, K. (2011) 'Four degrees and beyond: the potential for a global temperature increase of four degrees and its implications', *Philosophical Transactions of the Royal Society A* 369: 6–19.

Reilly, M. and Willenbockel, D. (2010) 'Managing uncertainty: a review of food system scenario analysis and modelling', *Philosophical Transactions of the Royal Society B* 365: 3049–63.

Robinson, S., van Meijl, H., Willenbockel, D., Valin, H., Fujimori, S., Masui, T., Sands, R., Wise, M., Calvin, K., Havlik, P., Mason d'Croz, D., Tabeau, A., Kavallari, A., Schmitz, C., Dietrich, J.P. and von Lampe, M. (2014) 'Comparing supply-side specifications in models of global agriculture and the food system', *Agricultural Economics* 45(1): 21–35.

Robinson, S., Willenbockel, D. and Strzepek, K. (2012) 'A dynamic general equilibrium analysis of adaptation to climate change in Ethiopia', *Review of Development Economics* 16: 489–502.

Rosegrant, M.W., Fernandez, M. and Sinha, A. (2009) 'Looking into the future for agriculture and AKST', in B.D. McIntyre, B.D. McIntyre, H.R. Herren, J. Wakhungu and R.T. Watson (eds), *International Assessment of Agricultural Science and Technology for Development (IAASTD): Global Report*, Washington, DC: Island Press, 307–76.

Rothman, D.S., Agard, J. and Alcamo, J. (2007) 'The future today', in UNEP *Global Environmental Outlook GEO 4: Environment for Development*, Nairobi: United Nations Environment Programme, 397–456.

Thornton, P.K., Jones, P.G., Ericksen, P.J. and Challinor, A.J. (2011) 'Agriculture and food systems in sub-Saharan Africa in a 4°C+ world', *Philosophical Transactions of the Royal Society A* 369: 117–36.

van der Mensbrugghe, D., Osorio-Rodarte, I., Burns, A. and Baffes, J. (2011) 'Macroeconomic environment and commodity markets: a longer-term outlook', in P. Conforti (ed.), *Looking Ahead in World Food and Agriculture: Perspectives to 2050*, Rome: Food and Agriculture Organization.

van Vuuren, D.P., Ochola, O. and Riha, S. (2009) 'Outlook on agricultural change and its drivers', in B.D. McIntyre et al. (eds), *International Assessment of Agricultural Science and Technology for Development (IAASTD): Global Report*, Washington, DC: Island Press, 255–305.

von Lampe, M., Willenbockel, D., Blanc, E., Cai, Y., Calvin, K., Fujimori, S., Hasegawa, T., Havlik, P., Kyle, P., Lotze-Campen, H., Mason d'Croz, D., Sands, R.D., Schmitz, C., Tabeau, A., Valin, H., van der Mensbrugghe, D. and van Meijl, H. (2014) 'Why do global long-term scenarios for agriculture differ? An overview of the AgMIP global economic model intercomparison', *Agricultural Economics* 45(1): 3–20.

von Lampe, M., Willenbockel, D. and Nelson, G.C. (2013) 'Overview and key findings from the global economic model comparison component of the Agricultural Intercomparison and Improvement Project (AgMIP)', 16th International Conference on Global Economic Analysis, Shanghai.

Willenbockel, D. (2009) *Global Energy and Environmental Scenarios: Implications for Development Policy*, DIE Discussion Paper No. 8/2009, Bonn: German Development Institute.

Willenbockel, D. (2012) *Extreme Weather Events and Crop Price Spikes in a Changing Climate: Illustrative Global Simulation Scenarios*, Oxfam Research Reports, Oxford: Oxfam International.

Wood, S., Ericksen, P., Stewart, B., Thornton, P. and Anderson, M. (2010) 'Lessons learned from international assessments', in J. Ingram, P. Ericksen and D. Liverman (eds), *Food Security and Global Economic Change*, London and Washington, DC: Earthscan, 46–62.

4

STRENGTHENING FOOD SECURITY THROUGH HUMAN RIGHTS

A moral and legal imperative and practical opportunity

Wenche Barth Eide[1]

Linking food security and the right to food – the 1996 World Food Summit and its follow-up

The right of all human beings to adequate food and to be free from hunger is well established in international human rights law (United Nations 1948; 1966). Nevertheless more than 800 million people are still food insecure and have too little to eat, and many more suffer from various forms of malnutrition. Most states recognize a responsibility to ensure food security for their people, but this is often addressed solely in terms of food production and trade, ignoring the fact that insufficient food supply is not the major issue.

The World Food Summit in 1996 was a milestone in making it abundantly clear that the main problem underlying hunger is the lack of access, not insufficient capacity for food production, reiterated and documented by numerous authors and fora in the years to follow. Nevertheless priority focus continues to be on increasing overall production of and trade in food while ignoring the critical aspect of access by all to the food they need. Both FIAN (Food First Information and Action Network, the largest international non-governmental organization (NGO) dealing with the right to food)[2] and the UN Special Rapporteur on the Right to Food[3] have emphasized that the question of access has more to do with the *modes* of food production and distribution than with levels of food production alone.

Hunger is caused by breakdowns or failures for particular groups or persons of the ways in which they procure their food. Amartya Sen (1982) mentions failures or deprivation of entitlements to *land* for one's own food production, lack of access to *work* to be able to earn money to buy food, lack of access through *trade* to be able to sell or barter food for other products, or lack of access to *transfers,* whether in the form of gifts or through social protection.

The realization of the right to adequate food varies with the level of commitment by governments, their available resources, their priorities and their political will. It also depends in large part on the degree of democratic mobilization by civil society, making active use of the whole range of human rights – civil, political, economic, social and cultural human rights. Many states are making great progress. Other states are lagging severely behind, particularly where internal conflicts and violence contribute to hampering the process towards responsive nation-building and accountability.

The need for a rights-based approach to development has been increasingly recognized during the last two decades, based on a broadening awareness that many contemporary development processes, including those justified in the name of eradicating hunger and malnutrition, impact very differently on different groups of people.

That *access* to food for all is the key factor in defining food security was clearly stated by the first World Food Summit hosted in Rome in 1996. The *World Food Summit Plan of Action* defined food security as follows:

> Food security exists when all people, at all times, have physical and economic access to sufficient, safe and nutritious food to meet their dietary needs and food preferences for an active and healthy life.
>
> *(FAO 1996: Para. 1)*

That food security, as defined in these terms, requires implementation of the right to food, was recognized by the Summit as stated in the opening words of the *Rome Declaration on World Food Security:*

> We, the Heads of State and Government . . . gathered at the World Food Summit . . . reaffirm the right of everyone to have access to safe and nutritious food, consistent with the right to adequate food and the fundamental right of everyone to be free from hunger.
>
> *(FAO 1996)*

This right of 'everyone', and the notion of 'all people' in the definition of food security above make the aim clear: *every human being* – man, woman and child – is entitled to physical and economic access[4] to food.

The right to food is contained in numerous international human rights declarations and conventions, above all Article 11 of the *International Covenant on Economic, Social and Cultural Rights* (ICESCR) adopted by the United Nations in 1966 (UN 1966) which is legally binding on all states parties[5] (160 as of December 2013). The Summit's reaffirmation of this right together with its people-focused concept of food security has been a major stimulus to international efforts to include the human rights dimensions in the food security discourse and practice. However, Objective 7.4 of the Summit's *Plan of Action* asked for a clarification of the content of the right to adequate food and the fundamental right of everyone to be free from hunger, with particular attention to the implementation and full and progressive realization of this right as a means of achieving food security for all. Such a clarification was made in

1999 by the Committee on Economic, Social and Cultural Rights (CESCR), whose mandate is to monitor the implementation of the rights contained in that covenant. The committee's General Comment No. 12 on the right to adequate food (GC12) provides the overall definition in terms that coincide closely with the definition of food security adopted by the World Food Summit:

> The right to adequate food is realized when every man, woman and child, alone or in community with others, have physical and economic access at all times to adequate food or means for its procurement. . . . The right to adequate food will have to be realized progressively. However, States have a core obligation to take the necessary action to mitigate and alleviate hunger as provided for in paragraph 2 of article 11, even in times of natural or other disasters.
>
> *(FAO 1996)*

Further details of this interpretation of the content of the right to food follow later in this chapter. The main point here is that the World Food Summit consolidated the intimate relationship between food security as defined by the world's leaders in 1996, and the right to food as a human right under international law.

While states remain the primary duty-bearers for realizing the right to food as discussed in the following section, civil society often plays an active role in demanding that states and authorities live up to their human rights obligations and responsibilities. Many civil society organizations are actively pursuing a rights-based approach to advocacy in agriculture and food security. In this context there is also a growing awareness of the conflicting interests and concerns held by many corporate investors in the agricultural and food sector, and producers of food on small plots as they strive for food security for their families. The tensions between interests in different food systems are many and call for a new common agenda that can facilitate the balancing of contrasting interests and strategies. The right to adequate food applied within a broader human rights perspective has the potential to serve that purpose.

Universal human rights and state obligations for implementation

The conceptual foundations of the right to food must be understood in the broader context of universal human rights as they have developed and been governed by the United Nations.[6] Universal human rights were made part of international law through the adoption of the *United Nations Charter* in 1945. While international law, prior to 1945, had been understood as a law of *co-existence* between sovereign states, the UN Charter transformed it into an international law of *cooperation* in solving common problems in economic, social, cultural and humanitarian affairs and in the promotion of human rights for all without discrimination (UN 1945: Article 1). The list of universal rights was set out in the *Universal Declaration of Human Rights* (UDHR) adopted by the UN General Assembly in 1948. Building on the UDHR, subsequent human rights conventions have elaborated in more detail both the rights contained in it and the corresponding obligations of member states that have ratified the conventions.

The right to food was included in the UDHR as part of an adequate standard of living (Article 25), and later specified in the ICESCR (UN 1966: Article 11.1 and 2). It is also contained explicitly or implicitly in other legally binding international instruments. Many institutions, functions and procedures have been created by the United Nations to protect and promote the human rights contained in these legal provisions.

Monitoring and reporting

States parties to ICESCR are obliged to report to the United Nations on the progress made in their implementation of the rights set out in the covenant. To monitor the compliance by states with their obligations, expert bodies have been established by the United Nations. Regarding the right to food, the relevant UN expert body is the Committee on Economic, Social and Cultural Rights (CESCR). The committee draws on the states' own reports and other sources, including reports from independent NGOs and national human rights institutions. Each state party meets separately with CESCR, which examines the progress made and the obstacles encountered in implementing the rights set out in the covenant, and recommends improvements and changes prior to the next round of reporting.

UN special rapporteurs on the right to food

Supplementing the work of these committees, the United Nations Human Rights Council – the main UN political body dealing with human rights – has also appointed 'special rapporteurs' on specific rights, including one on the right to food. They explore and report annually back to the Human Rights Council and the UN General Assembly on general issues pertaining to the special right of their mandate, and visit countries on invitation from the government concerned to examine their problems and progress in enhancing those rights. The rapporteurs' reports express their critique of shortcomings and praise for positive achievements made, together with recommendations for further improvement of the realization of the right concerned.

The mandate of the special rapporteur on the right to food was established in 2000 and has since 2007 been held by Professor Olivier De Schutter. He and his predecessor[7] have visited a range of countries and dealt with a number of thematic issues relevant to food security and the right to food.

Transparency

The monitoring process is based on full transparency of the progress and shortcomings of states. All country reports and CESCR's findings are accessible through the website of the Office of the UN High Commissioner for Human Rights. All special rapporteurs' reports are accessible on the website of the Office of the UN High Commissioner for Human Rights[8] or on the rapporteurs' own websites.[9]

The role of the specialized agencies and other UN bodies, including the UN Committee on World Food Security

More specific technical-political discussions take place in intergovernmental agencies, funds or programmes within the UN family.[10] For the right to food, the Food and Agriculture Organization of the UN (FAO) is the most important. In 2009 the intergovernmental FAO Committee on World Food Security (CFS) underwent a fundamental reform to make it more effective by including a wider group of stakeholders, including relevant UN agencies and civil society – all with the aim of increasing its ability to promote polices that reduce food insecurity. It is therefore now often called the UN Committee on World Food Security, or just CFS, with its secretariat based in FAO but including members from the three Rome-based agencies: FAO, IFAD (International Fund for Agricultural Development) and WFP (World Food Programme). Two major outcomes from CFS have a particular focus on human rights, a set of voluntary guidelines on responsible tenure (discussed below) and the recent *Global Framework for Food Security and Nutrition* (CFS 2013).

A special feature of CFS is its independent High Level Panel of Experts (HLPE), which prepares reports, based on themes selected by the CFS, on the state of the art of various development issues impacting on global and national food security. One of these dealt with social protection for food security in an explicit human rights perspective, to which we return later in the chapter.

Human rights-based approach to development

A 'Common Understanding on Human Rights-Based Approaches to Development Cooperation and Programming' was developed and adopted by the United Nations Development Group (UNDG) in 2003,[11] emphasizing that all programmes of development cooperation, policies and technical assistance should further the realization of human rights, which should guide all development cooperation and programming. Development cooperation should contribute to the capacities and resources of states to meet their obligations and of people to claim their rights. Development processes should be guided by the fundamental human rights principles such as participation, accountability, non-discrimination, transparency, human dignity, empowerment and the rule of law.

Rights-holders and duty-bearers: State obligations for human rights

Under international human rights law, human beings are the holders of rights, and the state is the primary duty-bearer. This means that the state must respect and ensure the rights set out in the conventions ratified by the state. On the other hand, every individual is expected to take, as far as possible, responsibility for her or himself and for those who are dependent on that person. It is these interlinkages between self-responsibility and the obligations that states have that constitute the human rights nexus.

Human rights include both *freedoms* and *entitlements.* One may think of the normative system of human rights as a framework of a *social contract,* where the starting base is that people are generally free to make their own choices and are expected to take care of their own needs as far as possible within the resources available to them and the capability they have for this. Their freedom of action in ensuring their livelihood should be limited only by measures required to protect the comparable freedoms for others. For those unable, for reasons beyond their control, to take care of their own needs, the social contract implied in human rights requires efforts from the state to create or improve the enabling conditions by which people can take care of their needs. Where such efforts fail or are insufficient, those in need should be directly assisted or provided for.

Based on a first study made for the United Nations on the right to adequate food in development (Eide, A. 1989), it is now generally recognized that state obligations for human rights fall into three categories, now regularly used by CESCR: the obligations to respect, to protect and to fulfil, the last being divided into facilitate and provide.

In generic terms, human rights comprise the freedoms of human beings to choose their own solutions and use their own resources as they themselves prefer. These must be *respected* by the state authorities unless they interfere unreasonably with the rights and entitlements of others and must be *protected* against harmful interference by third parties. When individuals cannot on their own manage to secure for themselves an adequate standard of living, the community as a whole as represented by the state must seek to *facilitate* the enjoyment of each right by creating the necessary enabling conditions, or directly to *provide* assistance when required.

For this to function without unnecessary tension and conflict there is a need for democratic governance ensuring effective participation by all in negotiating the necessary regulations and systems of assistance. It requires transparency in public affairs, general principles of equality and non-discrimination must be secured, and the state must be held accountable for its acts and omissions, while remedies must be available in case of alleged violations of the freedoms and entitlements that people should enjoy. Formal democracy is therefore not a sufficient safeguard. Unless social discrimination on grounds of race, caste, gender or ethnicity is effectively eliminated, marginal groups are often excluded and the democratic system can easily be captured by the elite or a given ethnic group. This is why prevention of discrimination and effective protection of economic and social rights are essential also in societies complying with the formal requirements of democracy.

It is through gradual development of reciprocal and interlinked rights and the corresponding institutional responsibilities that cohesive nations can be built. In our globalized world this can only operate within a broader framework of international cooperation underpinned by human rights, much of it negotiated or planned within activities and mandates of the various United Nations agencies and programmes. The main ones relevant to the management of food security and the right to food will be discussed below.

Through open political life in society, guided by the framework of human rights, the balancing of different rights can be achieved and reciprocal benefits can be obtained. While leaving people the freedom to choose their own approaches, every well-functioning state pursues and should pursue policies involving some regulations, permissions and prohibitions in economic and social life, to the extent necessary for ensuring the implementation of human rights. The purpose of human rights is to provide guidance to those regulations, to set limits to some and to require positive measures in other contexts. Modern human rights therefore constitute a comprehensive normative system including civil, political, economic, social and cultural rights.

Towards addressing the right to adequate food in context: Norms and content, guidelines and tools

As already noted, the 1996 World Food Summit sparked a new era in the 'right to food movement'. Different actors and fora contributed content to the very general legal provisions regarding the right to food and formulated guidelines and tools for implementing this and related rights.[12] Both intergovernmental and civil society actors thereby obtained a more holistic normative base on which to work, and several states with high food insecurity among population groups began to recognize a human rights-based approach to food security as an opportunity to fight food insecurity.

Normative content of the right to adequate food

In response to the World Food Summit's request in 1996 for further clarification of the right to food and the corresponding state obligations, the General Comment No. 12 (hereafter GC12) was adopted and launched in 1999 by the Committee on Economic, Social and Cultural Rights. Further to its definition of the right to adequate food (cited above in the introductory section) the committee states that the *core content* of the right to adequate food implies the following:

> The availability of food in a quantity and quality sufficient to satisfy the dietary needs of individuals, free from adverse substances, and acceptable within a given culture; the accessibility of such food in ways that are sustainable and that do not interfere with the enjoyment of other human rights.
>
> *(GC12: para. 8).*

Several concepts needed further clarification, such as 'adequacy', 'availability', 'access', and 'sustainability':

> The notion of *sustainability* is intrinsically linked to the notion of adequate food or food *security,* implying food being accessible for both present and

future generations. The precise meaning of 'adequacy' is to a large extent determined by prevailing social, economic, cultural, climatic, ecological and other conditions, while 'sustainability' incorporates the notion of long-term availability and accessibility.

(GC12: para. 7)

As the dietary and nutrition dimensions of food security are often overlooked in the contemporary food security debate, it is particularly useful that GC12 further describes the meaning of the terms implied in 'adequacy': *dietary needs,*[13] *free from adverse substances*[14] and *cultural or consumer acceptability.*[15]

Availability refers to the possibilities either for feeding oneself directly from productive land or other natural resources, or for well-functioning distribution, processing and market systems that can move food from the site of production to where it is needed in accordance with demand (GC12: para. 12). *Accessibility* encompasses both economic and physical accessibility:

Economic accessibility implies that personal or household financial costs associated with the acquisition of food for an adequate diet should be at a level such that the attainment and satisfaction of other basic needs are not threatened or compromised. Economic accessibility applies to any acquisition pattern or entitlement through which people procure their food and is a measure of the extent to which it is satisfactory for the enjoyment of the right to adequate food. Socially vulnerable groups such as landless persons and other particularly impoverished segments of the population may need attention through special programmes.

Physical accessibility implies that adequate food must be accessible to everyone, including physically vulnerable individuals, such as infants and young children, elderly people, the physically disabled, the terminally ill and persons with persistent medical problems, including the mentally ill. Victims of natural disasters, people living in disaster-prone areas and other specially disadvantaged groups may need special attention and sometimes priority consideration with respect to accessibility of food. A particular vulnerability is that of many indigenous population groups whose access to their ancestral lands may be threatened.

(GC12: para. 13)

These specifications are essential to a full-fledged understanding of the notion of food security. What the right to food adds to the discussion of food security, however, are the corresponding state obligations to make the right to food a reality.

State obligations for the implementation of the right to food

The right to food does not mean that the state shall feed its people. People will normally take care of their own needs, through their own efforts, either by self-production or access to the market. The role of the state is to promote and consolidate the conditions by which everyone can feed *themselves*. Like all other economic, social and cultural rights, the right to adequate food shall be *progressively* implemented. The state has a social responsibility to monitor developments concerning food security within their territory, and to take the appropriate steps to achieve food security for all, and do so at the earliest possible time.

The obligation to respect and protect the right to food

To *respect existing* access to adequate food requires states not to take any measures that result in preventing such access as already exists and need not change. Where people already produce and process their own food or part of it with reasonable success in line with their own preferred cultural food patterns, or have access to functioning markets to sell their surplus, or earn sufficient income from the formal or informal sector to ensure household food security on a stable basis, there may be no reason for the state to engage.

The obligation to *protect* people's right to adequate food is summarized by GC12 as requiring measures by the state to ensure that enterprises or other third parties do not deprive individuals of their access to adequate food by whatever entitlement and mode they wish to continue. Illustrations may be drawn from the whole gamut of threats and risks to people's entitlements – to land, forests and fishing territories, other productive assets including traditional seeds, or paid work as well as income-generating activities in the informal sector. Legislation on food safety protection is another aspect of the duty of the state to protect the right to food, requiring effective food legislation and inspection. There is also a clear role of the state and local government in protecting consumers, especially children and youth, from unethical marketing of inadequate high-energy, low-nutrient-dense junk foods which contribute to obesity and related chronic non-communicable diseases (NCD), often side-by-side with undernutrition in poor households (WHO 2013).

The obligation to fulfil the right to adequate food

According to GC12, states must proactively engage in activities intended to fulfil/facilitate people's access to and utilization of resources and means to ensure their livelihood, including food security. It is under this category that most economic and social development efforts can be grouped, as long as they enable people to procure food and other necessities. But do these efforts also reach the economically vulnerable, and what is expected from the state when they fail to do so?

To *fulfil/provide* is required whenever an individual or group is unable, for reasons beyond their control, to enjoy the right to adequate food by the means at their disposal. GC12 emphasizes that this obligation of assistance also applies towards persons who are victims of natural hazards or other disasters.

The situations and options for interventions by the state vary considerably under these broad categories. To stimulate and organize both analytical research work and

In: FAO Right to Food Toolbox
Tool 2: Methods to Monitor the Right to Adequate Food, Vol. II
http://www.fao.org/righttofood/knowledge-centre/right-to-food-methodological-toolbox/en/

Normative Principles/ FS Attributes / Categories of State obligations	**H o u s e h o l d F o o d S e c u r i t y**				
	Adequate food			**Sustainable _supply_ of adequate food**	**Stable _access_ to adequate food**
	Dietary adequate (quantity, nutritional quality)	Safe for human beings to eat	Culturally acceptable	Environmentally and economically sustainable food systems	Physical and economical access to food within a household's livelihood
Respect					
Protect					
Fulfil - Facilitate - Provide					

*Adapted from Oshaug A, Eide WB and Eide A (1994), Human rights : a normative basis for food and nutrition policies., *Food Policy*, 19: 491-516

FIGURE 4.1 The 'right to adequate food' matrix

Note: adapted from Oshaug et al. (1994).

central and local plans for food and nutrition security measures, one may conceptually juxtapose each category or level of obligation systematically with each of the broad attributes of the right to food, as illustrated in Figure 4.1.

GC12 emphasizes that violations of the covenant (ICESCR) occur when a state fails to ensure the satisfaction of the minimum essential level required to be free from hunger. In determining which actions or omissions amount to a violation of the right to food, the *inability* of a state party to comply must be distinguished from the *unwillingness* to do so.

On national implementation

The most appropriate ways and means of implementing the right to adequate food will inevitably vary significantly from one state to another:

> Every State will have a margin of discretion in choosing its own approaches, but the Covenant clearly requires that each State party take whatever steps are necessary to ensure that everyone is free from hunger and as soon as possible can enjoy the right to adequate food.

> *(GC12: para. 21)*

This will require the adoption of a national strategy to ensure food and nutrition security for all, based on human rights principles that define the objectives and the policies and corresponding benchmarks, as well as the resources needed for the most cost-effective way of meeting these.

From norms and content to guidelines for implementation

GC12, as the most authoritative interpretation to date of the right to food, is still formulated in rather general terms. The second World Food Summit in 2002 called for a set of practical guidelines for implementation. The 'Right to Food Guidelines'[16] (hereafter RtFG) were endorsed by the FAO Council in November 2004 following 18 months of negotiations among interested governments (FAO 2005; Oshaug 2005; Rae et al. 2007). The 19 guidelines constitute a range of activities by which the right to food can be put into practice, and seven practical steps and ways in which human rights principles should be followed in all fields related to food security.

GC12 and RtFG have stimulated extensive research on the right to food and numerous publications, some on food security in relation to specific themes.[17]

The FAO toolbox

FAO convened the 1996 and 2002 World Food Summits and followed up by establishing the FAO Right to Food Team, which has done a remarkable job in translating the GC12 and RtFG into a number of practical aids for food and nutrition security analyses and activities. They include among others an electronic interactive primer to the right to food and a 'Toolbox' aimed at the legislative and executive branches of governments in operationalizing measures towards the right to adequate food.[18] Based on this a series of ten more popular 'handbooks on the right to food' was launched on Human Rights Day 10 December 2013.[19]

Implementing the right to adequate food – selected issues and evolving examples

Applying a human rights approach to development efforts including food security implies the conscious, systematic and concrete integration of certain values and standards into policies, plans, programmes, priorities, processes, outputs and outcomes of an organization.

Two examples of further interpretation of obligations to respect, protect and fulfil the right to food

The case of governance of tenure

The 2004 Right to Food Guidelines recommended that states should respect and protect the rights of individuals with regard to resources such as land, water, forests, fisheries and livestock without any discrimination (RtF Guideline 8.1). States were called on to take measures to promote and protect the security of land tenure, especially with respect to women, poor and disadvantaged segments of society, through legislation that protects the full and equal right to own land and other property, including the right to inherit (RtF Guideline 8.9).

This was followed up in the landmark *Voluntary Guidelines on Responsible Governance of Tenure in Land, Fisheries and Forests for Food Security* (VGGT) (FAO-CFS 2012). These guidelines address the fundamental recognition that the livelihoods of many, particularly the rural poor, depend on secure and equitable access to and control over land, fisheries and forests that are their sources of food and shelter; the basis for social, cultural and religious practices; and a central factor in economic growth, as expressed in the preface to VGGT, which continues,

> Various forms of tenure systems have determined who can use which resources, whether based on written policies and laws or on unwritten customs and practices. But tenure systems increasingly face stress as the world's growing population requires food security, and as environmental degradation and climate change reduce the availability of land, fisheries and forests. Inadequate and insecure tenure rights increase vulnerability, hunger and poverty, and can lead to conflict and environmental degradation when competing users fight for control of these resources.
>
> *(FAO-CFS 2012)*

States are advised on how they can formulate responsible tenure policies based on their full obligations to '*recognize and respect all legitimate tenure holders and their rights . . .*';[20] '*safeguard legitimate tenure rights against threats and infringements . . .*';[21] and '*protect and facilitate people's access to resources through good tenure governance . . .*'.[22] Readers are referred to the document for details.[23]

The case of social protection

Not everybody holds tenures to be respected and protected, and some may have tenure to resources too small on their own to ensure a decent livelihood and food security. For them, *social protection* may be required to fulfil the right to food. Social protection was in the past rarely included in the global food security discourse. This has been changed by the fourth report prepared for the CFS by its independent High Level Panel of Experts: *Social Protection for Food Security* (HLPE 2012). It was launched in 2012 with recommendations to states and to CFS and is the first of these reports that fully develops the topic in an explicit human rights framework. Its primary recommendation is that 'Every country should strive to design and put in place a comprehensive and nationally owned social protection system that contributes to ensuring the realization of the right to adequate food for all' (Rec. 1).

Another recommendation to states makes it clear that human rights should be at the base of social protection: 'Social protection for food security should be underpinned by the human rights to food and social protection at every level, from governments signing up to global agreements, to national legislation and programme implementation' (Rec. 4).

The report recommends a 'twin-track' strategy for social protection systems to maximize their positive impacts on food security, by providing essential assistance in the short term and supporting livelihoods in the long term. The CFS itself should

> . . . actively encourage, monitor and report on the incorporation of the provisions on the right to adequate food and the right to social protection contained in the Universal Declaration on Human Rights, and in the corresponding international human rights conventions, into *national legislation and programmes* supported by an enforceable legal framework in all countries.
> *(HLPE 2012; emphasis as in text)*

Comment on the significance of the two documents

The two examples show that food security strategies increasingly draw explicitly on the international human rights framework and interpretations of the right to adequate food and related rights, such as the human right to social security, set out in Article 9 of ICESCR (UN 1966).[24] This reflects the dynamic development within the food security debate since the World Food Summit in 1996. A further example under elaboration, also strongly based in a human rights approach, is the *Draft Voluntary Guidelines for Securing Sustainable Small-scale Fisheries in the Context of Food Security and Poverty Eradication,* building on both the Right to Food Guidelines and those in Governance of Tenure.

Emerging examples of implementation at a country level

When seeking examples at a country level, one must acknowledge that it has been both conceptually and methodologically difficult to find a good way to systematize how state performance for the right to adequate food can best be described and documented. As mentioned in the introduction, human rights can only be realized 'progressively and to the maximum of available resources'. Thus one must select human rights/right to food-sensitive characteristics or parameters by which to judge, where the necessary information is accessible, whether performance of a certain state is going in the *right direction* – or the reverse.

In August 2013 the UN Special Rapporteur on the Right to Food, Olivier De Schutter, submitted an interim report to the UN Human Rights Council and General Assembly providing his judgments of progress with the right to food realization over the last ten years (De Schutter 2013). He outlines an innovative framework within which to evaluate progress in different states at three levels:

- First, to consider the right to food as a self-standing right that imposes special obligations on states to *respect, protect and fulfil* this right, each of which can be assessed accordingly.

- Second, to examine whether the right to food has encouraged the transformation of social welfare benefits under governmental food security schemes into *legal entitlements*, thus moving away from charity.
- And third, to check the extent to which states have adopted *national strategies* for the progressive realization of components of the right to food in the longer term perspective.

De Schutter notes progress at each of these levels, brought about by the interplay of courts, parliaments, governments, national human rights institutions, civil society and social movements. Key examples are those of Brazil and India. Brazil with its development of a rights-based multi-sectoral food security advocacy and promotion has, in part through effective cooperation between government and civil society, created a 'Zero Hunger' programme which has yielded results in terms of effects on reducing malnutrition rates. India, with its Mahatma Gandhi Rural Employment Guarantee Act, has also shown positive impacts on infant malnutrition (Nair et al. 2013). Further, in July 2013 the government issued the National Food Security Ordinance (Government of India 2013) based on a legislative bill initiated in 2011. This aims to ensure access to food for two thirds of the population of India through a variety of programmes that will hereafter be considered legal entitlements, making it unlikely that they will be removed with changing political paradigms and governments (De Schutter 2013).

According to both De Schutter and FAO, a growing number of right to food-based policies and programmes have been introduced and legislative and constitutional changes made. These changes in policy and law have placed emphasis on mechanisms for participation and accountability. According to Rosales, a member of the FAO Right to Food Team, this has accelerated the implementation of legal and policy measures that contribute to the realization of the right to food. De Schutter provides among others the following examples as summarized by Rosales (more details in De Schutter's report):

> South Africa, Kenya, Mexico, Ecuador, Bolivia, the Ivory Coast and Niger have given direct constitutional protection to the right to food, while reform processes are underway in El Salvador, Nigeria, and Zambia.
>
> Right to food framework laws, often taking the shape of 'Food and Nutrition Security' laws, have been adopted in Argentina, Guatemala, Ecuador, Brazil, Venezuela, Colombia, Nicaragua, and Honduras, with several other Latin American countries in the process of adopting similar measures.
>
> Countries including Uganda, Malawi, Mozambique, Senegal and Mali have adopted, or are in the process of adopting, framework legislation for agriculture, food and nutrition that enshrines rights-based principles of entitlements and access to food.
>
> The South African High Court ordered a revision of the Marine Living Resources Act and the creation of the Small-Scale Fishers Policy to ensure the socio-economic rights of small-scale fishers (2012).

Increasing litigation and Court rules for the protection of the right to food had been made in The African Commission on Human and Peoples' Rights and the ECOWAS Court of Justice ECOWAS; India and Nepal Supreme Courts.

Regional policies that are human rights based are being implemented in the CARICOM countries and Portuguese speaking countries in Africa.

(Rosales 2013)

More examples and information can be found at the FAO's right to food website.[25]

Conflicting interests and counterforces to a rights-based approach to food security

While the human rights aspect of food security has been more widely recognized during the last decade, its implementation faces conflicting interests and values.

Sometimes the obstacles to a human rights-based approach stem from self-serving governments or corrupt regimes encouraging investments that yield revenues favouring themselves or a relatively small circle of the elite. More significant are conflicting development theories with widely different policy implications for governments regarding the scope and nature of regulations and allocation of resources.

At a very general level this is seen in the debates between advocates for rapid economic growth, even when the livelihoods of many people may be threatened, confronting those that argue in favour of a more socially just development with greater prospects for long-term sustainability and better social cohesion.

Choosing development paradigms

A salient example of contrasting views can be seen in the recent publications by two leading Indian economists and the debates that have followed their publication; they clash over economic policy paradigms and development priorities for India. Jagdish Bagwathi argues in favour of rapid economic growth driven by market forces and unrestricted trade, and with a relatively passive role of the government in economic life by abstaining from protective regulations and distributive measures (Bagwathi and Panagaryia 2013). Nobel laureate Amartya Sen and his co-author Jean Drèze take a very different view. They argue that India's main problems lie in the lack of attention paid to the essential needs of the people, especially of the poor, and often of women. They point to major failures to foster participatory growth and to make good use of the public resources generated by economic growth to enhance people's living conditions, and insist that sustainable growth is possible only when combined with devoting resources to remove illiteracy, ill health, undernutrition and other deprivations (Drèze and Sen 2013).

A significant recent step by the government of India is the 'right to food' bill described in the previous section, although it is too early to say how effective its impact will be on the enjoyment of the right to food for the most vulnerable population and age groups.

The same conflicting economic paradigms for development have characterized many debates around rights-based approaches over the years. They should be critically analysed to clarify which interests and values are underlying the different positions taken, and who will be the beneficiaries of and losers from the choices made.

The case of food and nutrition security and the role of business

Food security approaches based on a rapid growth paradigm encourage an open door for large-scale modern and high-tech agriculture. What impact this has on small producers whose livelihoods are at stake is then seen as a secondary concern. The main actors involved are large transnational agricultural and food corporations, together now often referred to as 'Big Food'. Ethics of corporations vary. Civil society networks have for long demonstrated that public food security policies intertwined with such corporate activities can sometimes directly contribute to creating hunger; they therefore advocate the need to find alternative policies (RTFN-Watch 2013). For the same reason the web-based journal *World Nutrition* in 2013 decided to establish a permanent column called 'WN Big Food Watch' aimed to serve as a neutral observation base to open the doors of corporations and facilitate public discussion of the activities of food companies and their implications for public health (World Nutrition 2013: 466).

Questions concerning business and human rights have received increasing attention in recent years, and have led to the adoption, by the United Nations Human Rights Council, of the *Guiding Principles on Business and Human Rights: Implementing the United Nations 'Protect, Respect and Remedy' Framework* – or 'the Ruggie principles' (UN Human Rights Council 2011).

The essence of that framework is that all states – both the home state and the host states of business enterprises – have a duty to protect the human rights of every person that risks being negatively impacted by the activities of business entities. The enterprises are themselves required to respect the human rights of everyone within their area of operation. Remedies shall be made available for persons whose human rights have been violated by the activity of the corporation.

While more and more corporations recognize their ethical responsibilities and adopt their own voluntary codes of social conduct, a full-fledged recognition of their human rights obligations is not yet in place. It is likely to emerge, however, as part of the growing public-private partnerships, including in the food arena. Among the central issues will be the availability of effective remedies where the right to adequate food has been violated. One must also take into account that while corporate executives may have a sense of social responsibility as regards the use of land and other resources, they may be pushed by their shareholders whose concern is with rapid economic return on investments rather than prevention or reduction of hunger (Allaire 2013).

Summary and conclusions: Achievements made and obstacles encountered

The right to food as an essential component of universal human rights was included in the post–World War II approach to a new or more cooperative world order, and spelt out in the Universal Declaration of Human Rights adopted in 1948. Following the end of the Cold War in the 1990s, the World Food Summit in 1996 significantly accelerated the attention to the recognition and implementation of the right to food for all.

Since then, great advances have been made, as described in this chapter. The content of the right to food and the corresponding state obligations have been clarified. Practical guidelines have been adopted, some of them focusing on the realization in general of the right to food in the framework of food security, and others on the modalities or requirements by which people can achieve their food security. Particularly important here are the Voluntary Guidelines on Responsible Governance of Tenure (VGGT). For others there is a growing attention to issues such as minimum wages and social protection, partly to facilitate transitions to new forms of livelihoods.

This chapter has also shown, drawing in part on the final report of the United Nations Special Rapporteur on the Right to Food, that great advancements are now being made in many countries at the constitutional, legal and practical levels.

But there is nothing to guarantee that, even if the legal and institutional systems are in place, a human rights approach to food security would result in better targeted, more efficient and sustainable human-centred policies and programmes in the long run – evidence for that is still to come.

Success or failure depends to a large extent on political priorities and conflicting values. The international system of human rights, while having made enormous progress during the last decades, does not have a powerful system of enforcement. Governments decide what priorities they want to make. Leaving aside the obstacles caused by corruption, governments are faced with quite contrasting development ideologies and related theories. Some argue for very rapid economic growth at all costs. Some governments seek to justify this by asserting that growth in the long run may lead to a trickle-down that will also benefit those at the bottom.

They may then conveniently forget the famous saying by one of the greatest economists that ever lived, John Maynard Keynes: 'This *long run* is a misleading guide to current affairs. *In the long run* we are all dead. Economists set themselves too easy, too useless a task if in tempestuous seasons they can only tell us that when the storm is long past the ocean is flat again' (Keynes 1924).

Efforts to justify unregulated and undistributed growth by reference to long-range future trickle-down effects are particularly brutal in relation to children and to women of reproductive age. Severe hunger and malnutrition during childhood and pregnancy does not wait long to take its toll.

Human rights, including the right to food, are the language of justice. For governments that want to build an inclusive and socially cohesive nation, the implementation of the right to food is an essential element. Fortunately it is now clearly recognized as such in the international discourse on food security. Good and accountable governance can only exist when principles of non-discrimination are recognized and when different interest groups are able to participate and give voice to their needs, and where there is transparency in all public matters.

The specificity of a rights-based approach is its legal anchorage in international and regional human rights law, progressively also in domestic law. This gives a particular leverage to NGOs and community-based organizations to claim the rights on behalf of disaffected people, and to demand the inclusion of their freedoms and entitlements in relevant policies. The human rights principles help to steer the concretization of the obligations to respect, protect and fulfil the rights into context-relevant policies and programmes.

The real problem of world hunger and food insecurity has neither in the past nor today been caused by a lack of world capacity to produce enough food to meet effective demand. Consider the following figures: in December 1948, when the Universal Declaration of Human Rights was adopted, the world population was approximately 2.5 billion. By November 2013 when this chapter was written, the world population was approaching 7.2 billion. FAO has estimated that in 2012, around 870 million people suffered from chronic undernourishment. Assuming that in 1948 there were also several hundreds of millions of hungry undernourished people, this means that the world food producers have since 1948 been able to expand their capacity to feed at least an additional 4.7 billion persons. There is no convincing ground to assume that the world today does not have the capacity to produce enough food also for those 870 million who still have too little to eat.

But as noted above, the modes of food production as well as its distribution are the keys to the elimination of hunger and malnutrition. The right to food and its clarification and elaboration described in this chapter provide the directions that this should take.

The ambitions may easily be set unrealistically high in our world of continuing tensions between vested interests, contradictions and conflicts, whether in the fight over resources or in protracted crises arising for other reasons. But human rights are given greater attention in the international agenda, and that opportunity must not be lost. The right to food is firmly established in international human rights law and increasingly in domestic legislation and practice in more and more countries. This process will undoubtedly continue. More and more examples are emerging of case law, framework laws and ordinances, and administrative or policy plans. These give grassroots food security activists and human rights defenders new openings to act (see also CIDSE et al. 2013). However, there will probably always have to be compromises between the views and wants of various interest groups. They must all accept that the world and societies are changing. There will have to be gives and takes. What a human rights approach to food security through the right to food can do is to move the balance steadily in the favour of the right to food also for the marginalized and vulnerable population groups.

Notes

1 I am grateful to Asbjørn Eide for his support by reading several versions of this chapter and contributing with criticisms and constructive ideas.
2 Home page: <www.fian.org> (accessed 16 December 2013).
3 Home page: <www.srfood.org> (accessed 16 December 2013).
4 Following the WFS, FAO proposed the addition of 'social', thus 'physical, economic and social access'; both alternatives are used.
5 Those states that have ratified a given international binding convention.
6 There are also regional human rights charters and institutions to manage them (European, African and Inter-American).
7 Professor Jean Ziegler.
8 <www.ohchr.org>.
9 See Professor De Schutter's webpage at <www.srfood.org> or that of his predecessor Professor Jean Ziegler (2000–7) at <www.righttofood.org>.
10 (Reference to UN organizational chart and interactive list of bodies of the UN family.)
11 See more at <http://hrbaportal.org/the-human-rights-based-approach-to-development-cooperation-towards-a-common-understanding-among-un-agencies#sthash.Mauli0Pz.dpuf>.
12 Academic work to give content to the right to adequate food started in the early 1980s and had, together with growing normative work by NGOs, considerable influence on the outcomes of the 1996 World Food Summit and the GC12.
13 Implies that the diet as a whole contains a mix of nutrients for physical and mental growth, development and maintenance, and physical activity that is in compliance with human physiological needs at all stages throughout the life cycle and according to gender and occupation. Measures may therefore need to be taken to maintain, adapt or strengthen dietary diversity and appropriate consumption and feeding patterns, including breast-feeding, while ensuring that changes in availability and access to food supply as a minimum do not negatively affect dietary composition and intake. (Paragraph 9)
14 The phrase '*free from adverse substances*' sets requirements for food safety and for a range of protective measures by both public and private means to prevent contamination of foodstuffs through adulteration and/or through bad environmental hygiene or inappropriate handling at different stages throughout the food chain; care must also be taken to identify and avoid or destroy naturally occurring toxins. (Paragraph 10)
15 Implies the need also to take into account, as far as possible, perceived non-nutrient-based values attached to food and food consumption and informed consumer concerns regarding the nature of accessible food supplies. (Paragraph 11)
16 Full name: *Voluntary Guidelines to support the progressive realization of the right to adequate food in the context of national food security*. In practice they are only referred to as 'Right to Food Guidelines'.
17 Major books since 2005 include Kent (2005); Eide, W.B. and Kracht (2005; 2007); Haugen (2007); Engh (2008); Ziegler et al. (2011); De Schutter and Cordes (2011); Hospes and van der Meulen (2009).
18 Both are accessible at <www.fao.org/righttofood>.
19 For information, see <www.fao.org/righttofood/knowledge-centre/right-to-food-handbooks/en/>.
20 Full text: '1. Recognize and respect all legitimate tenure right holders and their rights. They should take reasonable measures to identify, record and respect legitimate tenure right holders and their rights, whether formally recorded or not; to refrain from infringement of tenure rights of others; and to meet the duties associated with tenure rights.'
21 Full text: '2. Safeguard legitimate tenure rights against threats and infringements. They should protect tenure right holders against the arbitrary loss of their tenure rights, including forced evictions that are inconsistent with their existing obligations under national and international law.'

22 Full text: '3. Promote and facilitate the enjoyment of legitimate tenure rights. They should take active measures to promote and facilitate the full realization of tenure rights or the making of transactions with the rights, such as ensuring that services are accessible to all.'
23 <www.fao.org/nr/tenure/voluntary-Guidelines/en/> (accessed 01 November 2013).
24 ICECSR Article 9: 'The States Parties to the present Covenant recognize the right of everyone to social security, including social insurance.'
25 <www.fao.org/righttofood/>.

References

Allaire, Y. (2013) 'The future of corporate governance', Blog Nov 2013, World Economic Forum, Geneva. Online posting. Available <http://forumblog.org/2013/11/the-future-of-corporate-governance/#disqus_thread> (accessed 16 December 2013).

Bagwathi, J. and Panagaryia, A. (2013) *Why Growth Matters: How Economic Growth in India Reduced Poverty and the Lessons for Other Developing Countries,* United States: Public Affairs/Perseus Books Group.

CFS (2013) *Global Strategic Framework for Food Security and Nutrition,* 2nd version. Committee on World Food Security. Online. Available <www.fao.org/fileadmin/templates/cfs/Docs1213/gsf/GSF_Version_2_EN.pdf> (accessed 17 December 2013).

CIDSE (International Alliance of Catholic Development Agencies), IUF (International Union of Food, Agricultural, Hotel, Restaurant, Catering, Tobacco and Allied Workers' Associations), La Vía Campesina and FIAN International (2013) *Using the Global Strategic Framework for Food Security and Nutrition to Promote and Defend the People's Right to Adequate Food. A Manual for Social Movements and Civil Society Organizations.* Online. Available <http://viacampesina.org/downloads/pdf/en/GSF-Manual_en.pdf> (accessed 16 December 2013).

De Schutter, O. (2013) 'Assessing a decade of right to food progress', Report to the UN General Assembly, 7 August 2013, UN Doc. A/68/288. Online. Available <www.srfood.org/images/stories/pdf/officialreports/20131025_rtf_en.pdf> (accessed 16 December 2013).

De Schutter, O. and Cordes, K.Y. (eds) (2011) *Accounting for Hunger: The Right to Food in the Era of Globalization,* Studies in International Law, Oxford: Hart Publishing.

Drèze, J. and Sen, A. (2013) *An Uncertain Glory: India and its Contradictions,* London: Penguin Books Ltd.; and Princeton: Princeton University Press.

Eide, A. (1989) *The Right to Food as a Human Right,* Final report to the UN Sub-Commission on Prevention of Discrimination and Promotion of the Rights of Minorities, by Asbjørn Eide, Special Rapporteur, UN Doc.E/CN.4/Sub.2/1987/23; subsequently published in 1989 as UN Study in Human Rights No.1, Geneva and New York.

Eide, W.B. and Kracht, U. (2005) *Food and Human Rights in Development, Volume I: Legal and Institutional Dimensions and Selected Issues,* Antwerp and Oxford: Intersentia.

Eide, W.B. and Kracht, U. (2007) *Food and Human Rights in Development, Volume II: Food and Human Rights in Development. Evolving Issues and Emerging Applications,* Antwerp and Oxford: Intersentia.

Engh, I.E. (2008) *Developing Capacity to Realize Socio-Economic Rights: The Right to Food in the Context of HIV/AIDS in South Africa and Uganda,* Antwerp and Oxford: Intersentia.

FAO (1996) *Rome Declaration on World Food Security and World Food Summit Plan of Action.* UN Food and Agriculture Organization. Online. Available <www.fao.org/docrep/003/w3613e/w3613e00.htm> (accessed 16 December 2013).

FAO (2005) *Voluntary Guidelines to Support the Progressive Realization of the Right to Adequate Food in the Context of National Food Security.* UN Food and Agriculture Organization.

Online. Available <www.fao.org/docrep/009/y7937e/y7937e00.htm> (accessed 11 February 2014).

FAO-CFS (2012) *Voluntary Guidelines on the Responsible Governance of Tenure of Land, Fisheries and Forests in the Context of National Food Security.* UN Food and Agriculture Organization. Online. Available <www.fao.org/docrep/016/i2801e/i2801e.pdf> (accessed 01 November 2013).

Government of India (2013) 'The National Food Security Ordinance', *The Gazette of India, 5 July,* Ministry of Law and Justice. Online. Available <www.prsindia.org/uploads/media/Ordinances/Food%20Security%20Ordinance%202013.pdf> (accessed 16 December 2013).

Haugen, H.M. (2007) *The Right to Food and the TRIPS Agreement: With a Particular Emphasis on Developing Countries' Measures for Food Production and Distribution,* Leiden: Martinus Nijhoff Publishers.

HLPE (2012) *Social Protection for Food Security,* Report No. 4 by the High Level Panel of Experts on Food Security and Nutrition of the UN Committee on World Food Security. Online. Available <www.fao.org/fileadmin/user_upload/hlpe/hlpe_documents/HLPE_Reports/HLPE-Report-4-Social_protection_for_food_security-June_2012.pdf> (accessed 17 December 2013).

Hospes, O. and Van der Meulen, B. (eds) (2009) *Fed Up with the Right to Food? The Netherlands' Policies and Practices Regarding the Human Right to Adequate Food,* Wageningen: Wageningen Academic Press.

Kent, G. (2005) *Freedom from Want: The Human Right to Adequate Food,* Washington, DC: Georgetown University Press. Online. Available <http://press.georgetown.edu/book/georgetown/freedom-want>.

Keynes, J.M. (1924) *A Tract on Monetary Reform,* Ch. 3, London: McMillan.

Nair, M., Ariana, P., Ohuma, E.O., Gray, R., De Stavola, B. (2013) *Effect of the Mahatma Gandhi National Rural Employment Guarantee Act (MGNREGA) on Malnutrition of Infants in Rajasthan, India: A Mixed Methods Study, PLoS ONE* 8(9): e75089. doi:10.1371/journal.pone.0075089

Oshaug, A. (2005) 'Developing voluntary guidelines for implementing the right to adequate food: anatomy of an intergovernmental process', in W.B. Eide and U. Kracht (eds), *Food and Human Rights in Development, Volume I: Legal and institutional dimensions and selected issues,* Ch. 12, pp. 259–84, Antwerp and Oxford: Intersentia.

Oshaug, A., Eide, W.B. and Eide, A. (1994) 'Human rights: a normative basis for food and nutrition policies', *Food Policy,* 19: 491–516.

Rae, I., Thomas, J. and Vidar, M. (2007) 'History and implications for FAO of the guidelines on the right to adequate food', in W.B. Eide and U. Kracht (eds), *Food and Human Rights in Development, Volume II: Evolving Issues and Emerging Applications,* Ch. 17, pp. 457–88, Antwerp and Oxford: Intersentia.

Rosales, M. (2013) 'Re. rights-based approaches to food security in protracted crises', FAO Food Security and Nutrition Forum. Online posting. Available <www.fao.org/fsnforum/protracted-crises/re-rights-based-approaches-food-security-protracted-crises-13> (accessed 16 December 2013).

RTFN-Watch (2013) *Alternatives and Resistance to Policies that Generate Hunger,* Right to Food and Nutrition Watch. Online. Available <www.rtfn-watch.org/fileadmin/media/rtfn-watch.org/ENGLISH/pdf/Watch_2013/Watch_2013_PDFs/Watch_2013_eng_WEB_final.pdf> (accessed 10 February 2014).

Sen, A. (1982) *Poverty and Famines: An Essay on Entitlement and Deprivation,* Oxford: Oxford University Press.

United Nations (1945) *Charter of the United Nations,* 24 October 1945, 1 UNTS XVI. Available <www.refworld.org/docid/3ae6b3930.html> (accessed 11 February 2014).

United Nations (1948) *Universal Declaration on Human Rights.* Available <http://daccess-dds-ny.un.org/doc/RESOLUTION/GEN/NR0/043/88/IMG/NR004388.pdf?OpenElement> (accessed 16 December 2013).

United Nations (1966) *International Covenant on Economic, Social and Cultural Rights* (ICESCR). Available <www.ohchr.org/EN/ProfessionalInterest/Pages/CESCR.aspx> (accessed 16 December 2013).

United Nations CESCR (1999) *General Comment No. 12: The Right to Adequate Food (Art. 11 of the Covenant),* 12 May 1999. United Nations Committee on Economic, Social and Cultural Rights (CESCR). Online. Available <www.unhchr.ch/tbs/doc.nsf/0/3d02758c707031d58025677f003b73b9> (accessed 11 February 2014).

United Nations Human Rights Council (2011) *Guiding Principles on Business and Human Rights: Implementing the United Nations 'Protect, Respect and Remedy' Framework,* Report of the Special Representative of the Secretary-General on the issue of human rights and transnational corporations and other business enterprises, John Ruggie, UN Doc. A/HRC/17/31. Online. Available <www.business-humanrights.org/media/documents/ruggie/ruggie-guiding-principles-21-mar-2011.pdf>. (accessed 16 December 2013).

WHO (2013) *Obesity and Overweight,* Fact sheet No. 311, updated March 2013. World Health Organization. Available <www.who.int/mediacentre/factsheets/fs311/en/> (accessed 02 March 2014).

World Nutrition (2013) 'Big food watch', Editor's note, *World Nutrition,* 4(7): 466.

Ziegler, J., Golay, C., Mahon, C. and Way, S.A. (2011) *The Fight for the Right to Food: Lessons Learned,* Basingstoke, UK: Palgrave Macmillan.

5

LARGE SCALE LAND ACQUISITIONS – CHALLENGES, CONFLICTS AND PARTIAL SOLUTIONS IN AN AGRO-INVESTMENT LIFE CYCLE PERSPECTIVE

Michael Brüntrup

The phenomenon of recent Large Scale Land Acquisitions (LSLAs) for agriculture in developing countries possibly more than any other food security related topic has made it to the headlines of the international press in recent years, pioneered by alarming reports of the non-governmental organization (NGO) GRAIN (2008). In academic circles, attention to LSLAs has risen dramatically, too, first – because of a lack of empirical data – more on a conceptual and theoretical basis but more recently also a wave of case studies is flooding scientific conferences and journals, e.g. Futures Agriculture 2011 and 2012 conferences in Sussex[1] and Cornell,[2] several annual World Bank 'Land and Poverty' conferences in Washington,[3] the *Journal of Peasant Studies,*[4] *Canadian Journal of Development Studies,*[5] *Globalizations,*[6] *Development and Change*[7] and others.

In the meantime, the disquieting news is receding. While previous reports went as high as 227 million hectares (OXFAM 2011), the numbers reported now mostly vary between 20 and 60 million hectares (ha). Reporting on land grabbing is a highly contentious issue, with some stakeholders having an interest in exaggerating the numbers, while others seeking to hide them (Scoones et al. 2013; Edelman 2013). These disputes also affect the precise definition of an LSLA (country type, minimum size, origin of investors, new investment or takeover). Arguably the most carefully compiled data collection on LSLAs is the Land Matrix, which looks at local and international deals over 500 hectares in poor and middle income countries but excludes takeovers. After its relaunch in June 2013, it now presents data on about 1,200 deals for an area of about 58 million ha, of which 942 (36 million ha) concluded, 183 (12 million ha) were intended and 75 (7 million ha) failed. The geographical distribution is shown in Figure 5.1. Almost half of the area is found in Sub-Saharan Africa (SSA). It should be noted that even many of the projects in the present lists did and do not get off the ground. In East Africa, for instance the implementation is between 0.3 and 3 per cent of projected area according to the

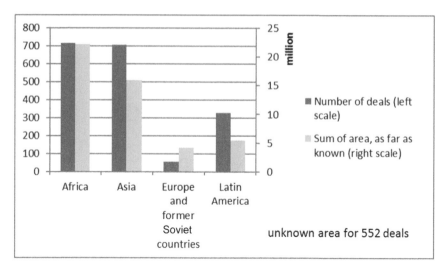

FIGURE 5.1 Distribution of deals and area of LSLAs by region

Source: author's calculation based on Land Matrix (2014) for 1,253 deals with a known area.

Land Matrix data. In Ghana in 2012 none of the many Jatropha projects (making up more than 50 per cent of all LSLAs in the Land Matrix) were still operational though they had not yet formally failed (Brüntrup et al. 2013a), and also elsewhere in SSA most large scale Jatropha projects – a major driver of LSLA in SSA – have been abandoned (Charles 2012).

Yet, there are still good reasons to consider LSLAs in a forward-looking book on food security: first, there are still enough ongoing and emerging cases to call it an important phenomenon; second, even where LSLAs have been abandoned, they are still affecting the rural communities concerned; and third, and possibly most important, the very underlying reasons for the emergence of LSLAs have not yet disappeared and lead to a continuation or interest of various stakeholders in establishing such projects, in the future possibly with better understanding and concepts than in the past. This text attempts to contribute to such better concepts, not in the sense that projects are successfully implemented in the interest of investors but that they create better situations for the affected rural populations at large.

In the following section, it is first briefly recalled what reasons have pushed and continue to push for LSLAs. Next, potential and observed positive and particularly negative impacts are reviewed – the latter being the issues which have to be tackled to make LSLAs acceptable or, if tackling is not possible, should lead to their rejection. The particular perspective of this text is to review not only the land acquisition but to have a more holistic perspective and look at the (stylized) life cycle of an LSLA as part of a longer term agro-investment project. This may not necessarily be the case – land acquisition can also happen (and certainly does) as speculative investment to profit from rising land prices only. In a macro-economic perspective, this may not even be considered as investment since productive capacity does not rise. However, more typically the acquisition is part of a project cycle which starts from planning and is followed by negotiations with stakeholders to the land and the investment, project implementation, and later adaptations as experiences are gained and/or internal or external factors change. These later phases of large scale agro-investments (LSAIs) and land exploitation, of accompanying investments in local processing and second round effects in rural areas, are still rarely considered in the literature.

One reason for this gap is that the recent wave is too young to have produced 'mature' investments. However, it is possible to gain a more complete picture of the long-term consequences of LSAIs, by looking at the fate of earlier investments and some hypothetical considerations. This is what the life cycle perspective in this text aims at. It must be acknowledged that some new factors and circumstances make new agro-investments different from old ones: greater land pressure nowadays increases costs for resettlement; less authoritarian regimes means higher human rights and legal standards, better connection of local rural populations with the rest of the world and stronger civil society organizations with improved negotiation skills and power *vis-à-vis* investors; better integrated markets mean more competition, and more demand for decent jobs in rural areas especially by the younger generation making new investments. Yet, many lessons may be drawn from older investments. Based on all observations and considerations, some ways forward as to how to tackle the phenomenon of LSLAs at the global, national and local level are provided in the conclusions.

Factors explaining the existence of LSLAs and their likely future endurance

LSLAs are not a new phenomenon in developing countries. They have been a motivation of Europeans colonialists and capitalists for centuries, land being one of the main productive assets in pre- and early industrialized times. In Latin America, the Caribbean, Southeast Asia and SSA a wave of LSLAs took place with colonization, in particular where colonialists settled, where a favourable combination of abundant land and (frequently forced) labour was available, and where products could be cultivated with large demand in the motherlands. Sugar, rubber, tea, coffee, bananas, palm oil, sisal, spices and timber are some of the typical plantation products of that period (Dinham and Hines 1984). Land and other formal laws were often shaped in a way by colonial powers to legalize these acquisitions (Wily 2011) – a reminder that rule of law may not be a good guide in such constellations of fundamentally unequal power.

In Latin America, Central and Southeast Asia these plantations showed mixed fate at best in terms of efficiency, but often survived because of political economy factors – they were an important asset of the old and new elites. In some countries, land reforms distributed a smaller (Latin America) or larger (East Asia) part of these lands (Dorner and Thiesenhusen 1990; Dorner 1992). In some countries with large frontier forest areas and vibrant capitalist regimes, plantations were even expanded (e.g. Thailand, Malaysia, Indonesia, Argentina, Brazil).

In SSA, in contrast, plantations had widely disappeared, and only a few new ones were established with the old plantation and some new crops, such as cashew, ornamental plants or fruits. Expropriation, coupled first to decolonization, then mismanagement on state farms and finally dismantlement of estates during structural adjustment programmes (SAPs), was widely observed (Brandt and Brüntrup 2012). In many poor and newly independent countries, the investment climate for large foreign investors was weak. A few exceptions are Ivory Coast and some highly remunerative land uses such as for specialty crops in Kenya or Senegal (Dinham and Hines 1984). In addition, increasing know-how of local commercial farmers (often trained on plantations) provided an alternative to estate production to procure raw material for international markets. In several cases it also emerged that with appropriate knowledge dissemination and under fair competition rules, smallholders were more competitive than plantations anyway (Deininger and Byerlee 2012). More generally, agriculture in SSA was exploited to facilitate industrialization; export cash crops were taxed to generate revenues; and food prices were kept low to please urban consumers and keep wages low (Heidhues and Obare 2011; Anderson and Brückner 2012; Brandt and Brüntrup 2012). Thus, several political and economic factors in SSA drove farms in general and large farms in particular down. Successful surviving LSAIs such as for sugar cane, tea, coffee, palm oil or vegetables, flowers and fruit are typically high value, export oriented and labour intensive but at the same time capital intensive and technically difficult to manage – thus market niches are not easy for smallholders to access.

So, what has changed in recent years to explain the recent boom of LSLAs? Three recent waves of LSLAs indicate some of the drivers pushing for such investments, many of which are of a structural nature.

The first wave of investors came in the mid-2000s with the biofuel boom of the 2000s (FAO 2008). About 50 per cent of worldwide notifications of LSLAs and almost 70 per cent in SSA were once attributed to biofuels (Anseeuw et al. 2012). Minor biofuel sectors had emerged early in the 20th century, and the two oil price crises in the 1970s had temporarily stimulated these further, but only Brazil (and to a lesser extent Malawi) had continued these programs and were using substantial amounts of biofuels for transport. In the mid-2000s, the European Union (EU) and the USA massively increased public support for biofuels, in particular through compulsory blending mandates for transport fuels and tax incentives. Motivation included climate change mitigation through supposedly low-carbon biofuels, reduction of oversupply, incentives for agriculture and rural areas through new value chains, and increase of agricultural prices through the creation of new demand for agricultural products (FAO 2008; IEA 2009). Many developing countries followed these examples and declared their own biofuel strategies or policies (Clancy 2013). This general boom, and particularly the EU policy created expectations for a large and effective demand for biofuel feedstocks. Since biofuel value chains have important economies of scale in processing, large amounts of feedstock are needed and must be produced regularly and with reliable and certified product quality to satisfy the information and product requirements of policy makers, fuel producers, car industries and owners. Because of logistical and some crop specific reasons, under many circumstances large scale farming is the economically preferred option to produce biofuel feedstocks. Since most African, Caribbean and Pacific (ACP) countries have quota and tariff-free access to the EU,[8] many investors chose land-abundant African countries to produce biofuels (Matondi et al. 2011). In particular, *Jatropha curcas,* a bushy plant with fruits of high oil content well suited for biodiesel production, attracted investments (in addition, many smallholder projects were initiated to supply the industry for local, national and international consumption). Also sugar cane and palm oil gained prominent places in the plans of investors. Investors mostly came from the energy sector, with easy access to capital, processing technology and the oil sector, but no experience in farming.

The second wave of LSLAs came with the food price crisis in 2007–8, when almost all agricultural products experienced rapid and substantial price increases. The subsequent analyses of that boom convinced politicians, researchers and investors that the period of low and decreasing food and agricultural prices was over and a new area of scarcity and high prices had begun (OECD and FAO 2008; Helbling and Roache 2011; Schaffnit-Chatterjee 2012). Population growth, income increases and related demand increase for animal products in developing countries, land loss and soil degradation, limited water for irrigation, limited reserves of oil and fertilizers and decreasing productivity gains in agriculture aggravated by climate change since then have (re)gained wide attention. The Malthusian pessimism that population growth will outpace production increase is (again) the major paradigm, this time possibly more soundly based on potential global natural resource capacity analyses (Rockström et al. 2009; Running 2012). There is evidence that many LSLAs are made with a perspective to solve problems in home countries of the investors, not to address local needs (Anseeuw et al. 2012).

A third wave of LSLAs is emerging – the 'bio-economy' in which bio-based materials are increasingly substituting fossil oil-based (Kircher 2014) for reasons of scarcity, price and climate effects of fossil sources. In addition, biological processes can be used to actively extract carbon dioxide from the atmosphere. Through supporting forestry in so-called Reducing Emissions from Deforestation and Degradation Plus (REDD+) projects, this is already taking place (Larson et al. 2013) and contributes to LSLAs worldwide (Deininger and Byerlee 2011; Anseeuw et al. 2012). In future, soil sequestration could be another sink for carbon linked to land use, but more ambitious ideas involve the transformation of biological materials into industrial use, capture and long-term storage (Lovett 2008). This could further increase large scale demand for agricultural products and, thus, LSLAs.

According to the new Land Matrix data (2014), 11 per cent of LSLAs are for food crops, 22 per cent for flex crops and 34 per cent for multiple purposes.

As a consequence of present and future prospects for natural resource-based markets, investors worldwide are adding agriculture to their portfolios, with some nuances in motivation: countries with import deficits and fear of food scarcity such as most Middle East countries, some of which experienced shortage of supply during the food crisis, try to gain more supply reliance through government-led investment funds; investors in many food-import dependent countries, for instance South Korea or China, are encouraged by their governments to invest in food production abroad; investors with purely financial interests see high returns on production, may speculate with increasing land prices, and can consider land and agricultural production as a new asset class with a low correlation to other financial market products. It is not easy to get consistent information on investor types, origin and motivations. Depending on definition of LSLAs and regional orientation, one can find sources which see investors from western countries (Anseeuw et al. (2012) for SSA), eastern countries (Visser and Spoor (2011) for former soviet union countries), countries of the region (Borras Jr. et al. (2012) for Latin America) or of local origin (Deininger and Byerlee (2011) for selected SSA countries) as leading. One can further detect from case studies that there are many overlaps and interdependencies between investors, for instance farms being transferred from one investor to another, local investors seeking to create joint ventures with locals, locals seeking land for sale to larger investors to gain windfall profits (Hilhorst et al. 2011). 'The involvement of national elites in the rush for land is perhaps the single most important reason for the pervasive lack of transparency surrounding the deals' (Cotula 2013: 54). What is sure is that the very large international investments are the focus of international discussion, sometimes even taken as synonymous with LSLA or land grabbing.

The demand-driven waves of LSLAs are supported by a constant stream of innovations and changes in the nature of farming, food and bio-based commercialization: information technologies, biotechnology and technical progress in mechanization can improve productivity (satellite-based precision farming) and marketing (market information) and may reduce the need for labour-intensive activities. Stricter food and agricultural product standards and regulations all over the world enforce traceability and the vertical integration of agricultural value chains. Similarly, supermarketization of food sales encourages large quantities of high quality food at regular intervals with reliable delivery. Many of these market developments enforce or encourage technology, information and capital-intensive innovations and cause high transaction costs. Thus, although it has been said that smallholders are astonishingly competitive under many circumstances and for many crops, these trends facilitate the possibilities, the economics and the optimism about large scale farming *vis-à-vis* smallholders (Wiggins et al. 2010, Deininger and Byerlee 2012; Brüntrup 2012; Reardon et al. 2012; Collier and Dercon 2013).

Potential and actual impacts in a life cycle perspective

Much has been written about the – mainly negative – consequences of recent LSLAs. However, the authors have only been able to observe the very first steps of agro-industrial investments, particularly the land acquisition phase. But arguably none of the investments since the food price crisis has already reached its full production capacity since investments under typical conditions of LSLAs (clearing, levelling, variety testing, nursery establishment, irrigation) can take several years if not decades. Also most investments where first processing steps are intended to be carried out on site, for instance necessary for perishable products like sugarcane, palm oil or many fruits, have not yet materialized. But without this perspective a complete assessment of impacts on rural economies and communities is not possible. This lack of full picture is illustrated by the many reports which complain about the lack of promised development effects, gratefully ignoring that these only can materialize (if at all) with full production.

Thus, it is argued that only a full life cycle analysis can provide a complete picture of the impacts of LSLAs on rural development. This means that the perspective on LSLAs has to be completed (except in cases where land acquisition is merely speculative, with no further productive investment made) with one on LSAIs, which may or may not comprise processing devices. While full production may be the final goal of most LSLAs, the fate of many recent LSLAs has demonstrated that another important step is to be considered when assessing these investments – their failure and abandon. While they finally may disappear from the statistics, anecdotal evidence suggests that their agony is severely impacting rural communities, and even after their final end they have repercussions.

Thus, adding a life cycle perspective to the analysis of impacts, one can distinguish at least four clear phases: screening and negotiation, investment and establishment, exploitation, and failure and abandon. In each, typical issues arise that have repercussions on the investments and on food security.

Screening and negotiation phase

Screening and negotiation may be regarded as separate phases by some authors (e.g. BEFS-WG and FAO 2013). In fact, concluded land deals are much fewer than those announced in the international databases which are usually based on information in the media (see above; the ratio of announced to concluded deals is probably 3:1). One of the reasons for this striking difference is that the attention of the international databases is oriented towards those deals which accumulate land from existing smaller users and units. Here, many uncertainties exist for investors and land sellers (national or local authorities) which can stop a deal from being realized. This situation is fundamentally different from the acquisition of existing large farms, frequent in former Soviet countries or in the already cultivated regions of Latin America where large existing farms change owner, and uncertainty concerning the land acquisition is much higher (Visser and Spoor 2011; Borras Jr. et al. 2012). Another reason for the high rate of failure before the deal is actually concluded is that these pioneer regions attract more pioneering, no-name and adventurous investors who have not much to lose, while more formal, settled investors who have a reputation to lose avoid them. The strong international campaign against land grabbing may further lead to the withdrawal of the second type of investors in favour of the first one.

Typical countries and regions screened for LSLAs have low population density and high land availability, yet fertile soils (a higher yield gap does not attract LSLAs), good water availability and acceptable market access (Cotula

2009; Deininger and Byerlee 2011; Anseeuw et al. 2012; Lay and Nolte 2013; Osabuohien in progress). Governance and land governance are, unlike for normal investors, no repellent for land investors. A particularly problematic area is the notion of unproductive or idle land which is often used not only by governments and investors but also communities to describe land considered ready for land deals. However, serious land use research all over the world shows that non-use land hardly exists. Farmers in many tropical, traditional land use systems use fallow to restore soil fertility (swidden agriculture, Ruthenberg 1980). These fallow rotations can take many years, thus implying a large ratio of secondary vegetation and crop land. This land is not idle but restoring fertility, and in addition used for grazing animals as well as tree and wild products. Pastoralists also use open lands. What is certainly right is that these lands are underused, and most often in more than one respect: yields in these farming systems are lower than input-intensive cropping systems, with yield gaps typically being 80 per cent (World Bank 2007; Deininger and Byerlee 2011) on the fields cropped, and rotation is further reducing the landscape productivity.

The negotiation phase determines which conditions prevail for the deals which, after all, are private contracts, even if concluded with governments (Cotula 2011; Cuffaro and Hallam 2011; German et al. 2011). Typically, in many pioneering countries central governments have formal ownership of most or all land, but *de facto* local populations – or local elites – often have an important say, too. Many central governments conduct active road shows and investment policies to attract investors. In a few cases (e.g. Cambodia, Neef et al. 2013), central governments concede land directly to investors. In many more, in practice there are several steps involved where both central and local governments and traditional authorities are negotiating and have to sign land transactions (Anseeuw et al 2012; Cotula 2011; Nolte 2013). In many countries, Social and Environmental Impact Assessments (SEIAs) are requested, to be paid for and commissioned by the investor. In a few countries, traditional authorities have formal exclusive rights to negotiate communal lands on behalf of their constituencies (e.g. Ghana).

The typical outcomes in poor countries, as far as is known (most contracts are not known to the public), are leasing contracts of several decades up to 99 years (sales are less frequent) with very low leasing fees (Cotula 2011; Deininger and Byerlee 2011). Contract farming arrangements may be important parts of some investment projects, for individual households are often more profitable than the jobs created and thus an important bargaining object in LSLA projects (Oya 2012; Cotula 2013; Herrmann et al. 2013). Often, Corporate Social Responsibility (CSR) plans are incorporated or attached, for example to construct schools or wells (Brüntrup 2012; Cotula 2011; Cotula 2013), or that require cultivation of the land in a given time. However, all these agreements are of a private nature, and thus without minimum standards, except those prevailing in the country anyhow. Global and specific crop or label governance frameworks may influence the outcomes, but are fragmented and not compulsory (Brüntrup et al. 2013b). Since many of the critical countries were not prepared for the land rush, it is no wonder that most of the contracts which have come to be known to the public are considered weak for the communities (Cotula 2011).

A frequent phenomenon in this phase that is reported in many case studies is the lack of sound information for locally affected people (e.g. Cotula 2009; Vermeulen and Cotula 2010; De Schutter 2009; Nolte 2013). Many reasons are cited for this situation, including lack of good will from the investor's or seller's side, lack of communication traditions and channels, lack of preparedness for such deals in areas which have never attracted investors before, unclear guidelines about legal or socially accepted procedures to follow, clash of cultural norms, illiteracy, corruption, unclear ownership and dispute about who is entitled to be informed or decide on the deals, etc. While for indigenous people an international norm exists which stipulates free, prior and informed consent (De Schutter 2009), under many circumstances the norm lacks clear validity, for instance because people are not officially accepted as indigenous (in much of SSA) or land is not considered to be under indigenous land regime. Often, it is the local traditional leaders who make the deals (on behalf of their people) but do not consider wider consultation (see below). What is considered adequate information, for instance on ecological risks and profitability calculations, may also be disputed or considered confidential (Rosenblum and Maples 2009; Silici and Locke 2013). Finally, a large investment under pioneer conditions automatically has a large intrinsic portion of uncertainty.

Another general observation of the negotiation phase of LSLAs is the problem of poor (land) governance, in particular if coupled with corruption. Deininger and Byerlee (2011) found that LSLAs, contrary to other investments in developing countries, did not bother about governance issues in general and to the contrary seemed to search for low land governance quality in particular. More recent analysis with new data does not seem to support this finding (Lay and Nolte 2013). However, it is clear that in many of the poorer countries and also in the forest frontier regions of wealthier countries (where many of the LSLAs take place and the presence of the state is much weaker than in urban areas) there are huge problems with land governance in particular and governance in general. In rural areas and for land in particular, the institutional setting is characterized by a low presence of central government institutions and organizations, a prevalence of traditional rules, and overlapping governance. Under such circumstances, arguably the politically and economically strongest stakeholders will impose their interests more easily than in an environment of clear rules and strict enforcement. In a comparative literature study of 18 LSLAs, in seven cases coercion was exerted to sign the contract, and in many cases irregularities were reported on the process (Richards 2013). Protests or clashes were reported in nine cases. But only in three cases was the legality doubtful.

The impact of the loss of land is rarely well documented. Richards (2013) finds that five of 18 LSLAs use forests, and more than half productive agricultural land, but which share is actually under cultivation is not known. Compensation is paid in half of the cases and qualified as low, not covering all affected persons and with long delays, which is also reflected in other publications (Cotula 2013; Väth 2012).

These are the strongest evidences of impact on local food security, but quantification is not available.

A further yet under-researched topic is that water issues are not or not sufficiently dealt with in negotiations on LSLAs (Mehta et al. 2012). There are abundant non-used water resources particularly in SSA, and there is a need for increased and improved use worldwide from various points of view – stabilizing and increasing production and productivity, and mitigating the effects of climate change (FAO and NEPAD 2002; De Fraiture and Wichelns 2010). Governments and smallholder farmers barely have the financial, technical, managerial and administrative capacities to exploit these options on a larger scale – thus, private investors in principle are welcomed. However, water abstraction is a delicate issue for several reasons: water governance and water organizations are very weak in many rural areas; the implication of water governance for LSLA deals is often not mandatory; LSLA impacts on water are even more difficult to assess than on land use; and very often LSLAs involve new stakeholders in negotiations, which is not in the immediate interest of investors and pushy governments and elites. Often, bureaucratic competition exists between land and water administrations, or the relation is not (yet) clarified. Water problems are frequently already existing and slowly accelerating, but it is only with large investments that they become apparent, since investors have the means to step up unsustainable, unfair or harmful practices.

A very frequent finding is that weak groups of local populations are particularly neglected. These groups are squatters and migrants, women, pastoralists, and fishermen and water users downstream.

- Squatters and migrants often are considered as non-permanent and do not have the same quality of rights as the autochthonous population. Under (semi-) formal constitutional rights (the frequent state principle 'the land belongs to who cultivates it') this may be less problematic, but these rules usually are not much respected in rural areas, and they disrespect the social realities of a strong ethnic and clan/family affectation of land (Wily 2011; Anseeuw et al. 2012; Cotula 2013; Nolte 2013). In addition, the user principle does not hold if formal state land is occupied – a frequent constellation in LSLA disputes. According to traditional rules, migrants in land-abundant regions are often welcomed as workers for cash crop cultivation (which does include food crops for markets) and given some land for their own cultivation as part of remuneration. However, these rules tend to be inconsistent or imprecise when it comes to long-term residential rights and permanent ownership; they have emerged under land surplus conditions and are effective as long as this condition holds but are not adapted sufficiently to land-constrained conditions (Platteau 2008). LSLAs accelerate the condition of or feelings about land scarcity.

- Women do not have strong (if any) land rights and weak representations in many traditional land governance systems, often underlined by gender-biased social and religious principles such as patrilocal marriage and patrilinear inheritance (Chu 2011). However, they are often the main users of land-related resources such as forests, wild products and pasture for small animals, and they have the main responsibility for fetching drinking water. Even if formal (national) rules stipulate equal rights, they may be disadvantaged in reality. Thus, they may have little voice in negotiations about land concessions, and will suffer from squeezing land resources by being pushed onto even more marginal lands (Behrman et al. 2012; Anseeuw et al. 2012).

- Pastoralists are another group frequently overseen and disadvantaged. They use pastures including harvested plots on a temporal basis; they often do not have formal rights in land and have traditionally not been interested in it. Yet, the continuous squeezing of their pastures and migrant corridors is sometimes dramatically enhanced by LSLAs because of their size and often compact nature (Koopman 2012; Suleiman 2013; Anseeuw et al. 2012).
- In parallel with the negligence of water issues, fishermen are a typically overlooked stakeholder group in LSLAs, particularly if involving irrigation and if they are outside the project area. Again, their informal and formal rights and political representations are weak (Mehta et al. 2012; Brüntrup et al. 2013b).

The negotiation phase finally results in land leasing or sale contracts. These can lay the grounds for later food security risks of LSLAs at the local level. Serious biases in representation of local populations (particularly some weaker groups), weak negotiation skills of local representatives, and strong information and power asymmetries insinuate that these contracts are indeed often not favourable for many locals, but lack of transparency means that not much is known about the actual content of most contracts.

Investment and establishment phase

It is only during the investment and establishment phase that the real consequences of land loss become increasingly visible to local populations. Fundamentally, LSLAs constitute a risk for food security of local populations as long as they are overwhelmingly smallholders, live from subsistence farming for a larger portion of their food consumption, earn a large proportion of their cash income from surplus of cash crop production, gain a living as cheap and unskilled day workers on local farms, use common grazing areas for livestock, or generate additional income from the collection of forest and fallow products (see above).

However, in many LSLA cases it takes years between contract signature and first changes of land use because often investors only start to seriously assemble project funding after they have signed the contract. Deininger and Byerlee (2011) found that only 21 per cent of projects had started implementation or production, and the Land Matrix 32 per cent (Richards 2013). Banks and financial investors wait for land acquisition before they make funding available. Many more years can pass before the full exploitation of a plantation of several thousand or even tens of thousands of hectares; for instance a 220,000 ha palm oil and rubber plantation in Liberia was scheduled to invest 3.1 billion USD over 15 years (The Munden

Project 2012). With previous experimentation and nursery phases, this time can easily double. This is of course hampering the fast implementation of the LSAI, and may be a threat to the entire project, for instance if contract clauses stipulate a (short) period for implementation or for financial reasons. In a sample of 39 mature agro-investments, only 50 per cent cropped more than half of their total area (World Bank 2014).

Very often there is not sufficient experience with the type of farming (crop, land clearing, land use, weather and water regime) and other operations (transport, product qualities under local conditions, energy and process water access, quality and stability) (Tyler and Dixie 2013). Qualified labour and services are not sufficiently available (Brüntrup et al. 2013a). Hiring non-local or even expatriate staff is costly and creates social tensions. Even if knowledge about failed LSAI exists, it is rarely available publicly and easily but scattered with individual managers and failed investors and closed off in bureaucracies.

One particularity of this phase is that the works and services required during this period are very different from the later exploitation phase. Land may be cleared to a larger extent and infrastructure built, but until a certain minimum production capacity is established it does not make sense to create larger processing plants and outsource production (if these are part of the investment plan), which are often the most valuable parts of the project for local development since these create high farm incomes, higher quality jobs (compared to manual farming), service industry demand and second round rural non-farm development loops. This means that for a considerable time, the job expectations of the communities and affected people and the promises of investors (which are typically referring to full production) are unachievable. Thus, this phase is probably the most critical for local development and food security even in well-designed investments since the problematic aspects (loss of land and other natural resources) are increasingly experienced while the productive phase – providing jobs, outgrower contracts (if foreseen) and cash flow for CSR and tax payments – is not yet (fully) reached.

Another consequence of long gestation periods until full production is that investor funds are stressed. Delays in this period, which are very common and almost unavoidable for green field investments in rural areas of poor countries with dozens of unpredictable obstacles, can easily exhaust even well-funded investors' coffers. Funding over such a long period also faces risks from volatile capital markets, since it is far from certain that no financial or economic depression in home countries will happen during such a long time. In fact, many of the past LSAI projects in the boom years 2005–8 have failed because finances dried up in the economic and financial crisis starting in 2009 in many industrialized countries, just after the bio-fuel and food price crisis hit LSLA waves (The Munden Project 2012; Cotula 2013).

Exploitation phase

If an LSLA project finally enters into its full productive phase and starts to create regular production streams, revenues and jobs, this does not mean that problems are gone. There are many often underestimated risks in agricultural production, and capitalist style farming especially (i.e. not based on family labour, survival-oriented and flexible decision making and mainly household resources) has been shown in the past to be ill-equipped to deal with many of the typical challenges in agriculture. Agriculture itself is highly dependent on external factors such as weather, pests and diseases. Market risks are also pervasive in agriculture, induced by natural variability, changes in demand, stocks, input and output markets, speculation and, not least, by policies and politics (Wright 2011; Anderson and Nelgen 2012).

Operating in a social environment marked by strong traditions, informal rules, beliefs and values challenges managers coming from other institutional environments. But local managers may be too weak to impose decisions against strong local interests, for instance in hiring and firing that involves local elites and family ties. In addition, in a poor country large projects almost automatically attract the attention of local and even national politics. Good management thus needs many strengths in agriculture, industry, logistics, social and political issues and more – definitely not easy to find or to combine in countries with little or no tradition of rural private agricultural and industrial enterprises (Brüntrup et al. 2013a; Tyler and Dixie 2013).

In a review of 179 agricultural investment projects of the Commonwealth Fund for Commodities (CDC), '[forty-nine] per cent of the projects were classified as failures or moderate failures in financial terms. In 60 per cent of these cases, the major cause of failure was that the project concept was fatally flawed, for example wrong location, wrong crop, or overoptimistic planning assumptions. About one third of these were unknowable at the time of appraisal. One in five had the "bad luck" to be adversely affected by government policies (10 per cent), or closed down due to civil unrest (8 per cent), or suffered from a collapse in markets (2 per cent). About 20 per cent failed due to bad management.' (Tyler and Dixie 2013: 3). In another survey, 'around 45 per cent of investors were materially behind schedule or operating below capacity. About the same share were unprofitable at the time of survey' (World Bank 2014: 4). Whether the many private investment funds coming up (Silici and Locke 2013) will have similar logic and impacts is, however, not evident. Borras Jr. et al. (2012) conclude for Latin America, where older LSLAs are frequent, that 'in most cases, land deals in the region have not resulted in any immediate large-scale negative impact on food security of the host country (although we can surmise that exceptions probably include sub-national local cases where there were clear displacements of communities resulting in the disruption of food production, supply and access, as in the case of Colombia)'. For Africa and Asia, case studies frequently report food security problems at the local level for displaced families (e.g. Richards (2013) reviewing 18 individual cases, or FAO (2013) compiling nine country case studies, though often the real versus potential impacts and the sampling procedures to count affected people remain unclear), but on the national level such findings are less evident and less likely in view of the low level of implementation (see above). For a few countries, non-negligible increases in food and export crop production are reported for older investments (Uganda rice for local market, Ghana palm oil for local market and fruits and vegetables for export). Food security is mainly affected by job creation, though often local people profit less than immigrants.

Two other major problems for local communities and for food security are worth special mention. One is monitoring: in large agro-industrial projects, the fulfilment of contracts and the gathering of data which may support the need to change business and contracts are difficult for individual farmers, communities and even national governments to oversee. Elements necessary to understand the situation of the enterprise and its local partners may involve productivity, costs, prices, benefits, corporate social responsibility and public good provision, environmental effects, and conflicts and their resolution. Monitoring these issues is important to guarantee trust, keep rumours and false accusations low, and renegotiate prices and longer term changes to initial contracts (Cotula 2013). At the same time, monitoring must be organized without over-burdening the enterprise.

The other issue is dependency: large agro-enterprises may become the dominant player in a given rural area. The fact that there is a key limited resource (land) which excludes other users makes this risk much higher than in most urban investments. Once the investment established, large investors tend to have a monopoly/monopsony role and good potential to extort local partners in negotiations and conflicts (Cotula 2013).

However, there is also evidence that many LSAIs finally produce benefits for rural communities, including food security, though thorough quantitative studies are rare. For sugar, several studies show positive economic effects on local populations (both plantation workers and particularly outgrowers, if these are part of the scheme) compared to the miserable conditions of comparable populations (Kennedy 1989; OXFAM 2004; Herrmann et al. 2013). Food security is supported through improved access to food via higher and more stable cash incomes, better infrastructure and markets rather than through food production increase – however, production does not fall for outgrowers. Estates often provide housing, water supply, electricity, schools, hospitals and social security for their workers. Some of these services are also offered to wider local communities. Another, although less pertinent, candidate for proven development-friendly LSAIs is the export-oriented cut flower and vegetables industry. In Kenya, this sub-sector has created several hundred thousand jobs, both on-farm and off-farm, many of them for women (McCulloch and Ota 2002; Minot and Ngigi 2003). In Senegal, it has been shown that large scale vegetable farms are more poverty-reducing than small to medium sized farms (Maertens and Swinnen 2009). Palm oil and rubber plantations have also strongly contributed to positive socio-economic development, in particular in East Asia (Wakker et al. 2004; Hayami 2010; Deininger and Byerlee 2012). Private large scale plantations created local industry and jobs, while state plantations, in addition, were used to train workers who later established their own farms.

Interestingly, Tyler and Dixie (2013) find that agro-industrial investment projects of CDC produced more development successes than financial successes, often because first unsuccessful projects were refinanced, credits written off and/or converted into equity and previous investments considered 'sunk costs'. This means that public funds for these projects were indeed subsidizing development more than profits.

The overall effects of operational LSLAs and investment projects are, thus, composed of a large variety of effects on different groups of people in different periods. As Väth (2012: 19) summarizes: 'The effects of a land-based, large-scale investment on people who received compensation, on neighbouring communities, on permanent and casual workers, as well as on contract farmers . . . the main finding [is] that outcomes are predominantly mixed and vary from very negative to positive for different population groups.' She further details the following five categories of impacted populations for her in-depth analysis of older and more recent palm oil plantations in Ghana:

> (i) While neighbouring villages realise mixed outcomes linked to land loss on the one hand and infrastructural improvements as well as employment creation on the other hand (see Okumaning village), (ii) communities which are further away are negatively affected as spillover effects cannot be accessed due to geographic distance to the core of the investment area (see Aboabo village). Nevertheless, (iii) people who had to relocate or who are just in the course of resettlement turned out to be the worst off as the institutional environment is too weak to guarantee legal entitlements to "fair, prompt, and adequate" compensation (see Congo village). Moreover, (iv) a detailed assessment of the discussion with workers revealed that the positive outcomes linked to employment creation also disclose their shady side when it comes to the quality of jobs (see the workers). Lastly, (v) contract farmers are the greatest beneficiaries because they profit from long-term economic integration (see the outgrowers), but at the same time they are still suffering from land loss (see smallholders), which was highlighted by all sub-groups of the local population.
>
> *(Vath 2012)*

While the mix of actors (plus other, more marginal ones, see above) is more or less the same everywhere, the specific impacts on each of these groups may vary. One area which is hardly researched is the long-term effect of agro-industry on structural transformation in rural areas. From industrial countries, it is known that agro-processing industries are often the core of a broader non-farm industry of rural hubs and urban centers which a mere smallholder-based economy cannot create. In developing countries, spurious evidence tells the same story (e.g. Losch et al. 2012; Tersoo 2014), but there are very few studies which trace rural transformation over the longer term and determine the importance of certain forms of agro-investments. If the introductory analysis of the drivers of the present land rush is correct, then LSLAs will have an important role to play in future rural transformations since smallholders will not be able to satisfy some of the requirements of growing demand. But this may only be true if employment effects are large, if large farms are linked to a local processing industry and through further local linkages, e.g. service industries. New industrial policy may have important lessons to teach (Altenburg 2011; Yumkella et al. 2011).

Failure and abandon phase

An area which has not received much attention yet, but which will certainly attract more in the future, is the problem of failed projects. In the past, most failures concerned the dismantlement of colonial and state-controlled agro-enterprises, in particular during SAPs (Tyler and Dixie 2013). These failures provoked some harm, for instance in replacing services and input delivery, but also created new opportunities. Many new LSLAs fail even before major investments take place (see above) for various reasons including wrong assumptions, inexperienced investors, missing or failed production technology, or simply planned failure when investors are only interested in access to timber, not production (GRAIN 2013). Also well-established LSAIs could and will fail, as past experiences teach (see above). If they are not re-funded (see the CDC experience in Tyler and Dixie 2013), these failures will harm local economies and food security for a number of reasons:

- The land occupied by the project may not fall back to the previous sellers or landlords. In several African countries existing laws stipulate that the land, transferred from the village to an investor or to an intermediary public body who rents or sells it on to the investor, returns to central government once the project ends (e.g. Tanzania, Zambia – compare Anseeuw et al. 2012 and Nolte 2013).
- The services linked to the agro-enterprise (specialized research, extension, inputs, local infrastructure, ice, marketing, etc.) are abandoned and may generate big gaps in the service supply in a given region. Moreover, because the services are concentrated in one hand or linked to (and dependent on) the large scale enterprise (e.g. supply contracts as guarantees for credits), more affected individuals and the entire region are vulnerable.
- Even if land is redistributed, it will take a long time to reorganize productive resources and agricultural production. Skills or mixes of skills necessary to create new productive units will no longer be available. The longer the investment has existed, the more the old structures including infrastructure will have perished, and the more the reorganization will be difficult. The end of the collective large scale agricultural structures at the end of the Cold War in former Soviet countries has shown that after a generation of dependent farming, attitudes, skills, knowledge and structures to support individual farming have almost vanished. Also post-SAP experiences show that the private sector may take a very long time to fill gaps created by failing monopolists (see above).
- The deconstruction or removal of old structures or cleaning of dumping sites may be very costly. Since the failure of an agricultural project often has a longer gestation time during which problematic issues are neglected and accumulated

(such as investments in soil fertility which may keep production sustainable in the long run but can be neglected in the short run), it is likely that such issues abound.

Conclusions

There has been a worldwide boom of LSLAs in the last 5–10 years. Though its extent was probably speculative, there are several fundamental factors which insinuate that the phenomenon of LSLA and agro-industries will continue to play a role in developing countries: higher demand for agricultural products will continue (driven by population and economic growth in developing countries, urbanization, new consumer demand, higher oil and energy prices and the emerging bio-economy), and some trends favour larger units (standards and traceability, advances in production and communication technology, capital and technology requirements stemming from changing demands in processing and consumption, and value chain integration) because of the weakness of smallholders to deal with some of these trends.

How developing countries should deal with this pressure is one of the most hotly debated issues in contemporary agricultural development circles. This text has identified many problems of LSLA deals, and the literature is full of examples that these problems are widespread. Particularly marginal and less powerful rural groups (migrants, women, pastoralists, water users) risk suffering and being disadvantaged. Thus, countries in SSA could radically oppose such deals. But the low performance of its smallholders and the many difficulties of improving smallholder agriculture (e.g. Ellis 2005), the mentioned trends in favour of large scale agriculture and the cited case studies of successful land investments suggest that this can mean forgoing important opportunities, not only for rural development but also to serve the upcoming markets (more generally, e.g. Briones and Felipe 2013; Collier and Dercon 2013). It may be more realistic to try to manage at least some deals in the best way for the most affected rural people.

However, it is important to underline that LSLAs should be analysed in conjunction with later investments. Perspectives which only look on immediate land and natural resource issues – important as they may be in terms of immediate risks particularly for food security through local (subsistence and market) production – neglect the opportunities of such large investments. These opportunities for workers, outgrowers, and the wider community, through direct and particularly through indirect effects, include food security through non-farm diversification, job creation and access to food. It is not only unfair but also short-sighted to ignore long-term effects. But risks of failure also have to be included in the overall considerations. This chapter therefore has looked at the project life cycle of LSLAs and agro-investments and has distinguished four phases – screening and negotiation, investment, exploitation, and (possible) failure and abandon.

Whatever the analysis reveals, LSLAs will finally be made between willing sellers and willing buyers, since it is highly unlikely that there will global prohibitions of such deals. Three key questions as to how fair deals are to be achieved are 1) what measures can attenuate the bad outcomes of a large scale agro-investment deal and support the positive ones? 2) how can negotiations be steered to ensure that these measures are considered in the contracts (most large scale projects are based on private contracts) or that bad contracts are not concluded? 3) how can general principles and concluded contracts be enforced?

One important answer to these questions concerns land and water governance and ownership. These are key issues in LSLAs, both for investors (who suffer from unclear procedures and uncertain deals) and for rural populations (who are often neither asked nor consulted). Clearer rules on ownership and governance must improve these uncertainties, including what influence the non-owner stakeholders can exert. Human rights-based approaches may be the best way to protect against down-side risks, particularly for food security, and also in cases where national laws or customary rights may prove unable to protect these rights, while the opportunities for development and the practicality of the processes also have to be considered. International guidelines (FAO 2005; FAO 2012) may provide orientation and standards, but ultimately national regulations have to create the legal frameworks for responsible LSLAs and land governance more generally. Compensation rules must be an important part of these considerations; they must be sufficient to cover not only losses of assets and short-term values but also to protect against livelihood losses. Minor but widespread losses such as access to forests and fallow can be compensated with CSR community projects including public infrastructure. Water issues are often neglected and should be integrated into assessments.

Another part of the answer must be the wider characteristics of the investments, including technical viability, accountability, social and ecological impacts on and off-site, labour issues, rules for settlement of complaints and renegotiation, potential impacts on the wider rural economy, etc. LSLAs have considerable scope to adjust to local realities, and these should be exploited. Again, international guidelines can help in orientation – presently, for instance Principles for Responsible Agricultural Investment are negotiated under the Committee on World Food Security (CFS), and other bodies have also declared principles (e.g. OECD 2011; or safeguards of World Bank or the 'Equator Principles', 'a credit risk management framework for determining, assessing and managing environmental and social risk in project finance transactions', see Equator Principles (2014), and compare Brüntrup et al. 2013b) – but cannot substitute national rules and good advice of all parties. The characteristics of the investors are also important – economic intelligence could help to identify track records and reveal bad behaviour. For these issues, transparency and good information are essential (key words: free, prior and informed consent or consultation, FPIC), and they should be accessible to all stakeholders.[9]

With all guidelines and regulations, large power asymmetries between investors and rural populations, in particular weak groups, remain a fact in large scale deals. Means to attenuate these imbalances have to be found, for instance inclusion of international civil society organizations, compulsory use (and funding) of lawyers for rural populations, ombudsmen for weak stakeholders, etc.

Project failures and their consequences must be dealt with in contracts and under national laws. Insolvency management, timing, rules to transfer assets (particularly land to third parties), redistribution of land, payment of debts and reorganization of services should be clearly organized. In order to insure against the consequences of project failure, exit and rehabilitation funds can be established which are filled by levies from investors and other stakeholders.

At regional level, harmonization of laws and investment regimes can help to avoid a race to the bottom as regards social and environmental standards. Systematic communication can help to spread another country's experience. And regional integration can facilitate investment by providing similar product standards, by creating larger markets for products, by pooling scarce resources such as specialized technical expertise and services, and so on.

Notes

1 <www.future-agricultures.org/index.php?option=com_content&view=category& layout = blog&id = 1547&Itemid = 978> accessed 19 September 2013.
2 <www.future-agricultures.org/land/7669-call-for-papers-land-grabbing-conference-october-2012-usa#.UjpA2H9gScM> accessed 19 September 2013.
3 The NGO GRAIN (2008) first attracted broad attention to LSLAs, much more evidence has been collected and presented since then, e.g. conference 2012 <www.landandpoverty. com/> accessed 19 September 2013, and conference 2013 <http://econ.worldbank.org/ WBSITE/EXTERNAL/EXTDEC/EXTRESEARCH/EXTPROGRAMS/EXTIE/0,,co ntentMDK:23278099~pagePK:64168182~piPK:64168060~theSitePK:475520,00.html> accessed 19 September 2013.
4 *Journal of Peasant Studies,* 38(2) (2011); and 40(3) (2013).
5 *Canadian Journal of Development Studies,* 33(4) (2012).
6 *Globalizations,* 10(1) (2013).
7 *Development and Change,* 44(2) (2013).
8 Least Developed Countries (LDCs) under the Everything but Arms (EBA) initiative and most non-LDCs under (provisional) Economic Partnership Agreements (EPAs).
9 FPIC as free, prior and informed consent is established as a human rights principle to involve indigenous people and communities in decisions concerning their territories and resources (UN_REDD 2013); however, for LSLAs it is disputed whether and how to apply it.

References

Altenburg, T. (2011) *Industrial Policy in Developing Countries, Overview and Lessons from Seven Country Cases,* Bonn: German Development Institute.
Anderson, K. and Brückner, M. (2012) *Distortions to Agriculture and Economic Growth in Sub-Saharan Africa,* World Bank Policy Research Working Paper 6206, Washington, DC: World Bank.
Anderson, K. and Nelgen, S. (2012) 'Trade barrier volatility and agricultural price stabilization', *World Development,* 40(1): 36–48.
Anseeuw, W., Alden Wily, L., Cotula, L. and Taylor, M. (2012) *Land Rights and the Rush for Land: Findings of the Global Commercial Pressures on Land Research Project,* Rome: International Land Coalition.
BEFS-WG and FAO (2013) *Draft Guidelines for Sustainable Agricultural and Bioenergy Investment,* Bioenergy and Food Security Working Group of Sierra Leone and Food

and Agriculture Organization. Online. Available <www.fao.org/energy/39236–0c8648696dd1c2302a98a613391200e1e.pdf> (accessed 28 March 2014).

Behrman, J., Meinzen-Dick, R. and Quisumbing, A. (2012) 'The gender implications of large-scale land deals', *Journal of Peasant Studies,* 39(1): 49–79.

Borras Jr., S.M., Franco, J.C., Gomez, S., Kay, C. and Spoor, M. (2012) 'Land grabbing in Latin America and the Caribbean', *Journal of Peasant Studies* 39(3–4): 845–72.

Brandt, H. and Brüntrup, M. (2012) 'Post-colonial agricultural experiences in Sub-Saharan Africa', in Christoph, E., Kyd-Rebenburg, D. and Prammer, J. (eds), *Global Growing Casebook: Insights into African Agriculture,* 12–33, Vienna: Global Growing Campaign. Online. Available <http://global-growing.org/sites/default/files/GG_Casebook.pdf> (accessed 21 March 2014).

Briones, R. and Felipe, J. (2013) *Agriculture and Structural Transformation in Developing Asia: Review and Outlook,* Asian Development Bank Economics Working Paper Series 363, Manila: Asian Development Bank.

Brüntrup, M. (2012) 'Detrimental land grabbing or growth poles? Determinants and potential development effects of foreign direct land investments', *Technikfolgenabschätzung – Theorie und Praxis* 20(1): 28–37.

Brüntrup, M., Swetman, T., Michalscheck, M. and Asante, F. (2013a) 'Factors of success and failure of large agro-enterprises (production, processing and marketing). A pilot study in Ghana – results of case studies in the fruit, maize, and palm oil sub-sectors', *African Journal of Food, Agriculture, Nutrition and Development* (AJFAND), 13(5). Available <www.ajfand.net/Volume13/No5/Reprint-Factors%20of%20Success%20and%20Failure.pdf> (accessed 21 March 2014).

Brüntrup, M., Scheumann, W. and Berger, A. (2013b) 'Looking at the broader picture: instruments to tame large scale land and water acquisitions for rural development', Paper presented at the 2013 Law and Development Conference 'Legal and Development Implications of International Land Acquisitions', 31 May 2013, Kyoto, Japan. Online. Available <http://lawanddevelopment.net/img/2013papers/MichaelBruentrup-et-al.pdf> (accessed 9 February 2014).

Charles, D. (2012) 'How a biofuel dream called Jatropha came crashing down'. Online. Available <www.npr.org/blogs/thesalt/2012/08/22/159391553/how-a-biofuel-dream-called-jatropha-came-crashing-down> (accessed 21 March 2014).

Chu, J. (2011) 'Gender and "land grabbing" in Sub-Saharan Africa: women's land rights and customary land tenure', *Development,* 54(1): 35–9.

Clancy, J. (2013) *Biofuels and Rural Poverty,* Milton Park and New York: Routledge.

Collier, P. and Dercon, S. (2013) 'African agriculture in 50 years: smallholders in a rapidly changing world?' *World Development.* Online. Available <www.sciencedirect.com/science/article/pii/S0305750X13002131> (accessed 21 March 2014).

Cotula, L. (2009) *Land Grab or Development Opportunity? Agricultural Investment and International Land Deals in Africa,* London: International Institute for Environment and Development.

Cotula, L. (2011) *Land Deals in Africa: What Is in the Contracts?* London: International Institute for Environment and Development.

Cotula, L. (2013) *The Great African Land Grab? Agricultural Investments and the Global Food System,* London and New York: Zed Books.

Cuffaro, N. and Hallam, D. (2011) 'Land grabbing in developing countries: foreign investors, regulation and codes of conduct', Paper presented at the International Conference on Global Land Grabbing, 6–8 April 2011, Institute of Development Studies, University of Sussex. Online. Available <http://papers.ssrn.com/sol3/papers.cfm?abstract_id=1744204> (accessed 12 May 2013).

De Fraiture, C. and Wichelns, D. (2010) 'Satisfying future water demands for agriculture', *Agricultural Water Management* 97(4): 502–11.

De Schutter, O. (2009) *Large-scale Land Acquisitions and Leases: A Set of Core Principles and Measures to Address the Human Rights Challenge*, Briefing note, Geneva: UN Office of the High Commissioner for Human Rights.

Deininger, K. and Byerlee, D. (2011) *Rising Global Interest in Farmland: Can It Yield Sustainable and Equitable Benefits?* Washington, DC: World Bank.

Deininger, K. and Byerlee, D. (2012) 'The rise of large farms in land abundant countries: do they have a future?' *World Development* 40(4): 701–14.

Dinham, B. and Hines, C. (1984) *Agribusiness in Africa,* Trenton, NJ: Africa World Press.

Dorner, P. (1992) *Latin American Land Reforms: A Retrospective Analysis,* Madison, WI: University of Wisconsin Press.

Dorner, P. and Thiesenhusen, W.C. (1990) 'Selected land reforms in East and Southeast Asia: their origins and impacts', *Asian-Pacific Economic Literature* 4(1): 65–95.

Edelman, M. (2013) 'Messy hectares: questions about the epistemology of land grabbing data', *Journal of Peasant Studies,* 40(3): 485–501.

Ellis, F. (2005) 'Small-farms, livelihood diversification and rural–urban transitions: strategic issues in sub-Saharan Africa', Paper prepared for the Research Workshop on the Future of Small Farms, Withersdane Conference Centre, 26–29 June 2005, Wye, Kent, UK.

Equator Principles (2014) Online. Available <www.equator-principles.com/> (accessed 28 March 2014).

FAO (2005) *Voluntary Guidelines to Support the Progressive Realization of the Right to Adequate Food in the Context of National Food Security,* Rome: Food and Agriculture Organization (FAO).

FAO (2008) *The State of Food and Agriculture – Biofuels: Prospects, Risks and Opportunities,* Rome: Food and Agriculture Organization (FAO).

FAO (2012) *Voluntary Guidelines on the Responsible Governance of Tenure of Land, Fisheries and Forests in the Context of National Food Security,* Rome: Food and Agriculture Organization (FAO).

FAO (2013) *Trends and Impacts of Foreign Investment in Developing Country Agriculture – Evidence from Case Studies,* Rome: Food and Agriculture Organization (FAO).

FAO and NEPAD (2002) *Comprehensive Africa Agriculture Development Programme,* Rome: Food and Agriculture Organization (FAO) and New Partnership for Africa's Development (NEPAD).

German, L., Schoneveld, G. and Mwangi, E. (2011) *Contemporary Processes of Large-Scale Land Acquisition by Investors: Case Studies from Sub-Saharan Africa,* CIFOR Occasional Paper 68, Bogor, Indonesia: Centre for International Forestry Research (CIFOR).

GRAIN (2008) 'Seized: the 2008 landgrab for food and financial security'. Online. Available <www.grain.org/article/entries/93-seized-the-2008-landgrab-for-food-and-financial-security> (accessed 6 May 2013).

GRAIN (2013) 'The many faces of land grabbing.' Online. Available <www.grain.org/fr/article/entries/4908-ejolt-report-10-the-many-faces-of-land-grabbing-cases-from-africa-and-latin-america.pdf> (accessed 24 March 2014).

Hayami, Y. (2010) 'Plantation agriculture', in Pingali, P.L. and Evenson, R.E. (eds), *Handbook of Agricultural Economics,* 3305–22, North Holland: Elsevier.

Heidhues, F. and Obare, G. (2011) 'Lessons from structural adjustment programmes and their effects in Africa', *Quarterly Journal of International Agriculture,* 50(1): 55–64.

Helbling, T. and Roache, S. (2011) 'Rising prices on the menu', *Finance & Development* 48(1). Available <https://www.imf.org/external/pubs/ft/fandd/2011/03/helbling.htm> (accessed 9 February 2013).

Herrmann, R., Grote, U. and Brüntrup, M. (2013) 'Household welfare outcomes of large-scale agricultural investments: insights from sugarcane outgrower schemes and estate employment in Malawi', Paper presented at the Annual World Bank Conference on Land and Poverty, 8–11 April 2013, Washington, DC. Online. Available <www.conftool.com/landand poverty2013/index.php/Herrmann-368_paper.pdf?page=downloadPaper&filename= Herrmann-368_paper.pdf&form_id=368&form_version=final> (accessed 6 May 2013).

Hilhorst, T., Nelen, J. and Traoré, N. (2011) *Agrarian Change Below the Radar Screen: Rising Farmland Acquisitions by Domestic Investors in West Africa. Results from a Survey in Benin, Burkina Faso and Niger,* Amsterdam: Royal Tropical Institute (KIT) and SNV Netherlands Development Organization.

IEA (2009) *Bioenergy – A Sustainable and Reliable Energy Source,* Paris: International Energy Agency (IEA).

Kennedy, E.T. (1989) *The Effects of Sugarcane Production on Food Security, Health, and Nutrition in Kenya: A Longitudinal Analysis,* Washington, DC: International Food Policy Research Institute.

Kircher, M. (2014) 'The emerging bioeconomy: industrial drivers, global impact, and international strategies', *Industrial Biotechnology,* 10(1): 11–18.

Koopman, J. (2012) 'Land grabs, government, peasant and civil society activism in the Senegal River Valley', *Review of African Political Economy,* 39(134): 655–64.

Land Matrix (2014) Online. Available <www.landmatrix.org/en/get-the-detail/all/> (accessed 27 January 2014).

Larson, A.M., Brockhaus, M., Sunderlin, W.D., Duchelle, A., Babon, A., Dokken, T. and Huynh, T.B. (2013) 'Land tenure and REDD+: the good, the bad and the ugly', *Global Environmental Change,* 23(3): 678–89.

Lay, J. and Nolte, K. (2013) 'Determinants of large scale land acquisitions – evidence on success and failure from the Land Matrix', Presentation at the PEGNet Conference 2013: *How to Shape Environmentally and Socially Sustainable Economies in the Developing World – Global, Regional, and Local Solutions,* 17–18 October 2013, University of Copenhagen, Denmark.

Losch, B., Fréguin-Gresh, S. and White, E.T. (2012) *Structural Transformation and Rural Change Revisited: Challenges for Late Developing Countries in a Globalizing World,* Washington, DC: World Bank Publications.

Lovett, R. (2008) 'Burying biomass to fight climate change', *New Scientist Magazine,* 2654: 32–5.

Maertens, M. and Swinnen, J.F. (2009) 'Trade, standards, and poverty: evidence from Senegal', *World Development,* 37(1): 161–78.

Matondi, P.B., Havnevik, K., and Beyene, A. (2011) *Biofuels, Land Grabbing and Food Security in Africa,* London and New York: Zed Books.

McCulloch, N. and Ota, M. (2002) *Export Horticulture and Poverty in Kenya,* Vol. 174, Brighton and Sussex: Institute of Development Studies.

Mehta, M., Veldwisch, G.J. and Franco, J. (eds) (2012) 'Special issue: water grabbing? Focus on the (re)appropriation of finite water resources', *Water Alternatives,* 5(2): 193–207.

Minot, N. and Ngigi, M. (2003) 'Are horticultural exports a replicable success story? Evidence from Kenya and Côte d'Ivoire', Paper presented at the InWEnt, IFPRI, NEPAD, CTA conference Successes in African Agriculture, Pretoria.

Neef, A., Touch, S., and Chiengthong, J. (2013) 'The politics and ethics of land concessions in rural Cambodia', *Journal of Agricultural and Environmental Ethics,* 26(6): 1085–1103.

Nolte, K. (2013) *Large Scale Agricultural Investments under Poor Land Governance Systems: Actors and Institutions in the Case of Zambia,* GIGA Working Papers, 221. Online. Available <www.giga-hamburg.de/working papers> (accessed 24 March 2014).

OECD (2011) *Guidelines for Multinational Enterprises,* 2011 Edition, Paris: Organization for Economic Co-operation and Development (OECD).

OECD and FAO (2008) *Agricultural Outlook 2008–2017,* Paris: Organization for Economic Co-operation and Development (OECD); and Rome: Food and Agriculture Organization (FAO).

Osabuohien, E.S. (in progress) *Foreign Land Acquisitions in Nigeria: Forces from Above and Voices from Below* (unpublished manuscript), Bonn.

OXFAM (2004) *A Sweeter Future? The Potential for EU Sugar Reform to Contribute to Poverty Reduction in Southern Africa,* London: Oxfam.

OXFAM (2011) *Land and Power: The Growing Scandal Surrounding the New Wave of Investments in Land,* Oxfam: Oxford. Online. Available <http://policy-practice.oxfam.org.uk/publications/land-and-power-the-growing-scandal-surrounding-the-new-wave-of-investments-in-l-142858> (accessed 24 March 2014).

Oya, C. (2012) 'Contract farming in Sub-Saharan Africa: a survey of approaches, debates, and issues', *Journal of Agrarian Change,* 12(1): 1–33.

Platteau, J.P. (2008) 'The evolutionary theory of land rights as applied to sub-Saharan Africa: a critical assessment', *Development and Change,* 27(1): 29–86.

Reardon, T., Timmer, C.P. and Minten, B. (2012) 'Supermarket revolution in Asia and emerging development strategies to include small farmers', *Proceedings of the National Academy of Sciences,* 109(31): 12332–7.

Richards, M. (2013) *Social and Environmental Impacts of Agricultural Large-Scale Land Acquisitions in Africa – With a Focus on West and Central Africa,* Washington, DC: Rights and Resources Initiative.

Rockström, J. et al. (2009) 'A safe operating space for humanity', *Nature,* 461: 472–5.

Rosenblum, P. and Maples, S. (2009) *Contracts Confidential: Ending Secret Deals in the Extractive Industries,* Revenue Watch Institute: New York.

Running, S.W. (2012) 'A measurable planetary boundary for the biosphere', *Science,* 337(6101): 1458–9.

Ruthenberg, H. (1980) *Farming Systems in the Tropics,* 3rd edition, Oxford: Clarendon Press.

Schaffnit-Chatterjee, C. (2012) *Foreign investment in farmland. No low-hanging fruit,* Frankfurt am Main: Deutsche Bank Research. Online. Available <www.dbresearch.com/PROD/DBR_INTERNET_EN-PROD/PROD0000000000296807/Foreign+investment+in+farmland%3A+No+low-hanging+fruit.PDF> (accessed 9 February 2013).

Scoones, I., Hall, R., Borras, S.M., White, B. and Wolford, W. (2013) 'The politics of evidence: methodologies for understanding the global land rush', *Journal of Peasant Studies,* 40(3): 469–83.

Silici, L. and Locke, A. (2013) *Private Equity Investments and Agricultural Development in Africa: Opportunities and Challenges,* Washington, DC: International Food Policy Research Institute.

Sulieman, H.M. (2013) *Land Grabbing Along Livestock Migration Routes in Gadarif State, Sudan: Impacts on Pastoralism and the Environment,* LDPI Working Paper 19. Online. Available <http://r4d.dfid.gov.uk/pdf/outputs/Futureagriculture/LDPI_WP_19.pdf> (accessed 24 March 2014).

Tersoo, P. (2014) 'Agribusiness as a veritable tool for rural development in Nigeria', *International Letters of Social and Humanistic Sciences,* 3: 26–36.

The Munden Project (2012) *The Financial Risks of Insecure Land Tenure: An Investment View,* prepared for the Rights and Resources Initiative by the Munden Project.

Tyler, G. and Dixie, G. (2013) *Investing in Agribusiness: A Retrospective View of a Development Bank's Investments in Agribusiness in Africa and Southeast Asia and the Pacific,* Washington, DC: World Bank.

UN-REDD (2013) *Programme Guidelines on Free, Prior and Informed Consent,* Online. Available <http://www.un-redd.org/Launch_of_FPIC_Guidlines/tabid/105976/Default.aspx> (accessed 20 August 2014).

Väth, S. (2012) 'Gaining neighbours or disruptive factors – what happened when large scale land-based investment in the Ghanaian oil palm sector met the local population on the ground?' Paper presented at the International Conference on Global Land Grabbing II, October 17–19, Cornell University. Available <www.cornell-landproject.org/download/landgrab2012papers/Vath.pdf> (accessed 9 February 2013).

Vermeulen, S. and Cotula, L. (2010) 'Over the heads of local people: consultation, consent, and recompense in large-scale land deals for biofuels projects in Africa', *Journal of Peasant Studies,* 37(4): 899–916.

Visser, O. and Spoor, M. (2011) 'Land grabbing in post-Soviet Eurasia: the world's largest agricultural land reserves at stake', *Journal of Peasant Studies,* 38(2): 299–323.

Wakker, E., Watch, S. and Rozario, J.D. (2004) 'Greasy palms: the social and ecological impacts of large-scale oil palm plantation development in Southeast Asia', Friends of the Earth. Available <www.foe.co.uk/resource/reports/greasy_palms_impacts.pdf> (accessed 9 February 2013).

Wiggins, S., Kirsten, J. and Llambí, L. (2010) 'The future of small farms', *World Development,* 38(10): 1341–8.

Wily, L.A. (2011) '"The law is to blame": the vulnerable status of common property rights in Sub-Saharan Africa', *Development and Change,* 42(3): 733–57.

World Bank (2007) *Agriculture for Development – World Development Report 2008,* Washington, DC: World Bank.

World Bank (2014) *The Practice of Responsible Investment Principles in Larger Scale Agricultural Investments: Implications for Corporate Performance and Impact on Local Communities,* Agriculture and Environmental Services Discussion Paper 08, Washington, DC: World Bank.

Wright, B.D. (2011) 'The economics of grain price volatility', *Applied Economic Perspectives and Policy,* 33(1): 32–58.

Yumkella, K.K., Kormawa, P.M., Roepstorff, T.M. and Hawkins, A.M. (eds) (2011) *Agribusiness for Africa's Prosperity,* Vienna: UNIDO.

6

WATER COMPETITION, WATER GOVERNANCE AND FOOD SECURITY

Helle Munk Ravnborg

Earlier this year I crossed a river in Nicaragua which I have crossed so many times before during the last ten years. There are a few houses at each side of the river and obviously people living in these houses collect water in the river to water their crops and their animals. A few years back, a couple who had lived and worked in the rapidly growing town of Estelí, 40 minutes' drive from the river crossing, decided to retire and to renovate their house along the river. They installed a water tank and a pumping system to fill the tank. They use the water from the tank to water a small plantation of avocado trees and also crops of tomatoes and other vegetables grown for sale in Estelí. This time when I crossed, I counted several new pumps. In at least two cases, the owners of the pumps were what I would consider 'agricultural entrepreneurs' rather than farmers. They were business people living in Estelí who had rented a piece of land along the river, prepared the plot, laid out their pipes and installed their pumps along the river bank to pump out water to water crops of tomatoes and green peppers. If everything goes well, they may rent the land for another year; if not, they are likely to move their equipment to another location.

Water is crucial to food security. Not only is it an essential input for the cultivation of food crops and for livestock rearing, it also plays a key role in the processing of many crops as well as in cooking. To farmers around the world, access to water makes the difference between surviving and thriving through farming. That is why every day, farmers around the world make great efforts digging canals, negotiating water distribution agreements, or filling and carrying buckets of water long distances to water crops, perhaps through artisanal drip irrigation systems made from used and amended plastic bottles turned upside down. And that is why farmers in command of sufficient economic resources place polythene tubes over long distances to transport water by gravity from a distant water source to a crop, or invest in equipment to pump out water from the ground or from nearby rivers.

According to the *Comprehensive Assessment of Water Management in Agriculture* (Molden 2007), improving access to agricultural water constitutes a key element in strategies to reduce rural poverty. Smallholder farmers make up the majority of the world's rural poor. According to a recent report prepared by the High Level Panel of Experts on Food Security and Nutrition (HLPE 2013b), there is an estimated total of 500 million smallholder farms (up to two hectares) in the developing world, supporting 2 billion people, and yet these small farms produce up to 80 per cent of food supplies in many Asian and sub-Saharan countries (FAO 2012; HLPE 2013b, quoting Hazell 2011). These smallholder farms depend mainly on rainfall for production, making them vulnerable to droughts and erratic rainfall. As climate change progresses, causing still more unpredictable rainfall patterns with erratic onset of rain and prolonged and unexpected dry spells (Bates *et al.* 2008), this vulnerability will only increase. In this context, upgrading rain-fed systems through providing supplemental irrigation combined with efforts to improve rainwater and soil moisture management is key (Molden *et al.* 2007). The greatest gains will come from focussing on low-yielding areas, e.g. large parts of sub-Saharan Africa (Molden *et al.* 2007). While for developing countries as a whole, it is estimated that the application of supplemental or full irrigation would increase cereal production by 43 per cent above rain-fed levels, for southern Africa the application of supplemental or full irrigation could bring a more than a three-fold (332 per cent) increase in cereal production (Fischer *et al.* 2002: 90). Beyond boosting yields and cutting yield losses from dry spells, improving smallholder farmers' access to agricultural water would provide them with the water security they need to invest in other types of farm improvements, and allow them to engage with the market on a more stable and rewarding basis.

Yet farmers, and in particular food-producing smallholder farmers, face increasing competition for water both from other sectors and from within agriculture itself. In an attempt to prepare for such competition, developing countries around the world have reformed their legal and administrative water governance frameworks during recent decades with a particular emphasis on how, and under which conditions, to allocate water resources between different types of uses and users in society.

This chapter provides an overview of the competition for water which agriculture and in particular food production faces and of the contents of the recent wave of water governance reform which has swept across the developing world and which in many places entails a process of transformation of water rights. The chapter concludes with a discussion of the implications for water access for farming, in particular smallholder farming, and for food security.

Water competition

As freshwater demands from other sectors such as industry, mining, energy generation and municipal water uses are growing, competition for water intensifies in terms of quantity, timing and quality. In some parts, and specifically in some locations of the world, e.g. in the mountainous parts of Latin America or Asia, this competition is for a finite water resource available in that location. In other parts of the world, where water availability is constrained by a shortfall of investments in water development – i.e. by what is usually referred to as economic water scarcity as opposed to physical water scarcity – it is rather a competition over how to distribute the costs for further water development. Considering the African continent as a whole, only 5.5 per cent of the total available renewable freshwater resource was being withdrawn in 2000 (Molden *et al.* 2007).

With rapidly growing investments in the mining sector, water which used to be allocated for irrigation through complex canal systems in the Andean hillsides, is increasingly being allocated for – or captured by – mining companies. Reports document how the water, when discharged back into the system of streams, lakes and canals, is heavily contaminated, making it inappropriate for irrigation as well as for human and animal consumption (e.g. Bebbington and Williams 2008; Sosa and Zwarteveen 2012). In the wake of this scramble for mineral resources affecting not only Latin America but also many countries in sub-Saharan Africa and Asia, an increasing number of similar cases are being reported from other parts of the world – of intensified competition for water between mining companies, domestic water consumers, and farmers and livestock herders (e.g. DIIS and Rehder 2010; Phuong *et al.* 2010; Larsen and Mamosso 2014).

Although by no means excluding agricultural uses of water, the increasing freshwater claims made for hydro-power generation significantly alter water flows both upstream and downstream of dam sites. By doing so, and by offering possibilities for large-scale irrigation schemes, hydropower generation often also alters the agrarian structure, i.e. the farm sizes, the type of irrigation systems, the crop choices, and thus ultimately the composition of the farming population at large of the areas affected by hydro-power dam construction. Depending on the specific crop choices and the market destination for the produce, this may have important implications for food security locally, nationally and internationally. Moreover, the intensive dam construction in for instance Southeast Asia along the Mekong River and its tributaries significantly changes and in many cases blocks the river flow, thus impeding fish migration and thereby reducing fish stock on which many people depend for part of their income and for protein supply (e.g. Orr *et al.* 2012). Not surprisingly, infrastructure development, typically dam construction, has been identified as one of the factors most likely to be associated with transboundary water-related conflict (Wolf *et al.* 2003).

However, it is not only water demands from the non-agricultural sectors which are growing. The freshwater demands of the agricultural sector itself are also growing and changing in composition. First of all, the global population is projected to grow from its current size of just above 7 billion people to around 9 billion people in 2050. This – hopefully combined with a decreasing number of persons suffering from food shortage – will obviously require an increase in food production and thus cause a growing demand for agricultural water. However, more profoundly affecting agricultural water demands are the projected changes in diets, as a growing number of people shift from a diet primarily based on cereals, root crops and pulses to a diet in which meat and dairy products as well as vegetables constitute a still higher share. According to Hoekstra (2013), in industrialized countries it takes an average of 3,600 litres of water a day to produce the food required to cover the dietary needs of a person living on what he refers to as a primarily meat-based diet, compared with 2,300 litres a day to produce the dietary needs of a person living on a primarily plant-based diet. As incomes rise in Asia, Latin America and gradually also in sub-Saharan Africa, the number and proportion of people living on a diet in which meat and dairy products constitute an increasing share is expected to grow, although per capita meat consumption in developing countries is expected to remain below that of developed countries (Molden *et al.* 2007). As most of the increase in meat and dairy production will be met through mixed and industrial livestock systems which depend on feed rather than grazing, this will lead to significant increases in agricultural water demands. Thus, altogether it is expected that food production will have to grow by 70 to 90 per cent by 2050 to meet global food demands (Molden 2007). Without further improvements in water productivity, this would imply a doubling of the total amount of water used in crop production. As livestock production is increasingly geographically separated from the production of fodder, the growing demand for animal products will lead to increasing international trade in fodder. This, in turn, will lead to new demands for water in fodder-producing regions of the world to be consumed in the livestock-producing parts of the world and by the livestock-consuming segments of the world's population.

Overall, agriculture is by far the biggest sector when it comes to freshwater withdrawals, i.e. water pumped or lifted out of lakes, rivers and streams or from the ground, the so-called blue water (Hoekstra and Chapagain 2007). Although on a global scale, 80 per cent of the water used for agricultural production is 'green' water (i.e. water that comes directly from the rain and is stored as soil moisture), agricultural irrigation accounts for 70 per cent of total global freshwater withdrawals (Molden 2007; UN-Water 2013).[1]

This, obviously, makes agriculture the focus of attention when demands for water from other economic sectors such as industry, mining and energy generation grow, and ways to accommodate these demands are sought. Thus, in addition to investments in water development, efforts to increase agricultural water productivity – to enable the production of more food per drop of water – are being pursued. This includes efforts made through agricultural research, e.g. through crop breeding and the development of new crop husbandry and irrigation techniques, as well as efforts made through policy development and change to stimulate more prudent use of water for irrigation, for instance through the removal of farm subsidies for electricity and the introduction of water use fees. Such efforts in combination may succeed in a relative reduction in agricultural freshwater withdrawals for irrigation, but even in an optimistic scenario, freshwater withdrawals for agriculture are expected to increase by 13 per cent by 2050 (Molden *et al.* 2007).

Finally, as part of efforts to reduce greenhouse gas (GHG) emissions and, perhaps more importantly, to reduce dependence on fossil fuels which at least until recently seemed to be becoming short in supply, substantial efforts have been made to boost the production and use of biofuels based on commodities such as maize, sugar cane, soy and palm oil. As an example, the European Union has established the target that by 2020, 10 per cent of the energy used for transportation should come from renewable sources and so far, efforts to meet this target seem to imply increasing levels of substitution of fossil fuels with biofuels (HLPE 2013a). Approximately 60 per cent of EU vegetable oil production and 40 per cent of the US maize production is destined for biofuel production (HLPE 2011: 32; here quoted from Woodhouse 2012a: 778). Besides target setting, this development has been encouraged through the provision of public subsidies for the US and EU biofuel production to the tune of an estimated USD 10 million in 2006 and USD 8 million in 2009 (Woodhouse 2012a: 778).[2]

This change in the composition of transportation fuels has significant implications for future demands for agricultural water. In 2010, the water footprint – the volume of water used to produce a product, taking into account the volumes of water consumed and polluted in the different steps of the production process – of the total global amount of biofuels consumed corresponded to 90 billion m[3] of water (Van Lienden *et al.* 2010). At the global level, it is expected that the annual biofuel water footprint will grow more than three-fold, from its 2010 level of about 90 billion m[3] per year to 970 billion m[3] per year (Van Lienden *et al.* 2010) in 2030. That large quantities of water are used for the production of non-food crops is far from a new phenomenon. Cotton is today by far the most water consuming commodity on a global scale, being responsible for close to 7 per cent of the global water footprint, including not only agricultural but also industrial and municipal uses of water, or the staggering amount of 569 billion m[3] water. However, if projections for future biofuel consumptions come true, the water footprint of biofuel consumption will by 2030 exceed the current water footprint of cotton by 70 per cent.[3]

Considering only the blue water footprint, i.e. the amount of water drawn from the rivers, lakes and the ground for irrigation, in 2010, the global blue water footprint of biofuel consumption was estimated to be 42 billion m[3], corresponding to 4.5 per cent of the total global blue water footprint and 0.5 per cent of the estimated amount of available blue water. By comparison, cotton was estimated to account for a global blue water footprint of 167 billion m[3], i.e. four times higher than that of biofuel. By 2030, these proportions are expected to have changed drastically. With a total blue water footprint of 465 billion m[3] – almost three times higher than the blue water footprint currently set by cotton – biofuel consumption is expected to account for 9 per cent of the total 2030 global blue water footprint and consume 5.5 per cent of the estimated amount of available blue water.

The largest consumers of biofuel in 2030 are expected to be the EU, the US, China and Brazil and with the exception of Brazil, none of these places have sufficient water resources to meet the demand for blue water owing to their expected biofuel consumption (Van Lienden *et al.* 2010). This means that a growing part of the biofuel consumption in Europe, the US and China will be covered through imports of biofuel or of biofuel feedstock from Brazil but also increasingly from low-cost producer countries in Asia, sub-Saharan Africa, Central America and the Caribbean (Cotula *et al.* 2008). Thus, through the demands put on land which have spurred the recent concern with land grabbing, e.g. in sub-Saharan Africa and Southeast Asia (Cotula *et al.* 2008; Borras Jr. *et al.* 2010; HLPE 2013a), and on water implying a significant export of virtual water embedded in the biofuel (feedstock) industry (World Water Assessment Programme 2009; Borras Jr. *et al.* 2011; Kay and Franco 2012; Ravnborg 2012; Woodhouse, 2012a), growing biofuel consumption will also intensify the competition for water within the agricultural sector and thereby put food production and food security at risk locally, nationally and perhaps even internationally.

Water governance reform

Water governance may be understood as the processes through which decisions are being made, enforced, sanctioned and challenged on the development, allocation and the conditions of use of water resources at all levels of society. It involves the interactions between political, social, economic, legal, and administrative or regulatory institutions – statutory as well as customary – that determine how decisions are taken and how authority is exercised (Graham *et al.* 2003; Rogers and Hall 2003; Cleaver and Franks 2005; Merrey *et al.* 2007; Paavola 2007).

In response to – and in some cases perhaps anticipating – such growing competition and the emergence of new types of uses and users (such as hydro-electric corporations, mining and industrial companies, large-scale agricultural enterprises, urban water supply utilities, etc.) who need secure water access in addition to formally sanctioned water rights to enable and protect their investments, countries across the developing world have embarked upon a process of water governance reform. During the past decades, countries such as Bangladesh,[4] Chile,[5] Ghana,[6] Kenya,[7] Mexico,[8] Nicaragua,[9] Peru,[10] South Africa,[11] Tanzania,[12] Viet Nam[13] and Zambia,[14] just to name a few, have reformed their legal and regulatory framework for water governance (Aagaard and Ravnborg 2006; Boelens *et al.* 2012; Bruns *et al.* 2005; Derman and Ferguson 2003). In many cases, particularly in Latin America, the attempts to reform the water governance framework have been met with broad-based popular protest and have had to be halted on several occasions (e.g. Boelens *et al.* 2012; de Vos *et al.* 2006) or are still underway.

Despite variations, many of these reforms share a number of common features. First of all, in most cases the water reforms mark an attempt to bring water, and particularly water allocation, under statutory control. Constitutionally, water has been considered a public resource and part of the national patrimony even before the current wave of water reforms. However, the reality has been different in many places with water access being determined through complex and context-specific mixes of private individual or collective rights to water, based on land ownership (e.g. riparian rights), being a first-comer (e.g. prior appropriation rights or customary rights), social or cultural belonging (e.g. community-based rights), or simply based on having achieved water access through economic, political, social or violent means.

Particularly in the multi-ethnic context characterizing many developing countries, the co-existence of such systems of more or less formalized water rights, often drawing on different legal – and extra-legal – systems, tends to lead to the existence of mutually overlapping and thus frequently contested water rights. This ambiguity, it is often argued, hampers the ability of governments to ensure that wider social, environmental and economic policy objectives are reflected in the allocation of water resources and more specifically it is perceived to hamper incentives for water development and water-dependent investments for private and institutional investors. Hence, the recent wave of water reforms may be seen as an attempt to abolish the concept of 'private waters' based on, e.g. riparian or prior appropriation rights, and the diverse and often overlapping sets of rules and practices which hitherto have governed water allocation. Instead, many governments are re-declaring water as a national patrimony or property through the creation of a single water administration agency which, on behalf of the president or a minister with whom control over all water resources is vested, is responsible for issuing water *use* rights.[15]

The establishment of a system of administrative water use rights as the mechanism for allocating water for specific uses and among specific users is the second common feature characterizing this recent wave of water reforms. Based on hydrological assessments of available water resources and, at least ideally, also on socio-economic assessments of already existing water uses, the water administration agency is expected to allocate water and specify the conditions for its use through a system of administrative water use rights. Moreover, the aim is that the legal and administrative enforcement of these water use rights will provide sufficient security to attract and effectively protect investments and other user interests and to enable the transfer of water use rights between users and uses through legal and administrative mechanisms.

Depending on the country, the volume of water implied, the intended water use and duration of the water use right, these administrative water use rights are referred to as licenses, permits, authorizations or concessions. Typically, permits for using water for irrigation have a duration of 5–10 years while concessions to divert and store water in the case of construction of hydro-water dams have a duration of up to 40 years. Apart from the duration of the right and the volume of water to be used, an administrative water use right tends to specify the kind of use to which water abstraction is permitted; where the water use takes place; the volume and quality of discharge water; and the source to which used water is discharged. Failure to comply with these specifications or conditions as well as, in many cases, failure to use the water for which the right is obtained may lead to the water right being forfeited. Often a water use fee will have to be paid in return for the water use right.

Third, in many countries the legal framework for the granting of administrative water use rights establishes a ranking in terms of priority between different types of use to be followed in cases where demands for water exceed the resource available. Where this ranking exists, water use for human (domestic) consumption, including for rural and urban water utilities, takes precedence over other types of uses. This is in line with the United Nations declaration of water as a human right (UN 2010). Domestic consumption is followed by the use of water for the following: agricultural, livestock and forestry production; environmental purposes; energy production intended to meet public needs; industrial purposes; tourism; navigation, etc. However, the exact order of priority of these uses varies from country to country and is often a key focus of the public water debate. Thus, both in Chile (Bauer 2008) and South Africa,[16] one of the issues addressed as part of the revisions of the newly reformed water legislations was the priority assigned to environmental water use.

A fourth common feature of recent water governance reforms in developing countries is that some water uses are exempted from the obligation to obtain an administrative water right. Typically, this applies to uses involving small quantities of water, such as for domestic uses (human consumption, cleaning and personal hygiene) as well as for watering of a few household animals and the garden around the house. These exemptions are commonly referred to as *de minimis* exemptions. Depending on the country, small-scale irrigation may also fall within the category of uses which are exempted from the need to hold a formal administrative water use right. In Uganda, it was assessed that the 1995 Water Statute implies that only some 200 water users who use water for irrigation would require a formal water abstraction right (Hodgson 2004, quoting Garduño 2001), whereas in Nicaragua, the 2007 water law implies that everybody who wishes to use water for watering crops would be legally required to apply for either an authorization (if the irrigated area is below three hectares) or for a concession (if the irrigated area exceeds three hectares). Estimates made for the small Condega district (370 km², less than 0.5% per cent of the Nicaraguan territory) in the northern hillsides of Nicaragua suggest that approximately 400 farmers and agricultural entrepreneurs, such as the ones who had installed their pumps along one of the river crossings of the Estelí River, would need an authorization from the district authorities in order to legalize their current use of water for irrigation (Gómez and Ravnborg 2011).

This leads to the fifth and final common feature of the recent wave of water governance reform to be highlighted here. Many countries undergoing a process of water governance reform face the challenge of how to deal with existing water uses – and users – for which access to water hitherto has been legitimized with reference to, e.g. riparian rights, prior use rights, etc. Rather than neglecting such existing water uses, most water governance reform processes entail the attempt to regularize existing water uses by bringing them into the administrative water rights regime. In many developing countries, this has proven to be a considerable challenge, given their often poor means of communication and their thinly staffed and sparsely funded national and sub-national administrative agencies. In Mexico, for instance, the deadline for registering and formalizing existing water use had to be consecutively extended from an initial three-year period to a total of eight years combined with vigorous information campaigns in order to ensure a reasonably complete water use register (Garduño 2005b).

Water governance in the pursuit of water and food security

Water security is defined as 'the capacity of a population to safeguard sustainable access to adequate quantities of acceptable quality water for sustaining livelihoods, human well-being, and socio-economic development, for ensuring protection against water-borne pollution and water-related disasters, and for preserving eco-systems in a climate of peace and political stability' (UN-Water 2013).

In the face of growing competition for water, secure water rights are increasingly important to the achievement of water security. 'As with land,' the United Nations Development Programme (UNDP) *Human Development Report 2006* argued, 'secure rights to water can expand opportunities for poor people to escape poverty. Conversely, the absence of secure rights leaves people open to the risk that they will be unable to assert their claims in the face of competition' (UNDP 2006: 178).

At face value, the recent water governance reforms contain a number of potentially pro-poor and pro-smallholder farming provisions, such as the *de minimis* exemptions, the provisions for regularizing existing water uses and the provisions for obtaining less administratively demanding authorizations for small-scale irrigation *vis à vis* the more demanding water use concessions applying to large-scale irrigation.

However, while part of the ambition underlying the recent water reforms was to create a single and unified set of water use rights to be allocated on the basis of politically agreed social, environmental and economic criteria, emerging evidence suggests that one system of overlapping and mutually exclusive water rights is being replaced by another system or even hierarchy of overlapping water use rights. This is owing to two interrelated factors. First, while intended to facilitate adjustment to growing competition for water, efforts to include reproductive measures and promote equality in the allocation of water use rights have been virtually absent in the water reform processes (UNDP 2006: 178). Second, even if not capable of promoting equality in the allocation of water, the effectiveness of the provisions in the water reforms potentially capable of providing security for rural populations and small-scale farmers (with respect to their current water access) critically depends upon the administrative and regulatory capacity of the governance institutions responsible for allocating water use rights, and for regulating compliance with the conditions specified in the water use right allocation.

Despite provisions for regularizing existing water rights as part of the transition to a new legal and administrative water governance framework, efforts are only rarely contemplated for addressing inequalities in access to information. The consequences in terms of equity and water security for small-scale water uses may be alarming, as experiences from Chile in the wake of the 1981 Water Code illustrate. As described by Bauer (1997), no public information campaigns were undertaken following the enacting of the 1981 Water Code and many small-scale farmers were thus unaware of the code's new features and the procedures for applying for new water rights or regularizing old ones. Thus, by the time peasants and their organizations learned about the new procedures, available water rights in many areas had already been granted by the water administration or regularized by those more legally adept (Bauer 1997: 650). It was in view of such experiences that, in Mexico, the period allocated for registering and formalizing existing water use rights was extended from the initial three-year period to a total of eight years.

The legal frameworks have tended to be rather vague with respect to how to cater to the *de minimis* exempted water uses in practice and thus guarantee that the required amounts of water will actually be available. Although applying to minor uses, the combined volume of water required to honour these legally endorsed claims for water may, in many parts of developing countries, be considerable given the large number of individual users. Moreover, the fact that a large part of the water needed for *de minimis* exempted uses will be used for human (or animal) consumption, as well as for personal hygiene, requires that it meet certain water quality standards. These standards may be compromised by other uses related to the same water body, for instance through contamination from agricultural pesticides, or from agricultural processing (e.g. of coffee or dairy). Despite *de minimis* uses tending to precede such non-exempted water uses in terms of priority, the lack of legal and administrative clarity with respect to how to cater to these *de minimis* exempted uses – and users – may imply that they are overridden and thus lose out to water uses – and users – for which rights are granted through more clearly described procedures (e.g. through licenses and concessions) (Aagaard and Ravnborg 2006; Woodhouse 2012a).

Perhaps most alarming from a food security point of view, the lack of administrative and regulatory capacity which characterizes many developing countries may end up causing a partial and uneven implementation of the water governance reform, thereby legitimizing the dispossession of water from typically small-scale upstream farmers to large-scale downstream farmers and agricultural enterprises, as well as to other industrial uses. As noted by UNDP, 'As a rule of thumb, the importance of power in shaping outcomes from legislation is inversely related to regulatory capacity. Weak regulatory capacity increases the scope for exploitation of unequal relationships' (UNDP 2006: 182).

Small-scale farmers often found in hilly and more remote areas of difficult terrain would rarely be required to obtain a formal water use concession, because they only undertake either rain-fed farming or small-scale irrigation which is either exempted from the need for soliciting a formal water use permission or may only need e.g. an authorization provided by the local district authority or from a catchment committee. By contrast, larger scale farmers and agricultural entrepreneurs operating irrigation systems that are apt for mechanization on generally larger plots on the more accessible plains and in the valleys would, following the recent wave of water governance reforms, in most countries be required to apply for a permit or concession from a national water authority to use water for irrigation, e.g. during a five-year period. Moreover, in cases where agricultural investments are financed through national or, even more importantly, international development investment banks, which is the case for much of the current expansion of biofuel feedstock production (Woodhouse 2012b), legally endorsed access to the amounts of water needed during the loan period to make the investment viable is increasingly being put as a condition by the financing institution. A permit or concession granted by a statutory national authority, serves to document such legal access. In Nicaragua, such a concession would be solicited from the National Water Authority of Nicaragua, which started receiving such applications and issuing concessions during 2010, following the approval of the new water law and regulation in 2007 (Gómez *et al.* 2012). However, seven years down the road from the signing of the new water law, the procedure envisaged for authorizing the use of water for small-scale irrigation (less than three hectares) is still awaiting endorsement from the national water authority. This means that while large-scale farmers and agricultural enterprises along with other larger scale water users are gradually making headway in legalizing their water access, small-scale agricultural users are prevented from regularizing their water use and from obtaining secure water rights.

Given the critical importance of small-scale farming both for local and for national food supplies in many countries (HLPE 2013b), this appropriation and concentration of water is likely to have significant implications for food security. As the competition for water intensifies, the increasing demand for biofuel, fodder, etc. may not only imply the legitimization of the shifting of water out of small-scale farming and into large-scale farming, but also of a global redistribution of water from 'low production cost' areas (Cotula *et al.* 2008), i.e. poor developing countries with cheap land and labour and limited environmental law enforcement, to higher income countries, unable to satisfy their 'virtual blue water' consumption, i.e. their 'blue water footprint' from internal sources.

Implementation tends to be the Achilles' heel of the process of reforming water governance (Garduño 2005a) and in particular of determining its social, environmental and economic outcome (UNDP 2006). Making water governance work in support of local and national water and food security requires closing the implementation gap which currently characterizes the water governance situation in many developing countries that have recently reformed their water governance framework and are subject to significant international attention as potential future suppliers of food, fodder and fuel feedstock. This entails regularizing and effectively backing the water use rights for domestic and small-scale productive use whether through individual or collective water use rights or reserved allocations, ensuring that environmental needs and water quality standards are effectively met, that water use charges are set in a way which reflects societal and environmental objectives, and finally that the water governance framework enables and encourages citizens and authorities to execute their water-related rights and responsibilities.

Notes

1 In addition to the distinction between 'blue' and 'green' water, the water footprint assessments have also begun to estimate the 'grey' water footprint of production and consumption of different products, i.e. the amount of freshwater required to assimilate the load of pollutants based on existing water quality standards (Hoekstra *et al.* 2011).

2 For comparison, total net official development assistance (ODA) from members of the OECD's Development Assistance Committee (DAC) was close to USD 120 billion.
3 This and the following proportions are own calculations based on Van Lienden et al. (2010) and Mekonnen and Hoekstra (2011).
4 National Water Policy (1999) and National Water Act (2013).
5 Water Code (1981), revised in 2005.
6 Water Use Regulation (2001) and National Water Policy (2007).
7 Water Act (2002), currently under revision (2012 draft).
8 National Water Law (1992); amended in 2004.
9 National Water Law (2007).
10 National Water Law (2009).
11 National Water Policy (1997) and National Water Act (1998); amended in 2013.
12 National Water Policy (2002) and Water Resource Management Act (2009).
13 Water Resource Law (1998); revised in 2012.
14 National Water Policy (2010) and Water Resource Management Act (2011).
15 Bangladesh constitutes an exception in this respect as 'surface water on any private land shall remain with the owner of such land, and such rights to use the water shall [. . .] be continued to be enjoyed' (<www.warpo.gov.bd/pdf/WaterActEnglish.pdf>, consulted 22 January 2014).
16 Available at <www.lawsofsouthafrica.up.ac.za/index.php/current-legislation>, then navigate to water/national-water-act-36-of-1998/regulations-and-notices/36-of-1998-national-water-act-regs-gn-665-6-sep-2013-to-date.pdf, consulted 27 March 2014.

References

Aagaard, C.A. and Ravnborg, H.M. (2006) 'Water reform – implications for rural poor people's access to water', DIIS Brief, Copenhagen: Danish Institute for International Studies.

Bates, B.C., Kundzewicz, Z.W., Wu, S. and Palutikof, J.P. (eds) (2008) *Climate change and water,* Technical paper of the Intergovernmental Panel on Climate Change, Genève: IPCC.

Bauer, Carl J. (1997) 'Bringing water markets down to earth: the political economy of water rights in Chile, 1976–95', *World Development,* 25(5): 639–56.

Bauer, C.J. (2008) 'The experience of Chilean water markets', Paper presented at Expo 2008, Zaragoza, Spain.

Bebbington, A. and Williams, M. (2008) 'Water and mining conflicts in Peru', *Mountain Research and Development,* 28(3/4): 190–5.

Boelens, R., Duarte, B., Manosalvas, R., Mena, P., Roa Avendaño, T. and Vera, J. (2012) 'Contested territories: water rights and the struggles over indigenous livelihoods', *International Indigenous Policy Journal,* 3(3): Article 5.

Borras Jr., Saturino M., McMichael, P. and Scoones, I. (2010) 'The politics of biofuels, land and agrarian change: editors' introduction', *Journal of Peasant Studies,* 37(4): 575–92.

Borras Jr., S.M., Hall, R., Scoones, I., White, B. and Wolford, W. (2011) 'Towards a better understanding of global land grabbing: an editorial introduction', *Journal of Peasant Studies,* 38(2): 209–216.

Bruns, B.R., Ringler, C. and Meinzen-Dick, R. (eds) (2005) *Water Rights Reform: Lessons for Institutional Design,* Washington, DC: International Food Policy Research Institute.

Cleaver, F. and Franks, T. (2005) *Water governance and poverty: a framework for analysis,* BCID Research Paper No. 13, Bradford: Bradford Centre for International Development.

Cotula, L., Dyer, N. and Vermeulen, S. (2008) *Fuelling exclusion? The biofuels boom and poor people's access to land,* London: IIED and FAO.

de Vos, H., Boelens, R., and Bustamante, R. (2006). 'Formal law and local water control in the Andean region: a fiercely contested field', *International Journal of Water Resources Development,* 22(1): 37–48.

Derman, B. and Ferguson, A. (2003). 'Value of water: political ecology and water reform in southern Africa', *Human Organization*, 62(3): 277–88.

DIIS (Danish Institute for International Studies) and Rehder, S. (2010) *Competing for water: when new powerful actors emerge*, Video Report No. 1. [Motion picture]. Online. Available <www.diis.dk/water> (accessed 20 March 2014).

FAO (2012) *Smallholders and family farmers.* Online. Available <www.fao.org/fileadmin/templates/nr/sustainability_pathways/docs/Factsheet_SMALLHOLDERS.pdf> (accessed 9 March 2014).

Fischer, G., van Velthuizen, H., Shah, M. and Nachtergaele, F. (2002) *Global Agro-ecological Assessment for Agriculture in the 21st Century: Methodology and Results,* Laxenburg: International Institute for Applied Systems Analysis, and Rome: Food and Agriculture Organization of the United Nations.

Garduño, H. (2001) *Water Rights Administration: Experience, Issues and Guidelines,* Legislative Study No. 70, Rome: Food and Agriculture Organization of the United Nations.

Garduño, H. (2005a) 'Making water rights administration work', Paper presented at international workshop African Water Laws: Plural Legislative Frameworks for Rural Water Management in Africa, 26–28 January 2005, Johannesburg, South Africa. Online. Available <http://projects.nri.org/waterlaw/workshop.htm> (accessed 20 March 2014).

Garduño, H. (2005b) 'Lessons from implementing water rights in Mexico', in Bruns, B.R., Ringler, C. and Meinzen-Dick, R. (eds), *Water Rights Reform, Lessons for Institutional Design,* chapter 4, Washington, DC: International Food Policy Research Institute.

Gómez, L. and Ravnborg, H.M. (2011) *Power, inequality and water governance: the role of third party involvement in water-related conflict and cooperation,* CAPRi Working Paper No. 101, Washington, DC: International Food Policy Research Institute.

Gómez, L.I., Ravnborg, H.M., Paz Mena, T. and Rivas Herman, R. (2012) *Competencia por el agua en Nicaragua,* Cuaderno de Investigación 41, Managua: Instituto de Investigación y Desarrollo.

Graham, J, Amos, B. and Plumptree, T. (2003) 'Governance principles for protected areas in the 21st century', Discussion Paper, Institute on Governance in collaboration with Parks Canada and CIDA, Ottawa (Canada).

Hazell, P. (2011) 'Five big questions about five hundred million small farms', Keynote paper presented at IFAD Conference on New Directions for Smallholder Agriculture, 24–25 January 2011, Rome.

HLPE (2011) *Price volatility and food security,* Report by the High Level Panel of Experts on Food Security and Nutrition of the Committee on World Food Security, Rome.

HLPE (2013a) *Biofuels and food security,* Report by the High Level Panel of Experts on Food Security and Nutrition of the Committee on World Food Security, Rome.

HLPE (2013b) *Investing in smallholder agriculture for food security,* Report by the High Level Panel of Experts on Food Security and Nutrition of the Committee on World Food Security, Rome.

Hodgson, S. (2004) *Land and water – the rights interface,* FAO Legislative Study 84, Rome: Food and Agriculture Organization.

Hoekstra, A.Y. (2013) *The Water Footprint of Modern Consumer Society,* London: Routledge.

Hoekstra, A.Y. and Chapagain, A.K. (2007) 'Water footprints of nations: water use by people as a function of their consumption pattern', *Water and Resource Management,* 21: 35–48.

Hoekstra, A.Y., Chapagain, A.K., Aldaya, M.M. and Mekonnen, M.M. (2011) *The Water Footprint Assessment Manual: Setting the Global Standard,* London: Earthscan.

Kay, S. and Franco, J. (2012) *The Global Water Grab: A Primer,* Amsterdam: Transnational Institute.

Larsen, R.K. and Mamosso, C.A. (2014) 'Aid with blinkers: environmental governance of uranium mining in Niger', *World Development,* 56: 62–76.

Mekonnen, M.M. and Hoekstra, A.Y. (2011) *National water footprint accounts: the green, blue and grey water footprint of production and consumption, Volume 1: main report,* Value of Water Research Report Series No. 50, Delft: UNESCO-IHE Institute for Water Education.

Merrey, D.J., Meinzen-Dick, R., Mollinga, P. and Karar, E. (2007) 'Policy and institutional reform processes for sustainable agricultural water management: the art of the possible', in Molden, D. (ed.), *Water for Food, Water for Life: A Comprehensive Assessment of Water Management in Agriculture,* pp. 193–231, London: Earthscan and Colombo: International Water Management Institute.

Molden, D. (ed.) (2007) *Water for Food, Water for Life: A Comprehensive Assessment of Water Management in Agriculture,* London: Earthscan and Colombo: International Water Management Institute.

Molden, D., Frenken, K., Barker, R., de Fraiture, C., Mati, B., Svendsen, M., Sadoff, C. and Max Finlayson, C. (2007) 'Trends in water and agricultural development', in Molden, D. (ed.), *Water for Food, Water for Life: A Comprehensive Assessment of Water Management in Agriculture,* chapter 2, pp. 67–9, London: Earthscan and Colombo: International Water Management Institute.

Orr, S., Pittock, J., Chapagain, A. and Dumaresq, D. (2012) 'Dams on the Mekong River: lost fish protein and the implications for land and water resources', *Global Environmental Change,* 22(4): 925–932.

Paavola, J. (2007) 'Institutions and environmental governance: a reconceptualization', *Ecological Economics,* 63: 93–103.

Phuong, L.T.T., Huong, P.T.M. and Skielboe, T. (2010) 'The case of the Tong Chai lead mine, Con Cuong, Viet Nam', Competing for Water Case Study Report No. 9. Mimeo. Online. Available <http://subweb.diis.dk/graphics/Subweb/Water/Case%20studies/Vietnam%20Tong%20Chai%20OK2.pdf> (accessed 20 March 2014).

Ravnborg, H.M. (2012) 'Water for all? Four issues to be dealt with at Rio+20', DIIS Comment. Online. Available <http://en.diis.dk/home/diis+comments/diis+comments/diis+comments+-+engelsk/diis+comments+2012/water+for+all_+four+issues+to+be+dealt+with+at+rio_20> (accessed 20 March 2014).

Rogers, P. and Hall, A.W. (2003) *Effective water governance,* TEC Background Papers No. 7, Stockholm: Global Water Partnership/SIDA.

Sosa, M. and Zwarteveen, M. (2012) 'Exploring the politics of water grabbing: the case of large mining operations in the Peruvian Andes', *Water Alternatives,* 5(2): 360–75.

UN (2010) *United Nations Resolution 64/292: the human right to water and sanitation.* Available <www.un.org/ga/search/view_doc.asp?symbol = A/RES/64/292> (accessed 20 March 2014).

UNDP (2006) *Human Development Report 2006: Beyond Scarcity: Power, Poverty and the Global Water Crisis,* New York: Palgrave, Macmillan.

UN-Water (2013) *Water Security and the Global Water Agenda: A UN-Water Analytical Brief,* Ontario, Canada: United Nations University.

Van Lienden, A.R., Gerbens-Leenes, P.W., Hoekstra, A.Y. and Van Der Meer, T.H.H. (2010) *Biofuel scenarios in a water perspective: the global blue and green water footprint of road transport in 2030,* Value of Water Research Report Series No. 50, Delft: UNESCO-IHE Institute for Water Education.

Wolf, A.T., Yoffe, S.B. and Giordano, M. (2003) 'International waters: identifying basis at risk', *Water Policy,* 5: 29–60.

Woodhouse, P. (2012a) 'New investment, old challenges: land deals and the water constraint in African agriculture', *Journal of Peasant Studies,* 39(3/4): 777–94.

Woodhouse, P. (2012b) 'Foreign agricultural land acquisition and the visibility of water resource impacts in Sub-Saharan Africa', *Water Alternatives,* 5(2): 208–22.

World Water Assessment Programme (2009) *The United Nations World Water Development Report 3: Water in a Changing World,* Paris: UNESCO, and London: Earthscan.

7

FOOD INSECURITY IN FRAGILE STATES AND PROTRACTED CRISES

Adam Pain

Introduction

Interventions to address food insecurity have classically been a humanitarian response to a crisis in food availability triggered by a natural hazard or conflict. They have conventionally been envisaged as relatively short term interventions within a framework of disaster response and post-disaster recovery where humanitarian action has been phased out, as development efforts have been phased in. There has been an explicit duality whereby humanitarian action has been separated from 'normal' development activity, reflected both in the organizational structures of donor agencies and actors – humanitarian versus development activities – and in funding sources and flows. In some respects the former has been about providing food, whereas the latter has been about supporting agriculture and livelihoods.

However, long-running political instability in some countries and enduring weak governance in others – combined with an increasingly complex and heightened risk environment due to global economic instability, climate change, progressively severe consequences of natural hazards and a tightening of global food supplies – have led to rising food prices and price volatility. This in turn has brought about a general squeeze on people with low incomes (Heltberg et al. 2012; Hossain et al. 2013) and in some cases revolution and government overthrow. Humanitarian action has increasingly had to engage with enduring and recurrent food insecurity rather than episodic events and this has had a number of consequences.

First it has brought into question the purpose and instruments of intervention. This has led to a shift from a simple response to the symptoms of food insecurity through emergency food supplies to more ambitious attempts to address some of the proximate causes of food insecurity through interventions to provide broader livelihood support and even to try to reduce vulnerability to food insecurity. This in part has been driven by growing donor fatigue in relation to the repeated demand for emergency funding for food insecurity responses.

Second this has gradually driven a rethinking of conceptual frameworks in relation to building food security in contexts of enduring food insecurity and attempts to build more effective policy frameworks to deal with such circumstances. As will be seen, though, new analytical tools and conceptual frameworks do not necessarily lead to change in actual responses.

Third, and given the role and importance of humanitarian action in contexts of political instability, the humanitarian response has become bent to grander purposes. Donor interventions in relation to state building and military action in relation to securing the state have increasingly made instrumental use of humanitarian support for other policy objectives. The emergence of the stabilization agenda, for example, which has justified interventions in fragile states for the purpose of mitigating perceived security threats to the West (Collinson et al. 2010), has sought to use humanitarian aid for this purpose. This has posed major challenges to humanitarian principles and practice.

Finally there has been a long-running critique of humanitarianism itself and the internationalization of public welfare (De Waal 1997; Duffield 1991; Polman 2010). Arguments have been made suggesting that part of the problem is humanitarian aid itself (see Jaspars, this volume), which may have undermined any need to build a social contract between the political elite and their constituents to deter famine and food insecurity.

This chapter begins by identifying the group of countries in which food insecurity and famine has come to be concentrated over the last decade and discusses how they might be characterized. It then moves on to investigate the underlying reasons for the persistence of food insecurity in such contexts and the links between food insecurity, conflict and the risk environment. It then focuses on the increasing evidence of urban food insecurity before examining humanitarian and state building responses to food insecurity and their evolving practices.

Food insecurity, fragile states and protracted crises

In 2010, the Food and Agricultural Organization (FAO) in its assessment of the state of food insecurity in the world (FAO and WFP 2010) noted that about 66 per cent of the estimated 925 million malnourished people were to be found in just seven countries – Bangladesh, China, the Democratic Republic of Congo (DRC), Ethiopia, India, Indonesia and Pakistan. Between them China and India accounted for 40 per cent of the total. However there was a smaller but significant population of undernourished people – estimated at more than 166 million (18 per cent of the malnourished total) found in 22 countries (see Table 7.1). These countries were noted to have a set of common symptoms of chronic and enduring food insecurity, which affected a large proportion of the population – from 14 to 69 per cent. These countries were also characterized by multiple disasters and conflict which had led to a breakdown in food systems and in which both formal and informal institutions functioned poorly. Common to these countries was the relatively high proportion of humanitarian assistance in the total overseas development aid (ODA) provided to them. The majority (17 of the 22) of these countries were to be found

TABLE 7.1 Countries in protracted crises

	Number of years with a disaster between 1996–2010	Per cent of humanitarian aid/total overseas development assistance (2000–08)	Percentage of population undernourished (2005–07)
Central Asia and Middle East			
Afghanistan	15	20	Na
Tajikistan	11	13	14.9
Iraq	15	11	Na
East Asia			
Democratic Republic of Korea	15	47	33
Caribbean			
Haiti	15	11	57
Africa			
Angola	12	30	41
Burundi	15	32	62
Central African Republic	8	13	40
Chad	9	23	37
Congo	13	22	15
Cote d–Ivoire	9	15	14
Democratic Republic of Congo	15	27	69
Eritrea	15	30	64
Ethiopia	15	21	41
Guinea	10	16	17
Kenya	12	14	31
Liberia	15	33	33
Sierra Leone	15	19	35
Somalia	15	64	32.8
Sudan	15	62	27
Uganda	14	10	21
Zimbabwe	10	31	30

Source: adapted from FAO and WFP 2010: table 1, p. 13 and table 2, p. 15.

in sub-Saharan Africa, one (Haiti) in the Caribbean, three in central Asia and the Middle East and one (the Democratic Republic of Korea) in the Far East.

FAO and WFP (2010) used the term 'protracted crises' to characterize these contexts of chronic food insecurity, a definition based on indicators or symptoms of the profile of food insecurity. Many of these countries are also labelled as 'complex emergencies' or 'fragile states'. The term 'fragility' encompasses a view of the state lacking capacity or legitimacy (World Bank 2011), thereby enhancing the risk of violence. Thus the definition depends on the policy stance and intended response taken towards them. A 'complex emergencies' definition invites a humanitarian response, a 'protracted crises' perspective argues for a longer term vision for humanitarian interventions, while a 'fragile states' terminology or even the phrase 'fragile and conflict affected states' (OECD 2011) situates chronic food insecurity in relation to the limits of state capacity to assure food security (Alinovi et al. 2007). But as the Asia Foundation (2013) has noted, large scale armed conflict can persist in strong states as well as weak ones, and sub-national conflicts affect over half of the countries in South and Southeast Asia although not necessarily with food security effects.

Thus although FAO and WFP's (2010) list of countries (and other authors had added to it) undoubtedly points to a series of commonalities in terms of symptoms that these countries exhibit with respect to food insecurity, there are also important differences among them. The food security crises of Ethiopia and DRC, for example, have had very different pathologies. Equally the Democratic Republic of Korea, a country that suffered a major famine in the 1990s and with over 33 per cent of its population estimated to be chronically undernourished, could hardly be called fragile. If anything it is a strong authoritarian state but, as has been seen with the recent collapse of authoritarian states in North Africa, it might not take more than a critical event to trigger collapse; in that sense it might be seen as brittle. Equally generic models that portray conflict states as having weak institutions have reflected a bias towards the formal state and have taken little account of other institutions, including other governance structures and markets, that populate the landscape. In Afghanistan for example (see Pain, this volume), markets have continued to function and ensure wheat supplies, even during conflict and the drought. The influx of NGOs into Afghanistan after September 2001, many of them drawing on African experience, led to an ill-founded drought crisis narrative and false assumptions of famine (Pinney and Ronchini 2007). In Somalia, the use of cash aid and its relative effectiveness as an instrument in response to the famine in 2011 pointed to well functioning financial institutions despite the apparent absence of the state (Leonard with Samantar 2011). These authors also pointed to enduring forms of informal local governance with state-like features.

The dynamics of change and the evolving risk environment: Food insecurity – a consequence or cause of conflict and instability?

Although the number of interstate and civil wars have declined (along with mortality rates) during the twentieth century (World Bank 2011), violence and conflict that is neither war nor peace nor specifically criminal or political (Roitman 2005) has persisted. This has given rise to enduring episodic but repeated violence and instability whereby countries and sub-national regions rarely graduate to a condition that can be described as post-conflict. While the underlying causes are structurally related to inequalities with respect to access to physical security, economic security and justice, the drivers of conflict are both internal to the country as well as externally linked to globalization.

At the same time there has been an increasing incidence of disasters triggered by extreme climate events. In the period from 1971 to 2003 alone they are reported to have increased fourfold (cited in Harvey et al. 2010: 7). Although the precise links between climate change and disaster incidence remain unclear, the general expectation is that the incidence of climate induced disasters is likely to increase. These will be superimposed over longer term effects of climate change which are likely in many places to make agriculture more risky. The coming together of the effects of a tightening of global food supplies, linked to both shifting supply and demand factors, greater food price instability and a global financial crisis since 2008 have further compounded the risk environment. Fragile and conflict affected states have experienced, as discussed with respect to Nepal (see Pain et al. this volume), a convergence of multiple crises which have compounded the effects on food insecurity. This gives rise to new questions as to the nature of the complex and dynamic links between food insecurity and conflict.

There is a recent body of literature that has sought to generalize about the causal relations between them; see for example FAO:

> a deeper analysis of the relationship between protracted crises and food security outcomes shows that changes in income, government effectiveness, control of corruption and the number of years in crisis are significantly related to the proportion of the population who are undernourished. These factors, plus education, are also all significantly related to a country's Global Hunger Index.
>
> *(FAO and WFP 2010: 16)*

The above quotation exemplifies this attention to correlating a range of factors with food security outcomes. However, little can be said about causality, and the factors identified (for example government effectiveness, control of corruption etc.) invite major questions as to how they can be robustly calibrated and assessed across countries.

There is certainly evidence and argument that can be brought to bear on the proposition that conflict can be a cause of food insecurity and famine. As Macrae and Zwi (1992) have argued with respect to Africa, the deliberate targeting of food production, consumption and distribution as an instrument of war played an important role in the pre-1992 famines in Angola, Ethiopia, Mozambique, Somalia and the Sudan. The Soviet Union was responsible for a massive destruction of rural infrastructure in Afghanistan during the 1980s. More contemporary conflicts such as those in DRC can also be seen as food wars because food can be used as a weapon; food systems can be destroyed during the course of a conflict (as shown by the Taliban on the Shomali plains outside Kabul in Afghanistan during the 1990s); and food insecurity persists as a legacy of the conflict. The conflict in Bosnia and Herzegovina during the early 1990s led to the loss of an estimated 55 per cent of cultivated agricultural land and half the number of farm animals in that region (Pinstrup-Andersen 2012).

But equally the case has been made that food insecurity can be a cause of violence, for which there is both historical and more recent evidence. The food price rises of 2007–8 are reported to have precipitated food protests and riots in 60 countries (Zaman et al. 2008) and, most notably in the cases of Egypt and Tunisia, are said to have contributed to the overthrow of their presidents. However, as Brinkman and Hendrix (2011) note, although there has been theorizing about the effects of food insecurity on economic and social grievance, the causal arguments lack detailed empirical evidence.

There is no doubt that food insecurity and poverty can also give rise to conflict and violence. Intergroup competition for resources, including access to cash crops and the means to produce and profit from them, can give rise directly to conflict, although the conditions or combination of conditions that trigger such violence may vary. In Ethiopia, drought was arguably the key trigger for violence in 1973–4, while price increases for rice in Indonesia (1998) or coffee in Rwanda (1994) have been suggested as the proximate causes of a violent response (Messer and Cohen 2006). But equally some of the countries that have had enduring conflict over the last twenty years such as Angola, DRC, Papua New Guinea (PNG) and Sierra Leone have been some of the most resource rich. These examples highlight the 'Dutch disease' wherein opportunities to exploit other valuable natural resources lead to both declining attention to agriculture and increased conflict over these highly profitable resources (Collier 2007).

There are severe limitations to the extent to which generalized statements about the links between food security and conflict are useful since the linkages are usually highly context specific. Most conflicts are complex and have deep causes. These can include competition for land and water in contexts where agriculture makes a major contribution to livelihoods. Such competition has been compounded by an increasingly risky growing environment brought about by climate change and an increased frequency of extreme weather events resulting in irregular rainfall, droughts, floods and extreme temperature fluctuations. Increasing scarcity of resources, particularly of water, is likely to exacerbate conflicts both between upstream and downstream farmers as well as between countries. But the fact that resource scarcity can lead to conflict is a reflection of power relations and the lack of peaceful means to resolve such conflict, reflecting governance practices and performance.

Accounts of country-specific food insecurity events often tend to be superficial and address proximate causes at best. As Majid and McDowell (2012) show in relation to the 2011 famine in Somalia, deep understanding of the context is needed to tease out some of the underlying determinants of the famine. They argue that the normative account that the famine was a consequence of drought and crop failure and lack of access to resources due to the Al-Shabaab is not well supported by the evidence. They also contend that the underlying causes of vulnerability were more social and political than natural. Further account, as Levine (2012) notes, also has to be taken of the role of international action in 2006 in provoking the crisis in Somalia given the stance of the international community towards the transitional government. On the one hand international actors adopted the stance of neutral humanitarians but on the other their practice was partisan and political.

Urban food insecurity

Much of the discussion on food insecurity has had an inherent rural bias. Thus the FAO and WFP's paper (2010) on food insecurity in protracted crises implicitly focuses on the rural dimensions in its analysis and suggested responses. The displacement of rural populations to urban areas as a result of conflict and other disasters has long been recognized (Pantuliano et al. 2012), raising important questions of urban food security. But only more recently has the scale of vulnerability of urban populations in fragile and conflict affected countries received more focussed attention. This recognition crystallized in responses to urban siege (Sarajevo) and most recently in relation to the earthquake of 12 January 2010 that struck near Port-au-Prince in Haiti.

Urban food insecurity is in many respects a consequence of growing rural food insecurity and conflict related displacement. Over 40 per cent of Haiti's population has been urbanized as a direct consequence of a flight from deep rural poverty and food insecurity over the previous decades. Similarly the rapid growth of Mogadishu in Somalia, Khartoum in the Sudan and Kabul in Afghanistan all reflect a long history of drought and conflict in these three countries. Mogadishu for example in the mid-1970s was a city of 0.5 million; in 1986 the population had grown to 1.2 million and since 2000 it has grown further, although the estimates of its size vary (Grünewald, 2012). Its growth rate has fluctuated since 1990, driven by a first bout of conflict induced displacement and reinforced by further periods of conflict. Since the early spring of 2011 a mass displacement to the cities has occurred as a result of drought. Khartoum has grown from a population of 0.25 million in 1956 to 3.3 million in 1990, and the most recent census of 2008 reported over 5.25 million, although this is disputed (Pantuliano et al. 2011). In 2011 over 2 million of this population (nearly 40 per cent) were estimated to be internally displaced persons (IDPs) uprooted by a series of severe droughts and civil conflicts (Pantuliano et al. 2011 and see Sudan chapter), although numbers may have declined since then. Kabul's population has been estimated to have doubled since 2001, when it was approximately 2 million (Metcalfe et al. 2012), driven by returning refugees from Pakistan, internal displacement from ongoing conflict within the country and the effects of drought and overall decline in the rural economy.

The rapid growth of city populations in these countries has in some cases generated specific acute humanitarian crises. Although the cities became a relative sanctuary for many rural Afghans during the 1980s, from 1992 to 1996 during the conflict over Kabul between the warring Mujahedeen factions it became a direct battleground with acute consequences for its inhabitants, many of whom fled (Metcalfe et al. 2012). Similarly Mogadishu has been a site of major conflict between warring factions that have divided the city between them, giving rise both to acute risk to life for the civilian population and major problems of access to food (Grünewald 2012).

More generally the rapid growth of such cities has generated chronic livelihood and food insecurity for both the large populations that have been displaced there as well as the pre-existing poor. Indeed, one of the problems that humanitarian agencies have constantly faced given the diffuse urban settlement of IDPs is separating clearly, for programming purposes, the IDPs from the already urban poor. Both populations have blended into each other, given the long history of displacement to the city and the fact that they face very similar life challenges (Haysom 2013). In many cases IDPs depend on host families for their survival. The growth of the displaced populations has in particular revealed the weak and problematic governance of many of these cities, where many of the national level political conflicts are played out in a more intense and localized arena by powerful actors. For this reason the IDP populations in many cases are seen to be a political problem and challenge, either isolated where possible to IDP camps on the margins of the city as in Khartoum (Pantuliano et al. 2011) and subject to control or marginalized through patterns of deliberate exclusion.

Many of these cities have long outgrown the basic provision of public infrastructure in relation to water supply, road, power, health care and education that they might have been endowed with three or four decades ago. War and neglect have aggravated this breakdown in provision. For example, the water supply in Mogadishu, (Grünewald 2012: S109) a city located in an area of natural water scarcity, has depended on underground reserves. These have become overdrawn and contaminated because of the breakdown of clean water supply and sewage systems during the many years of conflict. Kabul and Khartoum face similar problems (Metcalfe et al. 2012, Pantuliano et al. 2011).

Equally critical have been questions of land and settlement with the influx of displaced people being forced to settle in slums, often on the periphery of the cities and often in unsafe places (e.g. on hillsides, in low lying areas subject to floods or areas subject to pollution) and with tenure insecurity. Such settlements have been subject to demolition as in Khartoum (Pantuliano et al. 2011) or the IDPs have been discriminated against with respect to claims to and access to services, as city leaders have attempted directly or indirectly to discourage such settlement.

Compounding the general lack of security, safety and access to basic public services has been the high degree of economic insecurity for IDPs and the resident poor. Many of these cities in conflict countries have evidenced economic growth driven by a range of factors. This growth has ameliorated the unemployment situation to some extent, but the highly skewed nature of this growth has also cemented many aspects of social exclusion. In the case of Khartoum, oil exports have driven the economy, feeding in to property development and land speculation (Pantuliano et al. 2011). In Kabul there has been a building boom driven both by the spillover effects of the level of aid and military spending since 2001, including a rental market for expatriates and the recycling of profits made from the opium economy. This has contributed to a land grab of public land by the political elite and powerful. The growth of these unruly city economies in which powerful non-state actors are engaged has led to a more general rise in violence and crime which has contributed to the general sense of unsafe conditions (Pantuliano et al. 2012: S8). While organized crime linked to powerful figures is one element of the climate of violence, as Pantuliano et al. put it (2012: S10), 'the phenomenon of urban violence is now increasingly linked to the discrimination and marginalization of certain groups, including displaced communities, in under-resourced or poorly serviced urban areas.'

At the heart of this violence is the competition for resources and scarce employment in the informal sectors of the city economies between the displaced populations and the existing or host communities. In some cases, as in Khartoum, the government has attempted to regulate movement of IDPs and their access to employment through the requirement for work permits (Pantuliano et al. 2011). Much of the employment is in the informal sector in casual semi-skilled work, and often on a part time or irregular basis, with better opportunities in the centre of the city in comparison with its margins where many of the IDP are located. There is a gendered dimension to this in that women are increasingly playing a more significant role as the wage earner through diverse activities including food sales, the brewing and sale of alcohol, and prostitution. But even where scarce casual employment can be found, average daily wages in Khartoum are barely sufficient to meet daily subsistence needs (Pantuliano et al. 2011: 15). While the evidence suggests that the levels of malnutrition in Khartoum are lower than elsewhere in Sudan, food insecurity is viewed as a 'hidden feature' of urbanization because of the unsafe environment and quality of the diet (Pantuliano et al. 2011: 15).

Employment opportunities for the poor in Mogadishu, both for host and IDP populations, are confined in general to begging and to irregular casual labour: men working as porters or garbage collectors, women undertaking domestic jobs or petty trading of food items and children performing small scale jobs. While the government in Khartoum has had a long term policy of subsidizing bread prices in the city (see Sudan chapter), Mogadishu on account of the patterns of violence and conflict within the city itself has suffered severe shortages of food supply and a high volatility in prices linked to speculation (Grünewald 2012: S118), thus potentially creating greater levels of food insecurity. But as Grünewald notes, under such conditions estimates of both IDP population sizes and malnutrition rates are at the very best 'rough extrapolations of limited knowledge and patchy information'.

For Kabul, as Metcalfe et al. (2012) found, the same conditions of uncertain and irregular work in the informal economy characterize the employment opportunities for the internally displaced. There is also a marked seasonal dimension to employment opportunities, with a sharp decline during the winter period when more income is needed to cover expenditures in relation to heating costs. A recent study by WFP (2013) indicated that of 8,886 households in 61 informal camps for IDPs in Kabul, 25 per cent of the households held no food stocks at all and another 23 per cent held stocks for up to three days. The food consumption score was estimated to be poor for over 92 per cent of households.

In summary it is clear that major cities in fragile and conflict affected countries exhibit specific food insecurity challenges generated by a range of factors that are considerably more complex than in rural areas. These include not only the displacement of significant rural populations into the cities but the collision of these populations with a range of drivers including exclusionary patterns of governance, limited public goods availability and economic insecurity.

Humanitarian responses to food insecurity

The congruence of the food price spikes, rising fuel prices and financial crises that came together in 2008 and which have contributed to the increased risk environment have had global effects. The complexity and scale of these shocks have been unprecedented and as Heltberg et al. (2012) show, they have had a severe impact on poor people in general. This escalation in the global risk environment has overlain a more general increase in the number of disasters triggered by extreme climate events and protracted conflicts giving rise to major humanitarian disasters. Accordingly, the long drawn-out crises of Sudan, DRC, Iraq and Afghanistan continue to draw more than 50 per cent of international humanitarian aid.

In parallel with the evolving landscape of disasters, there has been an evolution in the instruments of humanitarian response, with an apparent shift from food aid to food assistance. However, given the fact that different donors and agencies are not consistent in their definition of food assistance and food aid (Harvey et al. 2010), the trends are not easy to untangle. In principle a shift from food aid to food assistance implies a move from the giving of food directly to those in need to the use of a wider set of instruments (e.g. food aid in kind, cash assistance, production and market support) to help people reliably secure food. However, such instruments can be used for purposes other than food aid. Food assistance can also be used to support transitions out of humanitarian need, i.e. within a framework for longer term development purposes, but again practices vary between donors. The consequences are that long term trends in food aid or food assistance are difficult to assess (Harvey et al. 2010).

In 2008 about 54 per cent of the 2.6 million tons of emergency food aid that was delivered went to just five countries – Ethiopia, Sudan, Somalia, Zimbabwe and Afghanistan. In the past much of the food aid was drawn from surpluses in the West, but increasingly humanitarian agencies have been procuring food more locally and regionally. The US continues to take the leading in providing food aid, with other major contributions coming from the EU, Canada and Japan. However, governments that are not members of the donor assistance community (DAC) or

Organisation for Economic Co-operation and Development (OECD) are increasingly funding food aid or contributing, as in the case of South Sudan, India and Kenya, to food aid in their countries.

A key driver behind the shift in terminology from food aid to food assistance, whatever the substance of that shift means in practice, has been the recognition that repeated food aid responses to long-running crises may have at best failed to address the root causes of the crises or at worst actually contributed to them (see Jaspars, this volume). There has therefore been an increasing interest in seeking to connect humanitarian responses to transitions out of crises and longer term development goals. FAO for example has developed what it calls its 'twin track' approach to explicitly link relief to longer term development processes (FAO 2003).

Food aid and food assistance have increasingly been seen as elements of a broader package of policy measures (often labelled as social protection) that are designed to reduce the risk of households becoming food insecure. However, as Ellis et al. (2009) make clear, social protection in its protective and preventative role has an anticipatory function in seeking to minimize food security risks caused by unexpected shocks. A social protection approach may have a role in fragile contexts under, in a sense, conditions of normal or predictable vulnerability to food insecurity. Where the state or government is part of the problem, and particularly where the crises are chronic, social protection provided by non-government actors may be one approach to use. But with more catastrophic and unpredictable failures of food security, whether due to disaster or conflict, the role of social protection may be limited. It has to be recognized that protracted crises have in many ways become the greater challenge than the short acute crises.

Over the past decades a set of assessment tools has been developed to provide early warning of impending food security crises. These tools combine various measures of assessment of production, prices and household food security. However as the recent drought in Somalia showed (Levine et al. 2011), even the availability of such early warning assessment (based on technical monitoring rainfall, prices, etc.) did not trigger a humanitarian response until it was almost too late.

Moreover, as a recent Feinstein Center study has made clear (Maxwell and Stobaugh 2012) in its analysis of what drives responses to humanitarian crises, the mandates of agencies essentially predetermine the sorts of analyses or needs assessments that they will do in the first place. Thus a mandate for nutritional support will lead to a focus on nutritional assessment. This in turn can often lead to a predictable set of interventions to respond to the need, often driven by the resources that the agencies know donors will be willing to provide. In sum, as Levine and Chastre (2004) noted with respect to an assessment of emergency food security interventions in the Horn of Africa, many of the responses were essentially pre-packaged, divorced from evidence or analysis; and they had little impact on food security. Developing a response analysis framework to more clearly link needs assessments to response design is still a work in progress.

There is a further dimension to humanitarian engagement in fragile and conflict affected countries and this relates to the instrumental use of humanitarian action, with food aid as a central component, for broader and political stabilization goals. This has come to the fore in Afghanistan and Iraq (Collinson et al. 2010) but is part of a wider practice. Stabilization reflects an agenda of the West in seeking to mitigate what are perceived to be threats to Western security posed by conflict countries. Accordingly, humanitarian agendas have been bent to the aims of stabilization and its role in transformative peace building. For example, while food assistance intended to deliver improved health outcomes in relation to nutrition is only part of the humanitarian response to post-conflict peace building, the influence of funders to direct such activities can be significant. Thus in 2007, 50 per cent of USAID's assistance programme in Afghanistan was focussed on just four provinces which had the greatest level of insurgent activity (Wilder and Gordon 2009), although not necessarily the greater levels of food insecurity. As Collinson et al. (2010) note, given the politics and divergent interests of the major Western players in state reconstruction, a principled humanitarianism that prioritizes response in relation to need may be severely constrained and action limited to short term relief because of insecurity and limited access.

More recently the emerging evidence (Jackson and Aynte, 2013) of humanitarian agencies making payments to Al-Shabaab in Somalia in order to access households at risk of famine has illustrated the double risk such agencies face. On the one hand there is a need to negotiate with armed groups in order to reach those at risk. On the other, Western governments, through labelling groups such as the Al-Shabaab as terrorists, make it even more difficult for humanitarian actors to negotiate because doing so may put them at risk of breaching counter-terrorism laws.

State building, food security and agricultural transformations

Many fragile states are heavily dependent on world markets for food imports (World Bank 2011) as seen in the cases of Sudan, Afghanistan and Nepal discussed in this book. This made them not only vulnerable to food insecurity during the food price spikes of 2008–10, but the long term effects of climate change, particularly in relation to water supply, are likely to increase rather than reduce the risks to national food supplies. As has been widely noted, the share of agriculture in development assistance in general has been under long term decline from a peak of about 17 per cent in the 1980s to about 5 per cent in the period from 2006 to 2008 (World Bank 2011: 230). Much of the aid to conflict affected countries has been relatively short term and humanitarian in nature, with a limited focus on longer term development. According to the World Bank (2011) only about 18 per cent of spending on agriculture over the last three decades has gone to fragile and conflict affected countries. The case has been made therefore that greater effort is needed within these countries to develop agriculture and improve domestic supply to reduce food security risks.

Much of the thinking on agricultural development draws from the evidence of history and patterns of agrarian transformation that have happened in the past. The World Development Report 2008 (World Bank 2008) for example has explicitly argued for an evolutionary model in its characterization of three worlds – agriculture based countries, transforming countries and urbanized countries. This assumes that the patterns of transformation seen in the past could be reproduced again through market driven agriculture and that an agriculture-led transformation of the past is possible now. Such an approach to transformation has been strongly evident in the agricultural policies of post-2001 Afghanistan (see Pain, this volume) and in Nepal (see Pain et al., this volume).

But there are several grounds on which to be doubtful as to whether new agricultural transformations that follow the path of past ones are possible now even in non-fragile or conflict settings. For a start, as Dercon has recently observed (2013), there has been a long-running tendency to overstate the role that agriculture can play in growth and poverty reduction. He argues with respect to Africa that given the heterogeneity in economic contexts, in an open economy the drivers of economic growth may be very different from those in a closed economy. Indeed the recent drivers of growth that have been seen in African economies have been more natural resource driven (e.g. mining) than agriculture driven. Accordingly, Dercon (2013) suggests that the potential for agriculture to play a key role in growth and poverty reduction is highly context specific. Ellis (2010) has been equally doubtful of the capability of agriculture in sub-Saharan Africa to drive growth and poverty reduction, arguing that much more attention needs to be given to rural–urban interdependencies. As Ellis put it:

> [R]ural poverty in SSA will sustainably decline only when people leave agriculture to participate in the growth of other sectors thus creating urban demand for food that serves to ensure higher and more stable incomes for those farmers that remain behind.
>
> *(2010: 48)*

Further, there is evidence that the pathways of successful agricultural transformation in the recent past, as seen in the successful developmental Asian states (see chapters on South Korea and Viet Nam), have followed policy routes distinctly different from the market driven agriculture model of the World Bank. Land reform, cheap credit and investment in public goods, combined with tariff barriers to defend an expansion of domestic industry and export-led manufacturing, were the pathways to development in Viet Nam and Taiwan (Studwell 2013). As others have noted (Dorward et al. 2004), India's green revolution–driven transformation in the 1960s and 70s was on the back of a long history of investment in public goods, subsidized credit, guaranteed prices and rising urban demand – conditions which simply do not exist in most fragile and conflict affected countries. But note should also be taken of the incompleteness of India's agrarian transformation, resulting in, on the one hand, a stagnant agriculture in many parts of the country which has become a poverty trap, and on the other, an industrial development that is disconnected from agriculture and draws its financing from elsewhere (Lerche et al. 2013).

India's stalled agrarian transformation is illustrative in many ways of the challenges that countries that have yet to undergo agrarian and demographic transitions face (Losch et al. 2012). Drawing attention to the challenges that many sub-Saharan African countries face as late developers in the context of a global open economy and the constraints of climate change, Losch et al. (2012) question the viability of the historical pathways of structural transformation, arguing that the market driven route may be feasible for a minority of farmers but not the majority. As Losch et al. (2012: 8) observe, even for Kenya's widely cited horticultural success story, which is the second largest commodity export of the country, less than 50,000 of Kenya's more than 3.5 million farm households are engaged in it. For the rest of Kenya's farmers, as with a majority of sub-Saharan farmers, given the needs of subsistence and the risks of markets, crop production is stubbornly centred around staple production for self-subsistence. Furthermore, the poorer the household the higher the proportion of farm production that is likely to be consumed. It follows that a focus on staple production as a step in integrating agriculture back into broader rural development processes may be a more secure route for late developing countries than the evolutionary model offered by the agriculture for development model (World Bank 2008).

But the policy prescriptions offered by Losch et al. (2012) implicitly require a level of state competence and social commitment that is largely absent from many fragile and conflict affected countries. Demands for rebuilding internal capacities and designing comprehensive development strategies for rural development are long term aspirations. But two important lessons can be drawn from the study. First a focus on market driven agriculture in fragile and conflict affected countries is at best likely to meet the aspirations of larger, better resourced and better located farmers and will neglect the subsistence needs of most. Further, an assumption that markets can work for the poor assumes a level playing field assured by the state. There is little reason to believe in contexts where political governance is so problematic that the governance of the economic marketplace will be any less so. Indeed all the evidence on the performance of markets under conditions of conflict (Roitman 2005) shows them to be characterized by exclusionary rent-seeking practices. The evidence on the development of commercial agriculture in Sudan (see Sudan chapter) also points to the limits of such development meeting food security needs.

The second lesson that can be drawn relates to the importance of achieving food security as a policy objective in state building efforts. In this respect the lessons from Afghanistan are clear. As a late developer in all respects, with a significant dependence on market supplied food imports to meet food security needs for a population with limited opportunities to move out of the rural economy, meeting food security needs should have been a primary objective. But the failure of the reconstruction effort to prioritize food security over agricultural market driven development was one of many serious missteps in policy. One outcome of this was the shift into opium poppy production, although for many this has not redressed the rising level of food insecurity since 2001. As Ha Choon Chang suggests (2009: 5), '[T]he issue of national food security needs to be taken very seriously when the country is at low levels of economic development', and there are high risks of food insecurity and malnutrition. If the capacity to provide basic physical security is an essential first step in state building (Department for International Development 2010), then assuring basic food security as the Southeast Asian states did in their trajectory of development is a second step that follows close behind. The case for doing so under conditions of rising market and climate induced risks is now even greater.

All this presupposes of course that 'the state is necessary to ensure human welfare and that its secure establishment requires some kind of social contract' (Leonard with Samantar 2011). If as Sudan (see Jaspars, this volume) illustrates, it is possible to have economic growth and rising prosperity at the same time as chronic food insecurity, the possibilities for building a social contract under such conditions are limited. Indeed for collapsed states such as Somalia, as Leonard with Samantar (2011) argues, one may have to recognize that the state-centric approach to reconstruction is part of the problem rather than the solution.

Conclusions

Under conditions of globalization and a rising risk environment, fragile states and protracted crises are unlikely to undergo positive transformations either quickly or easily. Chronic food insecurity is likely to persist. There can be economic growth in rentier economies or those endowed with natural resources such as oil, but growth from which the poor and marginal benefit requires a set of basic preconditions. The first condition is that of security, but security requires a political settlement. The liberal model of state building combined with its short time horizons has proved incapable of creating the conditions for political settlement, not least for reasons of conflicting interests and agendas by international actors. There are clear dangers of humanitarian practice – despite principles of neutrality, impartiality and independence – such as being influenced by donors' strategic objectives and thus becoming part of the problem. The humanitarian needs are likely to grow. So are the risks of external interventions subverting the potential for a more gradual creation of a domestic political order that may not necessarily be liberal but may well be the best that can be achieved under the circumstances.

References

Alinovi, L., Hemrich, G. and Russo, L. (eds) (2007) *Beyond relief: Food Security in Protracted Crises,* Rugby, UK: Practical Action.

Asia Foundation (2013) *Aid to Subnational Conflict Areas: Synthesis Report,* San Francisco: The Asia Foundation.

Brinkman, H-J. and Hendrix, C. S. (2011) *Food Insecurity and Violent Conflict: Causes, Consequences, and Addressing the Challenges,* Occasional Paper No. 24, Rome: World Food Programme.

Chang, H.-C. (2009) 'Rethinking public policy in agriculture: Lessons from history, distant and recent', *Journal of Peasant Studies,* 36(3): 477–515.

Collier, P. (2007) *The Bottom Billion: Why the Poorest Countries Are Failing and What Can Be Done About It,* Oxford: Oxford University Press.

Collinson, S., Elhawary, S. and Muggah, R. (2010) 'States of fragility: stabilisation and its implications for humanitarian action', *Disasters,* 34(S3): S275–S296.

Department for International Development (2010) *The Politics of Poverty: Elites, Citizens and States. Findings from Ten Years of DFID-funded Research on Governance and Fragile States 2001–2010, A Synthesis Paper,* London: Department for International Development.

Dercon, S. (2013) 'Agriculture and development: revisiting the policy narratives', *Agriculture Economics,* 44(S1): 183–7.

De Waal, A. (1997) *Famine Crimes: Politics and the Disaster Relief Industry in Africa,* Indiana: James Currey.

Dorward, J., Kydd, J., Morrison, J. and Urey, I. (2004) 'A policy agenda for pro-poor agricultural growth', *World Development,* 32(1): 73–89.

Duffield, M. (1991) *The Internationalisation of Public Welfare: Conflict and Reform of the Donor/ NGO Safety Net,* The Hague: Institute of Social Studies.

Ellis, F. (2010) 'Strategic dimensions of rural poverty reduction in sub-Saharan Africa', in Harriss-White, B. and Heyer, J. (eds), *The Comparative Political Economy of Development: Africa and South Asia,* pp. 47–63, London: Routledge.

Ellis, F., Devereux, S. and White, P. (2009) *Social Protection in Africa,* Cheltenham: Edward Elgar.

FAO (2003) *Anti-Hunger Programme. A Twin Track Approach to Hunger Reduction: Priorities for National and International Action,* Rome: Food and Agricultural Organisation.

FAO and WFP (2010) *The State of Food Insecurity in the World: Addressing Food Insecurity in Protracted Crises,* Rome: Food and Agricultural Organisation and World Food Programme.

Grünewald, F. (2012) 'Aid in a city at war: the case of Mogadishu, Somalia', *Disasters,* 36(S1): S105–S125.

Harvey, P., Proudlock, K., Clay, E., Riley, B. and Jaspars, S. (2010) *Food Aid and Food Assistance in Emergency and Transitional Contexts: A Review of Current Thinking,* London: Humanitarian Policy Group, Overseas Development Institute.

Haysom, S. (2013) *Sanctuary in the City? Urban Displacement and Vulnerability. Final Report,* HPG Report 33, Humanitarian Policy Group, London: Overseas Development Institute.

Heltberg, R., Hossain, N. and Reva, A. (eds) (2012) *Living Through Crises: How the Food, Fuel and Financial Shocks Affect the Poor,* Washington, DC: The World Bank.

Hossain, N., King, R. and Kelbert, A. (2013) *Squeezed: Life in a Time of Food Price Volatility, Year 1 Results,* Joint Agency Research Report, Brighton and Oxford: Institute of Development Studies and Oxfam.

Jackson, A. and Aynte, A. (2013) *Talking to the Other Side: Humanitarian Negotiations with Al-Shabaab in Somalia,* Humanitarian Policy Group, London: Overseas Development Institute.

Leonard, D. K. with Samantar, M. S. (2011) 'What does the Somali experience teach us about the Social Contract and the State?', *Development and Change,* 42(2): 559–84.

Lerche, J., Shah, A. and Harriss-White, B. (2013) 'Introduction: Agrarian questions and left politics in India', *Journal of Agrarian Change,* 13(3): 337–50.

Levine, S. (2012) 'Livelihoods in protracted crises', Paper presented to the High-Level Expert Forum on Food Insecurity in Protracted Crises, 13–14 September 2012, Rome.

Levine, S. and Chastre, C. (2004) *Missing the Point: An Analysis of Food Security Interventions in the Great Lakes,* HPN Network Paper 47, Humanitarian Practice Network, London: Overseas Development Institute.

Levine, S., with Crosskey, A. and Abdinoor, M. (2011) *System failure? Revisiting the problems of timely response to crises in the Horn of Africa*, Humanitarian Practice Network Paper 71, London, Overseas Development Institute.

Losch, B., Fréquin-Gresch. S. and White, E. T. (2012) *Structural Transformation and Rural Change Revisited: Challenges for Late Developing Countries in a Globalizing World*, Washington, DC: Agence Française de Développement and World Bank.

Macrae, J. and Zwi, A. B. (1992) 'Food as an instrument of war in contemporary African famines: a review of the evidence', *Disasters*, 16(4): 299–321.

Majid, N. and McDowell, S. (2012) 'Hidden dimensions of the Somalia famine', *Global Food Security*, 1: 36–42.

Maxwell, D. and Stobaugh, H. (2012) *Response Analysis: What Drives Program Choice?* Boston: Feinstein International Center.

Messer, E. and Cohen, M. J. (2006) *Conflict, Food Insecurity and Globalization*, Food Consumption and Nutrition Division, Discussion Paper 206, Washington, DC: International Food Policy Research Institute.

Metcalfe, V. and Haysom, S. with Martin, E. (2012) *Sanctuary in the City? Urban Displacement and Vulnerability in Kabul*, Humanitarian Policy Group Working Paper, London: Overseas Development Institute.

OECD (2011) *Busan Partnership for Effective Development Co-operation*, Fourth High Level Forum on Aid Effectiveness, Busan: Republic of Korea. Online. Available <www.oecd.org/dac/effectiveness/49650173.pdf> (accessed 17 December 2013).

Pantuliano, S., Assai, M., Elnaiem, B. A., McElhinney, H. and Schwab, M. with Elzein, Y. and Ali, H.M.M. (2011) *City Limits: Urbanisation and Vulnerability in Sudan Khartoum Case Study*, London: Overseas Development Institute.

Pantuliano, S., Metcalfe, V., Haysom, S. and Davey, E. (2012) 'Urban vulnerability and displacement: a review of current issues', *Disasters*, 36(S1): S1–S22.

Pinney, A. and Ronchini, S. (2007) 'Food security in Afghanistan after 2001: from assessment to analysis and interpretation to response', in Pain, A. and Sutton, J. (eds), *Reconstructing Agriculture in Afghanistan*, chapter 6, pp. 119–64, London and Rome: Practical Action and the Food and Agriculture Organization.

Pinstrup-Andersen, P. (2012) 'Food security and human conflict'. Online. Available <http://smkern.com/pinstrup/wordpress/?p = 317> (accessed 2 February 2013).

Polman, L. (2010) *War Games: The Story of Aid and War in Modern Times*, New York: Viking.

Roitman, J. (2005) *Fiscal Disobedience: An Anthropology of Economic Regulation in Central Africa*, Princeton and Oxford: Princeton University Press.

Studwell, J. (2013) *How Asia Works: Success and Failure in the World's Most Dynamic Region*, London: Profile.

WFP (2013) 'Kabul 61 informal settlement (KIS) vulnerability to food insecurity', Internal presentation to World Food Programme, Kabul, 17 March 2013. Online. Available <http://afgarchive.humanitarianresponse.info/sites/default/files/WFP Presentation – KIS Vulnerability to Food Insecurity 28 Feb 2013.pdf> (accessed 17 December 2013).

Wilder, A. and Gordon, S. (2009) 'Money can't buy America love', *Foreign Policy*, 1 December. Online. Available <www.foreignpolicy.com/articles/2009/12/01/money_cant_buy_america_love> (accessed 17 December 2013).

World Bank (2008) *World Development Report 2008: Agriculture for Development*, Washington, DC: World Bank.

World Bank (2011) *World Development Report 2011: Conflict, Security and Development*, Washington, DC: World Bank.

Zaman, H., Delgado, C., Mitchell, D. and Revenga, A. (2008) 'Rising food prices: are there right policy choices?', *Development Outreach*, 10(3): 6–8.

PART II
Case studies

8

SUDAN'S PERMANENT FOOD EMERGENCY

A historical analysis of food aid, governance and political economy

Susanne Jaspars[1,2,3]

Introduction

This chapter examines the development of a permanent emergency in Sudan and the role and influence of food aid and food security policies and strategies. The focus is on food aid as part of the international development package until the late 1980s, and as the major component of international food security and humanitarian assistance to Sudan from the 1990s until the present time. Trends in food aid during the latter period can be seen to reflect trends in humanitarian assistance overall. The chapter covers fifty years of food aid and reviews historical changes in food aid policy and practice in relation to changes in Sudan's food security, governance and political economy.

Sudan, like many countries in the Horn of Africa, has received food aid since the 1960s and is one of the largest recipients of food aid. At the same time, Sudan experiences unacceptably high levels of acute malnutrition and food insecurity and has done so for a long time. Food aid, whether as a development or humanitarian tool, has therefore not been successful in addressing food insecurity in the longer term. To analyse the effect of food aid further, this chapter looks at food aid as a form of governance, rather than simply a commodity to be distributed. Present day food aid involves a range of activities, institutions and authorities. The process of providing food aid includes procurement, logistics, assessments, ration planning, targeting, distribution and monitoring (Jaspars and Young 1995: 2). A range of policies, principles, standards and guidelines have been developed to determine who should receive food aid, how much, when, how, and for what purpose. It is also an industry that employs thousands of people. As such, food aid can be seen as a technology or a form of governance involving the strategization of power. Food aid is bound up with the production of knowledge and its practical application through a range of techniques, and by a number of authorities and agencies concerned with shaping the behaviour and attitudes of food aid recipients. Even when food aid has nutritional or livelihood support goals, it is also associated with many unpredictable consequences, effects and outcomes (Dean 2010; Foucault 1991; Foucault 2007). This chapter examines the changes in food aid as a technology of governance over time.

Food aid as a technology of governance interacts with local forms of governance in different ways, the power effects of which change over time as both food aid policy and practice, governance in Sudan, and global politics also shift. Whilst food aid can be good at saving lives, it cannot address the structural causes of food insecurity. Instead, food aid has been a tool for managing populations, used by international actors, the Sudanese state and non-state actors (opposition movements, internally displaced person (IDP) leaders), although the nature and functioning of this relationship is different for each. Long-term food aid also raises questions about the role of food aid in structures of patronage and the extent to which it provides compensation for an exclusionary process of development, thus maintaining inequality between centre and periphery. At the local level, links between the structures and institutions of food aid and state or non-state governance systems in turn raise questions about the impartiality and neutrality of emergency food aid.

The chapter will start with an overview of the nature of the protracted crisis in Sudan, its historical origins and changes in food security in relation to three periods in Sudan. The first period is from independence to the mid-1980s – a period when assistance was mainly state-centred, development-oriented and dominated by Cold War politics. The second period is from the mid-1980s, when food security responses in Sudan were dominated by emergency food aid – in particular at the end of the Cold War and during the military coup by Colonel El-Bashir in Sudan – and the international non-governmental organizations (INGOs) that helped distribute it. The third period is the first decade of the 2000s, when Sudan's protracted crisis became widely acknowledged whilst at the same time it became a middle income country as a result of oil revenues. This decade is also characterized by the prolonged and ongoing conflict in Darfur, the peace agreement and subsequent secession of South Sudan. The focus is on Darfur rather than other parts of Sudan because the Darfur crisis led to one of the largest global food aid responses so far.

Country context

Sudan has experienced an emergency requiring external assistance every year since 1984 (FAO 2012a). In the past decade, this included the crisis in Darfur, which has been called the world's worst humanitarian crisis and in which Sudan was the top recipient of humanitarian aid (Development Initiatives 2011: 24–5; BBC 2004). It has also been amongst the top food aid recipients for over twenty years.[4] Acute food insecurity and malnutrition remain alarmingly high for a large proportion of Sudan's population. According to the World Food Programme (WFP), '[L]ocalized nutrition surveys across Sudan reflect that the nutrition situation has not improved in any significant way in the last 25 years' (WFP Sudan 2010a: 9). A national health survey done in 2010 shows a prevalence of acute malnutrition above the emergency threshold for Sudan as a whole (Federal Ministry of Health and Central Bureau of Statistics 2012; WHO et al. 2000).[5] Not surprisingly, Sudan is also on the Food and Agriculture Organization's (FAO) list of low-income food-deficit countries (FAO 2012b). This makes the food crisis in Sudan one of the world's largest and most protracted.

Also in first decade of the 2000s this long-term emergency was accompanied by the longest and strongest period of economic growth since independence, largely as a result of the advent of oil revenues and the peace agreement with South Sudan in 2005. Growing Middle Eastern and Asian investment in telecommunications, commercial farming and Khartoum real-estate, as well as the energy sector, made Sudan one of Africa's fastest growing economies during these ten years (World Bank 2009). The case of Sudan therefore suggests that economic growth and protracted food emergency are perfectly compatible in the longer term. Modern industry, the service sector and professions in Sudan are geographically highly unequally distributed, being mainly concentrated in the greater Khartoum region between the Niles. In contrast, people on the peripheries have experienced long-term economic, social and political marginalization and regular emergencies (World Bank 2009; National Population Council General Secretariat et al. 2010: 5 and 22). The state assumes limited responsibility for social welfare beyond Khartoum but this does not mean that Khartoum is somehow disconnected from its hinterland. Since independence, the state has invariably interpreted what are essentially humanitarian or welfare crises, from rapid urbanization to disasters, refugee influxes and internal displacement, primarily as security issues (Karadawi 1999; Duffield 2002c). Accordingly, much of the recent increased government spending resulting from economic growth went to defence, national security, public order and safety rather than education, health or sanitation (Patey 2010).

Inequality is a major contributing factor to food insecurity in Sudan, as structural causes of food insecurity relate to instability and inequity in the food system (Maxwell et al. 1990). Traditionally, development policies, both national and those supported by international agencies such as the World Bank and the International Monetary Fund (IMF), focussed on the introduction of modern commercial agriculture in central Sudan. This includes large irrigated farms along the Nile (for example the Gezira scheme introduced as early as the 1920s by the British colonial government) and rain-fed farms in eastern Sudan (Gedaref), South Kordofan and Blue Nile provinces. The expansion of large-scale farms in the 1970s was facilitated by the abolition of native administration, which made it easier for the state to appropriate land for large-scale commercial farming schemes (Johnson 2003). Many of these farms were owned by those with close links to central government but much of the labour pool came from western and South Sudan (Ali 1989). National development policies largely neglected subsistence farmers. These interventions therefore favoured rich over poor and core over periphery. Real incomes of the poorest groups in Sudan fell by nearly a quarter between 1965 and 1986

(Maxwell et al. 1990). Following a period of neglect when oil revenue was high, attempting to revive these agricultural schemes once again became a government priority in 2008 (Sudan Council of Ministers 2008). This uneven development favouring the central areas of Sudan has not only been part of the war economy, it has also incorporated food aid strategies along the way.

Sudan frequently experiences drought but it was not until the 1980s that this required an international response. In contrast to its neighbouring countries, Sudan was able to deal with the 1970s crisis through a combination of government responses and the availability of agricultural work in central Sudan (African Rights 1997b). However, with economic decline, a growing national debt, austerity measures and government denial of famine, drought in 1983 and 1984 led to a severe famine in Darfur, Kordofan and Red Hills – the peripheral areas of Sudan. These same provinces (now states) have experienced drought-related acute food insecurity or famine on a regular basis since that time. Central Sudan has also been at war with one or another of its peripheral regions since 1983, when conflict in the south resumed. Large-scale conflict in Darfur has now been ongoing for more than ten years (since 2003), and in June and September 2011 conflict started in the southern Kordofan and Blue Nile, respectively. These conflicts are accompanied by war strategies aimed at deliberately undermining the resource base of those believed to support the other side, resulting in destruction, death and displacement. Militia drawn from marginalized pastoral populations often fight alongside Sudan's army (Keen 1994; Duffield 1990; Tanner 2005). War also affects livelihoods indirectly, for example by restriction of movement and therefore limiting access to land, work and markets, and by disruption of public services (Young et al. 2005; Jaspars and O'Callaghan 2008). This resulted in severe famines in South Sudan in 1988 and 1998 and humanitarian crises in Darfur from 2003 onwards. Food aid has been manipulated by all parties in these conflicts in different ways. These same strategies are now being used in Blue Nile and southern Kordofan with a major difference in that the government has imposed severe restrictions on access by international agencies to conflict-affected people, and has developed its own response apparatus, including national NGOs and a national strategic grain reserve.

These restrictions are the culmination of a long line of government actions to limit and hamper the activities of international agencies while attempting to assert greater control over the distribution of food aid – and the populations to which it is given. Every period of large-scale United Nations (UN) and INGO activity responding to famine or humanitarian crisis has been followed by a period of restriction of INGO activity. Since the late 1980s, the Sudan government has promoted local NGOs and a greater role for the Sudan Red Crescent (SRC), as well as a shift from relief to development and thus closer links with government. A strategy of Sudanization of the aid sector, a term first used in the 1990s, was revived in 2009 after the expulsion of thirteen INGOs from Darfur. Since the 1980s, senior government officials have perceived INGOs as political tools of Western governments, as supporters of rebel movements, and as carrying out activities which are not transparent, for which they have no permission, and which are therefore considered as a threat to national security (Humanitarian Aid Commission 2009; Key informant 17, 2012).

Locally therefore, the current context with regard to food aid is one of ongoing violent conflict, displacement, acute food insecurity and threats to livelihoods. The UN Workplan for 2013 estimates almost 3.4 million people in Darfur to be in need (including 1.4 million IDPs in camps), 695,000 IDPs or severely conflict-affected people in South Kordofan and Blue Nile, 142,000 refugees and another 123,000 vulnerable people including returnees and people of South Sudanese origin (UN Sudan 2013: 28). Large quantities of food aid continue to be distributed in Sudan's peripheries, mainly by WFP but recently also by the Sudan government. Access to affected populations is almost entirely by local NGOs, committees or government departments with movements by the UN and INGOs being severely restricted as a result of insecurity and denial of access. In Darfur, security has deteriorated since 2005, with an increase in attacks and kidnappings of international agency staff since 2009, turning Darfur into one of the most violent contexts globally for aid workers (Stoddard et al. 2006; 2009). Nationally, Sudan is becoming more urban, with a rising middle class who have benefited from oil wealth but are increasingly restless in the face of renewed economic decline and austerity measures.[6] Internationally, the influence of the West is decreasing in Sudan. China and the Middle East are becoming more important players – as investors in oil exploration, as co-owners with the government of former state farms along the Nile and as providers of infrastructural aid. The remainder of this chapter examines how food aid, and the agencies that provide it, fit into a history of declining food security and protracted crisis.

Food aid, governance and political economy

The early days – state-centred aid

From the first days of independence, food aid has been a key component of Sudan's development strategy. Sudan accepted US food aid in 1958, which was offered as part of an American aid package as early as 1955. As this was programme food aid, objectives were not humanitarian but to encourage economic development, provide support for US farmers and promote US foreign policy (Riley 2004). Within Sudan, programme food aid was sold on the domestic market and starting in the mid-1960s, these funds were used to promote development projects, which at the time focussed on infrastructure such as dams and roads and providing support for irrigated agricultural schemes along the Nile.

The 1970s was a period of government optimism for development. The Addis Ababa agreement of 1972 brought peace to the south and the government initiated a number of ambitious development plans, including agricultural projects to turn Sudan into the 'breadbasket of the Middle East', oil exploration, digging the Jonglei canal in South Sudan, and a range of industrial projects (for example the Kenana sugar factory) (African Rights 1997a). It was a period of rapid expansion of commercial agriculture. Large-scale farmers in central Sudan were supported with subsidized credit, a favourable exchange rate for inputs, and a floor price support by the Agricultural Bank of Sudan (Maxwell et al. 1990). The economy began to decline in the late 1970s, however, and worsened further in the 1980s as national debt mounted and Sudan's professional classes left the country. A switch to export crops and a reduction in land cultivated when rainfall was low increased labourers' vulnerability (O'Brien 1985).

As Sudan's debts grew, it became the largest recipient of US aid. Food aid supplies to Sudan also grew in the mid to late 1970s (see Figure 8.1). This was for a number of reasons, including the rise in Sudan's geopolitical importance as the Nimeiri government turned away from socialism while left-leaning regimes were established in neighbouring Ethiopia and Libya. By 1976, Sudan was designated as the chief anchor of US policy in the Horn of Africa (Ayers 2010). Other reasons for the growth in food aid in the 1970s was an increase in refugees (from Ethiopia) and returnees following the Addis Ababa peace agreement (Karadawi 1999); both groups received emergency food aid.[7] Also WFP, established in 1962, implemented its first development project in Sudan. By the 1970s, therefore, Sudan received three types of food aid: programme food aid, emergency food aid and project food aid.

WFP's projects were mainly aimed at supporting the agricultural schemes and other commercial projects which involved labour-intensive works. WFP's first project in Sudan (and globally) was to assist in the resettlement of people whose land was flooded as a result of the Aswan High Dam in Egypt. They were resettled on a large mechanized agricultural scheme (Shaw 1967; WFP 1969).[8] Other early projects included resettlement of nomads or refugees and part payment of wages

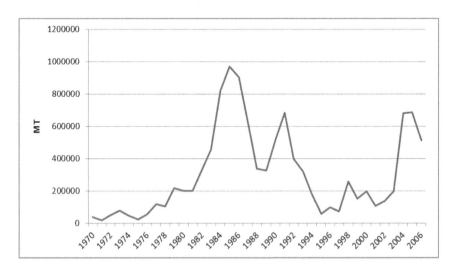

FIGURE 8.1 Food aid to Sudan

Note: The data for this graph was extracted from FAO's statistical database by combining data for Sudan on total cereals and total non-cereals for all donors and all years for which data are available.
Source: FAOSTAT (FAO 2012a).

or incentives on commercial government projects, such as forestry, gum Arabic or dairy schemes (WFP 1970; 1972; 1975; 1983; 1988). From WFP's evaluations it is clear that food aid was seen as a way of supporting government budgets. It may therefore have unintentionally supported the unequal development process that existed in Sudan at the time.

The most important food security intervention in the late 1970s and early 1980s, however, was a wheat subsidy, underwritten by food aid (Maxwell et al. 1990). The government used US programme food aid, mostly wheat, to provide a bread subsidy for the urban population. As the ability to produce or purchase wheat declined (local production was discouraged by the IMF), the contribution of wheat aid to meet consumption requirements rose from 29 per cent to 77 per cent between 1978–9 and 1986–7 (Hussain 1991). Cuts in the subsidies, for example in 1989, immediately led to civil unrest (Bickersteth 1990). Food subsidies can therefore be understood as a tool for containing potential social unrest in urban areas and reducing any threat to the state. The US in turn maintained wheat aid for as long as possible to help keep Nimeiri's regime in power as part of the politics of the Cold War (Hussain 1991).

Refugees, from the start, were seen as both an external and internal security threat. External, because involvement in political activities could antagonize neighbouring countries, and internal because, in the context of general economic decline, they were seen as a social and economic burden and blamed for an increase in crime and unemployment. Refugees were initially settled on agricultural schemes or lived in urban areas. However, in 1978 the police requested the restriction of their movements from settlements and evacuation from urban areas (Karadawi 1999: 104). State security, the army, and provincial authorities wanted refugees close to the Ethiopian border moved to camps to restrict the activities of political factions amongst them (Karadawi 1999: 118). Camps, and the categorization of the people concerned as 'refugees', also facilitated requests for international assistance and addressed both the need for security and for assistance. Whilst food aid was provided, and thus helped contain the security threat posed by refugees, the restriction of movement associated with the creation of camps meant refugees faced difficulties in accessing food. The failure of refugees in eastern Sudan to achieve self-reliance remains an issue, at the time of writing, for these reasons (Ambroso et al. 2011).

Refugees were also a key factor in the externalization of responsibility for welfare provision in Sudan. The UN and INGOs became key service providers to refugees in eastern Sudan beginning in the early 1980s and to returnees in southern Sudan following the Addis Ababa peace agreement of 1972. In South Sudan, they soon ended up replacing the state, as the new regional government was inexperienced and underfunded (Karadawi 1999; African Rights 1997a). INGOs, in contrast, were well financed and attracted the most educated people. Whole districts or sections of government were handed over to INGOs, thus forming an alternative form of governance (Tvedt 1994). These events contributed to a loss of domestic accountability for humanitarian action and signaled the beginning of international agency influence on government policy.

The 80s and 90s – emergency, people-centred aid and the community

New trends in food aid

The 1980s and 90s saw a dramatic change in the type of aid provided to Sudan. The famine in 1985 led to a huge increase in emergency food aid, which became the main type of food aid in Sudan (see Figure 8.1) (Benson and Clay 1986). In contrast, development aid to Sudan almost ceased in the early 1990s, including programme food aid (see Figure 8.2). Reasons for this included the military coup staged by Colonel El-Bashir and believed to be closely linked to the National Islamic Front (NIF) (Gallab 2008), little progress in resolving the war with South Sudan, and President El-Bashir's government supporting Iraq in the first Gulf War. Globally, food aid also shifted from programme to emergency food aid in the 1990s as a result of policy shifts in donor countries[9] (Clay and Stokke 2000: 31). From 1990 onwards, therefore, Western donors provided mainly humanitarian aid to Sudan – much of which was food aid, and emergency food aid became one of the main forms of food security support. In Sudan, the shift to emergency food aid also reflects a change in its geographical distribution. Whereas food aid up to 1985 was mainly concentrated in the centre, emergency food aid from 1985 onwards was largely distributed in its peripheries: Red Sea Hills, Darfur and Kordofan. WFP's food-for-work projects also shifted to the peripheries, and starting in 1994 it focussed on the hungry poor (WFP 1994).

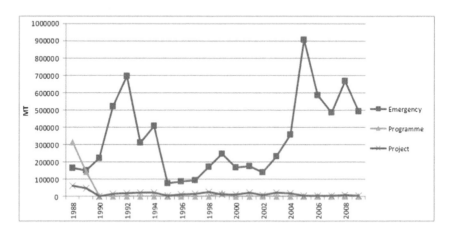

FIGURE 8.2 Trends in different types of food aid in Sudan

Note: The data for Sudan by food aid type was extracted from WFP's international food aid information database. The difference in the 2004–5 peak compared to Figure 8.1 is because FAO reports food aid on a split year (July–June) basis and WFP on a calendar year basis.

Source: WFP 2009a.

Trends in emergency food aid from 1990 onwards closely follow political developments in Sudan. Emergency food aid peaked in 1992, when Operation Lifeline Sudan (OLS) became fully active in providing aid to people affected by the conflict in South Sudan. OLS was a unique programme where the UN negotiated access to conflict-affected populations in a situation of ongoing conflict in South Sudan. The next peak was in 1998, which reflects response to the second Bahr El-Ghazal famine in South Sudan, and in 2005 food aid reached its peak in response to the Darfur crisis. The troughs are also revealing – 1988, the time of the first Bahr El-Ghazal famine, shows little humanitarian response. Still during the Cold War period, donors were reluctant to challenge the Sudanese government's conduct in the war with the south and its strategy of restricting food aid to particular groups or areas (Keen 1994). The drop in 1994 can be explained by donor, government and aid agency perceptions that the emergency in Sudan was over and consequently the promotion of a shift to more developmental relief. INGOs had a brief period of implementing community development or food security projects just after the 1985 famine, including village grain banks, agricultural extension, animal traction, and water harvesting as well as the establishment of village development committees, community-based organizations and professional associations (for example blacksmiths) to eventually manage these projects themselves (Jaspars 2010: 8–9). Although these continued on a small scale throughout the 1990s, the main tool available for development was emergency relief. These issues are discussed further below.

The 1980s and 90s saw further involvement of international agencies in Sudan, in response to famine in 1985, but in particular as part of OLS. These agencies worked directly with drought- or conflict-affected people at community level. This change in food aid from state-centred to people-centred support reflects a conceptual shift from state security to human security. Agencies introduced practices of targeting and community-based programming based on new concepts of vulnerability, entitlements and household food security. As one of the main forms of aid since the 1990s, the objectives of emergency food aid expanded to support livelihoods, coping strategies and later peace-building and protection. As a consequence, food aid as a technology of governance changed. From the 1960s to the mid-1980s, food aid had been mainly state-centred with even emergency food aid distributed through government structures. From the mid-1980s onwards, it penetrated deep into society and changed people's actions and power relations, and therefore as a technology of governance took on a different form. The remainder of this section illustrates this by examining some of the governance effects of the new assessment and targeting methods, the distribution of food aid in conflict, and how food aid may have unintentionally supported government policies and provided benefits to government and the private sector.

Food aid, politics and power in the 1990s

INGOs first tried to target food aid[10] in Sudan during the famines of the 1980s, but could not overcome the government bias towards urban populations (Keen

1991: 198–9; Borton and Shoham 1989). Most rural populations received only tiny quantities of food aid and some displaced and nomadic groups were excluded completely. Instead, they used their own 'coping strategies', including reducing food intake, eating wild foods and selling livestock (De Waal 1989). At the same time, however, merchants benefited from the low price of livestock and from controlling the price of grain. Famine was therefore not only associated with impoverishment and malnutrition but also accumulation of wealth (Bush 1988).

This trend of growing inequality continued in the 1990s, to become what Duffield calls the 'asset transfer economy', including transfer of land, livestock and labour from the politically weak to the strong through direct appropriation, market manipulation and control of food aid (Duffield 1994). During the 1990s, inflation eroded government salaries, local government budgets dwindled and corruption became endemic (African Rights 1997a: 16). Profitability from mechanized farming and other conventional activities also declined and a new economic regime emerged dependent on speculation, hoarding and using state office for personal gain (Duffield 1990). Food aid not only provided hard currency through local purchase and transport contracts, or could be diverted and sold at the local level, but restricting or delaying food aid supplies could yield benefits through raising grain and transport prices and thus greater profits for merchants and farmers, often from central Sudan. All food aid operations for Darfur in the late 1980s and 1990s suffered serious delays, starting with the Western Relief Operation in 1988 (Buchanan Smith 1989; Duffield 1990; Buchanan Smith and Davies 1995; DfID 1997; Tanner 2002). In its war with the south, the denial of relief to Bahr El-Ghazal was a key part of the government's counter-insurgency. However, it also benefited traders, military and farmers by maintaining high grain prices, increasing livestock sales and creating a source of cheap labour. It caused one of Sudan's most severe famines in 1988 (Keen 1994), and eventually led to the creation of OLS in 1989.

The creation of OLS established two very different working environments, particularly in terms of access to conflict-affected populations. Managed from Kenya, the OLS southern sector had relative ease of access and movement in South Sudan, mainly opposition-controlled areas, compared to the much more restricted environment in northern Sudan. The government had in effect temporarily ceded sovereignty of large parts of the south to the UN (Karim et al. 1996). This did not stop government concern about taxation and diversion of food aid by, and therefore support for, the Sudan People's Liberation Army (SPLA) and other southern movements. It could still restrict flights and aid convoys to the south. Aid agencies had similar concerns about the manipulation of food aid and introduced a number of strategies to promote the application of humanitarian principles of neutrality and impartiality, such as the ground rules to guide the conduct of aid agencies and rebel movements (Levine 1997). Agencies also included the 'do no harm' approach to minimize negative impacts on relations within or between groups or alternatively to promote peace by supporting positive relations (Anderson 1999). Senior government figures interviewed in 2012, however, all dated the politicization of food aid in Sudan to this period – the start of OLS. Note that this was a period when food aid was exclusively for humanitarian purposes, in contrast to the programme food aid in earlier years which was explicitly political. Despite the attempts of aid agencies to minimize the negative impacts of food aid, the Sudan government viewed INGOs as supporters of the SPLA and accused them of transporting weapons on OLS flights (Aboum et al. 1990; Karim et al. 1996).

In the north, President El-Bashir's National Salvation government brought in a form of government in 1989 which changed every aspect of society by putting Islamists in charge of key state institutions, and transforming all social, economic and administrative institutions into agents of Islamization (De Waal 2007b: 14–15). In many ways this was seen as an alternative, and perhaps a reaction, to the many years of failed development interventions from the West. The creation of an Islamist state was considered a civilizational project at the heart of which was the 'comprehensive call' – bringing all Sudanese into the project through a process of coercion, religious indoctrination, political mobilization and various forms of local jihad (Gallab 2008: 11). Since by now the country faced massive debts, part of the project was to be self-sufficient in food production and reduce dependence on Western food aid and food imports. The National Economic Salvation Programme and the Comprehensive National Strategy in 1990 and 1992 included investment in agriculture (such as expansion of the area under sorghum and wheat in the irrigated sector), liberalization of trade, privatization of state-owned enterprises, removal of food subsidies and the establishment of a national strategic reserve (Elbashir Mohamed and Elhaj Ahmed 2005). At the local level, subsidized food was distributed through popular committees: government committees established for monitoring basic services, for political mobilization, and for surveillance; but in reality these were closely linked to the ruling National Congress Party (NCP) and dominated by Islamists (Sidahmed 2011; Hamid n.d.). INGOs used these same committees to distribute food throughout the 1990s thus arguably enhancing their power.

The work of INGOs was highly controlled in northern Sudan in the 1990s. From 1993 onwards, they were required to sign a country agreement committing them to a move from relief to development, to limit the number of international staff, and to fill as many posts as possible with Sudanese staff (through the Ministry of Labour). They were also required to work in cooperation with government and national non-governmental partners, the latter involving a strategy of twinning with Sudanese NGOs. The OLS review argued that this would effectively make them extensions of the state (Karim et al. 1996: 49). Travel permits for international staff were introduced soon after the coup but permission to access displaced camps was particularly difficult to obtain (Karim et al. 1996: 97). Information on food security was tightly controlled. In light of the government's policy of food self-sufficiency, any mention of famine, malnutrition or mortality was now highly controversial and assessments had to be cleared at the ministerial level to ensure that information did not contradict the policy (Key informant 9, 2012; Key informant 26, 2012).

Government and aid agency thinking in fact converged on the need for transition from relief to development. In the mid-1990s, food aid objectives centred on promoting self-reliance by supplementing people's own food access strategies. New assessment methods gathered information on people's coping strategies and food and income sources, thus lowering estimates of food deficits, which also suited the political motivations of the government. Strategies for targeting food aid also became more prominent in the mid-90s. Food distribution was limited to certain times of the year, cutting rations to half or quarter size as well as targeting specific areas or vulnerable population groups. At the same time, however, acute malnutrition levels remained as high as those used to justify the earlier emergency operation (Karim et al. 1996: 112, 126–7). In South Darfur, rather than improving self-reliance, this strategy meant that displaced people from the south became integrated into the local economy as cheap and tied labour (Duffield 2002a; Macrae et al. 1997). In Sudan as a whole, mechanized farming continued to accelerate, as control over agriculture was crucial for the NIF (Coalition for International Justice 2006). It required an ever expanding pool of labour for which displaced populations were well-suited. In the OLS-managed south, food aid was reduced on the assumption that people coped but this led instead to a failure to pick up early signs of famine in 1998 (Seaman 1993: 27; Deng 2002: 35).

What is clear is that food aid by the end of the 90s had a new range of functions. The actions of NGOs, during the late 1980s can be seen as marking a shift in food aid as a technology of governance from geopolitical to biopolitical concerns. That is from an earlier association with support for state-based development to a more direct involvement at household and community levels, especially in relation to the increasing support for livelihoods and coping strategies. International aid agencies increasingly took responsibility for managing the lives and livelihoods of populations in Sudan's peripheries. The Sudan government perceived this as a threat to its own sovereign power, and increased its efforts to control INGOs, as well as adopting or using food aid strategies for its own political and economic purpose.

Darfur and the implications of Sudan's protracted crisis

Widespread violent conflict started in Darfur in 2003, associated with large-scale destruction and displacement. The humanitarian response beginning in 2004 was massive, including a food aid operation which was unprecedented in scale and reach, with distribution to over 3 million people in more than three hundred distribution points. Mortality and malnutrition levels were quickly brought down (Young 2007: S49). However, neither the scale of the operation nor the international attention could be sustained and in 2006 both nutrition and food security started to deteriorate again[11] (WFP et al. 2008; 2009). WFP reduced food rations from 2006 onwards, first because of inadequate funding (Young 2007), then for security reasons (WFP Sudan 2008), and later because some conflict-affected populations were considered able to meet part of their food needs themselves (WFP Sudan 2010b).

The Darfur conflict presents many of the familiar challenges of food distribution seen in South Sudan and conflict situations more generally, such as manipulation or diversion of food aid and boosting the authority of existing or new leaders (Mahoney et al. 2005). However, it was also radically different from past aid operations in Sudan. The quantities of food aid distributed were far higher than in the past; more than five times as much food aid was distributed in 2005 than in 1985[12] and almost eighty times as much as in 1997 (WFP 2006; DfID 1997). The distribution apparatus established by WFP was also unprecedented. It included massive support for the private transport sector, INGOs implementing partners, food relief committees (FRC), and from 2009, support for government ministries and Community Based Organization (CBO) to carry out assessments, food-for-recovery and voucher programmes. The scale of displacement resulted in a huge increase in urban centres where many displaced are likely to remain. Many of the camps were highly politicized and militarized, with their own shadow economies.

What was new on the part of international agencies was a focus on protection as this was the first emergency to be labelled a 'protection crisis'. This reflected international recognition of a collective responsibility to protect civilians caught up in conflict and the severity of violence in Darfur, involving widespread destruction, death and sexual violence (Pantuliano and O'Callaghan 2006: 6–7). In reality, the West was reluctant to respond with military or diplomatic action in Darfur (Williams and Bellamy 2005; Traub 2010). African Union (AU) and later UN-African Union Mission in Darfur (UNAMID) forces were unable to ensure protection for large sections of the population, thus leaving much of it in practice to humanitarian assistance agencies and to conflict-affected populations themselves.[13] Emergency food aid acquired a new role in promoting safety, helping prevent displacement, registering displaced populations and offering protection through WFP's widespread presence (Mahoney et al. 2005; Jaspars and O'Callaghan 2008; Young and Maxwell 2009). As registration for food aid was the only proof of IDP identity, food aid quickly became linked to claims for protection and political entitlements (such as claims for compensation for losses incurred during the war) (Young and Maxwell 2009).

In March 2009, the International Criminal Court (ICC) indicted President El-Bashir for genocide, crimes against humanity and war crimes and issued a warrant for his arrest. This immediately led the government to revoke the registration certificates of thirteen international and three national organizations, accusing them of involvement in activities which violated their humanitarian mandates and threatened national security, for instance by cooperating with the ICC, false reporting, and advocating with the international community and the UN Security Council for more pressure on Sudan (Humanitarian Aid Commission 2009: 4). Many of the agencies expelled were involved in protection, food aid and livelihoods activities. The president also announced a programme of Sudanization, reviving earlier efforts whereby international agencies were required to 'twin' with Sudanese counterparts. In contrast to the 1990s, when twinning was unsuccessful, local ministries or national or Islamic NGOs took over much of the work of the expelled INGOs. Even prior to 2009, insecurity and restricted access meant that many INGOs worked with local partners as a way of remotely managing the provision of assistance. For this reason, FRCs had already distributed food without supervision by INGOs on a number of occasions (Young and Maxwell 2009: 25). Access for international agencies continued to deteriorate from 2009 onwards, with fewer agencies working with international staff and travel permission to South Kordofan and Blue Nile almost impossible to obtain. More agencies were expelled in 2012, this time from eastern Sudan, and in July 2013 twenty United Nations High Commissioner for Refugees (UNHCR) staff were expelled, or rather, did not have their residence permits renewed. In January 2014, the government suspended the activities of the International Red Cross.

Sudanization closely resembles the recovery agendas of many international agencies, which include capacity building and working with local partners. Recovery approaches may include various forms of food security and livelihood support to promote self-reliance, and were promoted as early as 2005 – an unusual period of relative calm in Darfur. Early recovery was first incorporated into the UN Workplan in 2008, and included vocational training, income generation and agricultural support projects, amongst others (Jaspars and O'Callaghan 2008: 21). Further impetus for developmental approaches came with assertions that the Darfur conflict stemmed from an ecological crisis linked to climate change, including one from UN Secretary General Ban Ki Moon in 2007. According to this perspective, intervention to support economic development is even more important than political intervention (Ban Ki Moon 2007). These views are also reflected in UN documents, such as the UN's Darfur strategy of 2010, which identifies climate change as an underlying trend contributing to the conflict and recommends interventions in environmental management, livelihoods, education and strengthening local governance (UN Sudan 2010). The Sudanese government, in turn, has adopted the language of food crisis and climate change to attract investment in the revitalization of mechanized agriculture. Calling Darfur a climate change crisis also absolves the government from responsibility (Verhoeven 2011). The links between climate change and conflict have been disputed by a number of researchers, however, on the grounds that rainfall did not decline in Darfur immediately prior to the crisis and that similar long-term rainfall patterns did not cause conflict in other countries (Kevane and Gray 2008). Others argue that whilst a changing climate is no doubt a contributing factor, the cause of the conflict is the Sudan government, its failure to address environmental stress and its militarized response to political disputes (De Waal 2007a)

It can also be argued that recovery interventions are neither appropriate without ongoing food distribution nor feasible because the conflict is ongoing and has worsened over the past few years. Localized conflict increased after the Darfur Peace Agreement (DPA) in 2006, as only one of the rebel movements signed, and non-signatory movements and Arab militia splintered. Large-scale displacement following army and militia attacks continued with a new surge in violence in 2010 when the only Sudan Liberation Army (SLA) movement that signed the DPA resumed armed conflict against government forces. In the first six months of 2013, more than 300,000 people were newly displaced, highlighting the volatile and constantly changing nature of the conflict (UN OCHA 2013). Banditry, looting, informal taxes and transport costs have increased over time, which limits freedom of movement and therefore markets, trade and migration for work. Levels of acute malnutrition remain persistently high, particularly in North Darfur – with rates as high as 34 per cent (<-2 Z-scores weight-for-height) in the hungry season but often remaining well above the emergency threshold at times of the year when malnutrition is usually low (UNICEF 2012). More than 1.5 million displaced people remain in camps or settlements in the main population centres and 3 million received food aid in 2012 (UN Sudan 2013).

In 2013, WFP continued to distribute only partial rations in camps, and for rural populations only at certain times of the year or aimed to change its programme to food-for-recovery.[14] Rations to North Darfur were further reduced because of security restrictions on transport of goods particularly to rural areas (Key informant 53, 2013). Access to areas firmly within rebel control is extremely restricted. Food security assessments focus on consumption, dietary diversity, income and expenditure, and tell us little about the nature or causes of food insecurity or the risks that people face. This makes it difficult to provide assistance according to need. In addition, emergency levels of malnutrition no longer grab much attention or the reliability of the data is questioned. Most aid agencies working in food security are keen to use other food assistance modalities, such as vouchers and food-for-recovery, as well as more general rehabilitation programming. This is very much in line with government thinking, as its preface to the UN Workplan for 2013 states: 'As the humanitarian crisis in Darfur enters its tenth year, the Government of Sudan is determined to accelerate the shift away from aid dependency to self-sufficiency and sustainable livelihoods for all of its people' (UN Sudan 2013). It also facilitates the government's policy of IDP return and the dismantling of the camps (Government of Sudan 2010).

At first glance, the trajectory of the Darfur crisis suggests a return to the humanitarian politics of the Cold War. The current context of limited access to conflict-affected populations for international agencies now resembles the 1980s, when government could restrict humanitarian access without challenge from the donor community. There are a number of reasons why joint donor pressure on the government of Sudan did not seem possible during the past decade. First, the War on Terror (WoT) has fractured the donor community. The Sudanese government now has an ambiguous position as both an Islamist pariah state and an ally in the WoT as a provider of intelligence information to the West. Second, key donors also felt that tackling the government over its conduct in Darfur would jeopardize the peace agreement with South Sudan and, third, Chinese economic interests in Sudan have meant the watering down of security council resolutions (Traub 2010; Williams and Bellamy 2005). This prevented early intervention to the Darfur crisis, a meaningful response to the expulsion of INGOs, and serious challenge to the government's denial of access by INGOs and the UN to conflict-affected populations in South Kordofan and Blue Nile. Compared to its brief period of autonomy during the 'negotiated access' phase of the 1990s, the UN has also reverted to type, in the sense of being an auxiliary of the host government. Not only in Darfur, but also in a number of other disaster zones, the reassertion of state authority has limited and curtailed international humanitarian access (WFP 2009b).

In Sudan, however, it would be wrong to take the 'return to the Cold War' idea too literally. While the ability of the government to restrict humanitarian access with impunity resembles the 1980s, in the intervening years the state has developed its own 'non-governmental' infrastructure and ability to respond. The state's tendency to interpret and act on humanitarian crises as security threats is now complemented by a network of popular committees, national NGOs and its own strategic grain reserve.

Strategic grain reserve initiatives started in the late 1980s in response to the Western-dominated relief operation in 1985, but it was not until 2001 that this became an official government act. Oil revenue enabled the government to purchase large quantities of food for the strategic reserve – which according to government representatives reached 500,000 metric ton (MT) in the past two years. The objectives of the reserve include price stabilization, acting as a purchase and sale agent and provision of humanitarian assistance (Asfaw and Ibrahim 2008). Although the latter is not included as such in the act, it has recently been used in Darfur, South Kordofan and Blue Nile for distribution to emergency-affected populations. This is of course done according to the government's own political priorities and in government-held areas. It has also been used as a foreign policy tool to promote friendly relations with neighbouring countries such as Chad and Ethiopia (Key informant 50, 2013). Part of the food from the reserve is used to provide subsidies to modern poultry farms supplying Khartoum, Kassala and Gedaref to maintain meat at a low price for urban consumers. The price of wheat, and therefore bread, is kept low by preferential exchange rates for wheat imports. At the same time, the government has renewed its efforts to achieve food self-sufficiency, including through the agricultural revival strategy of 2008, and before this, a wheat self-sufficiency strategy. As before, much of the emphasis is on large-scale irrigated and mechanized farming in the centre. Investment now comes from China or the Arab states, often more concerned with their own food security than that of Sudan. Given the extent of food insecurity in Sudan's peripheries, it is likely, however, that Western food aid will be provided to Sudan for many years to come.

Conclusions

Despite more than fifty years of food aid and food security interventions, and a recent period of rapid economic growth, inequality in Sudan has grown. Food insecurity and malnutrition remain unacceptably high and Sudan's populations continue to experience disasters and humanitarian crises on a regular basis. This shows, first, that a food security response dominated by food aid is unable to address the structural causes of food insecurity and second, that economic growth and permanent food emergency can exist side by side. In the early years, food aid may have inadvertently contributed to an exclusionary development process, and since the 1980s it has helped address the consequences of political and economic marginalization of Sudan's peripheries.

This chapter shows that food aid is not only a commodity that can save lives but also that it functions as a form of governance, the nature of which has changed over time. When food aid was first introduced into Sudan it promoted a form of state-centred development, including the introduction of modern commercial agriculture. Food aid boosted government budgets, and projects were implemented through government structures. It was also a geopolitical tool to keep Sudan as a US ally against its left-leaning neighbours during the Cold War. Internally, it helped the Sudan government contain potential security threats from a growing urban and refugee population. Beginning in the mid-1980s emergency food aid increased and a number of Western INGOs came with it who, in parts of Sudan, took over the welfare functions of the state, and who worked at the level of the household or community. New techniques of assessment, targeting and distribution had the potential to change power relations and people's actions at the local level by linking food aid with new objectives like supporting livelihoods, self-reliance, peace-building or protection, and by interacting with local forms of governance. As such, international agencies gradually assumed responsibility for managing the lives and livelihoods of populations in Sudan's peripheries. These changes reflect a shift from geopolitics, or power over territory, to biopolitics, or the power over lives and livelihoods of populations by intervening in the processes that influence them.

The distribution of food aid during ongoing conflict, beginning in the early 1990s, raised the possibility that this could support rebel movements and prolong conflict. The Sudan government therefore began to view food aid as highly politicized during this time. Camps, whilst earlier seen as a convenient way of separating refugees from society for security and assistance purposes, are now seen as magnets for INGOs and as hives of support for the rebellion, and therefore to be avoided or disbanded as quickly as possible.

The government of Sudan has responded to this new 'politics of international food aid' by attempting to Sudanize the aid industry through the development of its own infrastructure of national NGOs, popular committees and a strategic grain reserve, and by restricting the actions of INGOs and UN agencies. This has enabled the government to more effectively deny access to international agencies and to direct food aid according to its own priorities. The government has also selectively adopted the language and strategies of Western agencies, for its own purpose. Reductions in food aid, associated with aims to move from relief to development, have suited government policies of self-sufficiency, the expansion of mechanized agriculture and, currently, IDP returns and the dismantling of the camps in Darfur. Food aid has enabled sections of the private sector to grow. What this chapter shows is that the Sudan government has learnt from fifty years of food aid and has responded accordingly. This is the *actually existing development* in Sudan, or the development that goes on indirectly as part of, or despite, the aid industry (Duffield 2002b: 160).

In contrast, the international aid industry has repeatedly shown itself to lack historical awareness or even an interest in its own past. Whilst the experience and research on food aid in Sudan has contributed hugely to international practice, an analysis of the past policies and practices within Sudan shows a lack of learning. Strategies like targeting and reducing food aid to encourage self-reliance are endlessly repeated but may have negative consequences in the absence of alternative ways to establish viable, dignified and safe livelihoods. International agencies have made important technical advances, such as food vouchers and other forms of livelihood support, but have consistently failed to examine the political effects of intended or unintended changes resulting from food aid. Learning from the past is an urgent task in countries which have received food aid for over fifty years, and is the first step in finding ways to address food insecurity in protracted crises.

Sudan's permanent emergency will no doubt continue to unfold. Khartoum is now home to a growing middle class sufficiently able to put its own food security on the political agenda and who therefore remain a government priority for access to subsidized food. At the same time, the desire of Sudan to be self-sufficient has to be considered in relation to the sale or leasing of land along the Nile by Middle Eastern and Asian countries seeking to improve their own food security. These tensions and dispossessions will be played out alongside continuing emergencies and insecurity in Sudan's peripheral regions. It seems inevitable that external food aid will continue to be required in Sudan.

Notes

1 Susanne Jaspars is a PhD candidate at Bristol University. The contents of this chapter are mainly based on her literature review but use some information gathered in interviews or documents gathered in Sudan in June 2012 and January 2013. The chapter uses text from draft chapters 4 and 5 of her PhD thesis, which at the time of publication has not been completed.
2 I would like to thank Mark Duffield, my PhD supervisor, for his valuable suggestions during discussions about the contents of this chapter, in particular relating to the existence of a dual economy in Sudan and the implications of the War on Terror and the re-assertion of state authority. This chapter is heavily influenced by his thinking.
3 Thanks also to Youssif El Tayeb for commenting on a draft of this chapter, and to editors for their suggestions for changes to earlier drafts.
4 Sudan has been in the top 10 per cent of food aid recipients in 19 of the past 23 years and in the top three recipient countries since 2004. I extracted data from WFP's food aid information system (WFP 2009a): *International Food Aid Information System*. Online. Available <www.wfp.org/fais> (accessed: November 2011) to identify the main food aid recipients globally.
5 The prevalence of wasting was 16.4 per cent (<-2 Z-scores for weight-for-height). Only Northern Province, Khartoum and Gezira were below the emergency threshold of 15 per cent, and the highest rates (>20 per cent) were found in Red Sea State (28.5 per cent), Sinnar (21.6 per cent), and North Darfur (21.6 per cent).
6 South Sudan became independent in July 2011, taking with it most of Sudan's oil. Production was shut down in January 2012 because of disputes over Sudan's pipeline charges.
7 Starting in the 1960s, Sudan both produced and received refugees. People from South Sudan fled to Ethiopia and the Congo, and Congolese, Chadians, Ugandans, and Ethiopians fled to Sudan.
8 Khasm El Girba in central Sudan.
9 Policy shifts included those in domestic agricultural policy (EU single market reduces influence of any one country on procurement, and consequent shift to financing local purchases), foreign policy (end of the Cold War), development (change to agricultural research and multi-sectoral rural development) and global trade. Climatic variability and global food prices also affect the availability of food aid.

10 Targeting is 'the process by which areas and populations are selected to receive a resource (emergency food aid) and then provided with it' for reasons of limited resources, the desire to concentrate on the worst affected areas and populations, and not to damage the local economy (Borton and Shoham 1989: 79).

11 With the exception of a temporary improvement in 2008.

12 Less than 100,000 MT was distributed in Darfur in 1985. In 2005 it was 565,000 MT. In 1997, SCF distributed 7,460 MT in Darfur in response to predictions of widespread humanitarian crisis.

13 Humanitarian protection programmes aim to promote the safety of civilians and include information gathering and advocacy and direct assistance. Direct assistance might include moving people to safe areas, tracing lost family members, health care for victims of sexual violence, replacing identity documents, proving information, or food and other assistance to help meet basic subsistence needs and prevent the need to engage in risky activities.

14 Similar to food-for-work.

References

Aboum, T., Chole, E., Manibe, K., Minear, L., Mohamed, A., Sebstad, J. and Weiss, T. (1990) *A Critical Review of Operation Lifeline Sudan: A Report to the Aid Agencies,* Humanitarianism and War Project, Boston: Tufts University.

African Rights (1997a) *Food and Power in Sudan: A Critique of Humanitarianism,* London: African Rights.

African Rights (1997b) 'Origins of the disaster', in African Rights (ed.), *Food and Power in Sudan: A Critique of Humanitarianism,* London: African Rights.

Ali, T.M. (1989) *The Cultivation of Hunger: State and Agriculture in Sudan,* Khartoum: Khartoum University Press.

Ambroso, G., Crisp, J. and Albert, N. (2011) *No Turning Back: A Review of UNHCR's Response to the Protracted Refugee Situation in Eastern Sudan,* Geneva: UNHCR Policy Development and Evaluation Service (PDES).

Anderson, M. (1999) *Do No Harm: How Aid Can Support Peace or War,* London: Lynne Rienner.

Asfaw, A. and Ibrahim, S. (2008) *The Strategic Reserve Corporation of Sudan: Learning from Best Practices,* Technical Discussion Paper 1, Khartoum: FAO and the Ministry of Agriculture and Forestry.

Ayers, A. (2010) 'Sudan's uncivil war: the global–historical constitution of political violence', *Review of African Political Economy,* 37(124): 153–71.

Ban Ki Moon (2007) *A Climate Culprit in Darfur.* Online. Available <www.washington-post.com/wp-dyn/content/article/2007/06/15/AR2007061501857.html> (accessed 8 August 2013).

BBC (2004) *Mass Rape Atrocity in West Sudan.* Online. Available <http://news.bbc.co.uk/1/hi/world/africa/3549325.stm> (accessed 9 March 2014).

Benson, C. and Clay, E. (1986) 'Food aid and food crisis in Sub-Saharan Africa: statistical trends and implications', *Disasters,* 10(4): 303–16.

Bickersteth, J.S. (1990) 'Donor dilemmas in food aid: the case of wheat in Sudan', *Food Policy,* 15(3): 218–26.

Borton, J. and Shoham, J. (1989) 'Experiences of non-governmental organizations in the targeting of emergency food aid: a report on a workshop on emergency food aid targeting at the School of Hygiene and Tropical Medicine, London, 4–6 January 1989', *Disasters,* 13(1): 77–93.

Buchanan Smith, M. (1989) *Evaluation of the Western Relief Operation 1987/88, Final Report,* Wokingham: MASDAR (UK) Ltd.

Buchanan Smith, M. and Davies, S. (1995) 'Chapter 5: Sudan', in *Famine Early Warning and Response – the Missing Link,* London: Intermediate Technology Publications.

Bush, R. (1988) 'Hunger in Sudan: the case of Darfur', *African Affairs,* 87(386): 5–23.

Clay, E. and Stokke, O. (2000) 'The changing role of food aid and finance for food', in Clay, E. and Stokke, O. (eds), *Food Aid and Human Security,* London: Frank Cass.

Coalition for International Justice (2006) *Soil and Oil: Dirty Business in Sudan,* Washington, DC: Coalition for International Justice.

De Waal, A. (1989) *Famine That Kills: Darfur, Sudan,* Oxford: Clarendon Press.

De Waal, A. (2007a) *Is Climate Change the Culprit for Darfur?* Online. Available <http://africanarguments.org/2007/06/25/is-climate-change-the-culprit-for-darfur/> (accessed 8 August 2013).

De Waal, A. (2007b) 'Sudan: the turbulent state', in De Waal, A. (ed.), *War in Darfur and the Search for Peace,* London: Justice Africa and Global Equity Initiative.

Dean, M. (2010) *Governmentality. Power and Rule in Modern Society,* 2nd edition, London: Sage Publications.

Deng, L. (2002) 'The Sudan famine of 1998: unfolding of the global dimension', *IDS Bulletin,* 33(4): 28–38.

Development Initiatives (2011) *Global Humanitarian Assistance Report 2011,* Wells: Development Initiatives.

DfID (1997) *Report on Sudan Emergency Food Distributions,* London: Department for International Development.

Duffield, M. (1990) 'From emergency to social security in Sudan – Part I: the problem', *Disasters,* 14(3): 187–203.

Duffield, M. (1994) 'The political economy of internal war: asset transfer, complex emergencies, and international aid', in Macrae, J. and Zwi, A. (eds), *War and Hunger: Rethinking International Responses to Complex Emergencies,* London: Zed Books.

Duffield, M. (2002a) 'Aid and complicity: the case of war displaced southerners in northern Sudan', *Journal of Modern African Studies,* 40(1): 83–104.

Duffield, M. (2002b) 'Social reconstruction and the radicalization of development: aid as a relation of global liberal governance', *Development and Change,* 33(5): 1049–71.

Duffield, M. (2002c) 'War as a network enterprise: the new security terrain and its implications', *Cultural Values,* 6: 153–65.

Elbashir Mohamed, A. and Elhaj Ahmed, A. (2005) *Study on Food Security Policies in Sudan,* Study prepared for the World Food Programme in Khartoum, Sudan, Khartoum: WFP.

FAO (2012a) *Global Information and Early Warning System.* Online. Available <www.fao.org/giews> (accessed March 2014).

FAO (2012b) *Low-Income Food-Deficit Countries (LIFDC) – List for 2012.* Online. Available <www.fao.org/countryprofiles/lifdc.asp> (accessed 2 March 2012)

Federal Ministry of Health and Central Bureau of Statistics (2012) *Sudan Household Health Survey 2010, National Report,* Khartoum: Government of Sudan.

Foucault, M. (1991) 'Governmentality', in Burchell, G., Gordon, C. and Miller, P. (eds), *The Foucault Effect: Studies in Governmentality,* London: Harvester Wheatsheaf.

Foucault, M. (2007) *Security, Territory, Population: Lectures at the Collège de France 1977–78,* trans. G. Burchell, Basingstoke and New York: Palgrave Macmillan.

Gallab, A. (2008) *The First Islamist Republic: Development and Disintegration of Islamism in the Sudan,* Aldershot: Ashgate.

Government of Sudan (2010) *Darfur: Towards New Strategy to Achieve Comprehensive Peace, Security and Development,* Khartoum: Government of Sudan.

Hamid, G. (n.d.). *Localizing the Local: Reflections on the Experience of Local Authorities in Sudan,* Riyadh: Arab Urban Development Institute.

Humanitarian Aid Commission (2009) *Humanitarian Situation in Sudan,* Khartoum: Ministry of Humanitarian Affairs.

Hussain, M.N. (1991) 'Food security and adjustment programmes: the conflict', in Maxwell, S. (ed.), *To Cure all Hunger: Food Policy and Food Security in Sudan,* London: Intermediate Technology Publications.

Jaspars, S. (2010) *Coping and Change in Protracted Conflict: The Role of Community Groups and Local Institutions in Addressing Food Insecurity and Threats to Livelihoods, A Case Study Based on the Experience of Practical Action in Darfur,* HPG Working Paper, London: Overseas Development Institute, Humanitarian Policy Group.

Jaspars, S. and O'Callaghan, S. (2008) *Challenging Choices. Protection and Livelihoods in Darfur: A Review of DRC's Programme in Eastern West Darfur,* HPG Working Paper, London: Overseas Development Institute, Humanitarian Policy Group.

Jaspars, S. and Young, H. (1995) *General Food Distribution in Emergencies: From Nutritional Needs to Political Priorities,* RRN Good Practice Review 3, London: Overseas Development Institute, Relief and Rehabilitation Network.

Johnson, D. (2003) *The Root Causes of Sudan's Civil Wars,* Oxford: James Curry.

Karadawi, A. (1999) *Refugee Policy in Sudan: 1967–1984,* Refugee and Forced Migration Studies Volume 1, New York and Oxford: Berghahn Books.

Karim, A., Duffield, M., Jaspars, S., Benini, A., Macrae, J., Bradbury, M., Johnson, D., Larbi, G. and Hendrie, B. (1996) *Operation Lifeline Sudan: A Review,* Unpublished.

Keen, D. (1991) 'Targeting emergency food aid: the case of Darfur in 1985', in Maxwell, S. (ed.), *To Cure all Hunger: Food Policy and Food Security in Sudan,* London: Intermediate Technology Publications.

Keen, D. (1994) *The Benefits of Famine: A Political Economy of Famine and Relief in South Western Sudan 1983–89,* Oxford: James Curry.

Kevane, M. and Gray, L. (2008) 'Darfur: rainfall and conflict', *Environmental Research Letters,* 3: 1–10.

Key informant 9 (2012) *Interview with long term aid worker.* By SJ on 21 June 2012, Khartoum.

Key informant 17 (2012) *Interview with former Humanitarian Aid Commissioner.* By SJ on 26 June 2012, Khartoum.

Key informant 26 (2012) *Interview with long term aid worker.* By SJ on 20 December 2012, London.

Key informant 50 (2013) *Interview with former government official.* By SJ on 2 February 2013, Khartoum.

Key informant 53 (2013) *Interview with WFP staff member.* By SJ on 5 February 2013, Khartoum.

Levine, I. (1997) *Promoting Humanitarian Principles: The Southern Sudan Experience,* RRN Network Paper 21, London: Overseas Development Institute.

Macrae, J., Bradbury, M., Jaspars, S., Johnson, D. and Duffield, M. (1997) 'Conflict, the continuum and chronic emergencies: a critical analysis of the scope for linking relief, rehabilitation and development planning in Sudan', *Disasters,* 21(3): 223–43.

Mahoney, L., Laughton, S. and Vincent, M. (2005) *WFP Protection of Civilians in Darfur,* World Food Programme, Unpublished.

Maxwell, S., Swift, J. and Buchanan Smith, M. (1990) 'Is food security targeting possible in Sub-Saharan Africa? Evidence from North Sudan', *IDS Bulletin,* 21(3): 52–61.

National Population Council General Secretariat, Ministry of Welfare and Social Security and Government of Sudan (2010) *Sudan Millennium Development Goals Progress Report 2010,* Khartoum: Government of Sudan.

O'Brien, J. (1985) 'Sowing the seeds of famine: the political economy of food deficits in Sudan', *Review of African Political Economy,* 33: 23–32.

Pantuliano, S. and O'Callaghan, S. (2006) *The 'Protection Crisis': A Review of Field Based Strategies for Humanitarian Protection in Darfur,* HPG Discussion Paper, London: Overseas Development Institute, Humanitarian Policy Group.

Patey, L. (2010) 'Crude days ahead? Oil and the resource curse in Sudan', *African Affairs,* 109(437): 617–36.

Riley, B. (2004) *US Food Aid Programmes, 1954–2004: A Background Paper,* Unpublished.

Seaman, J. (1993) 'Famine mortality in Africa', *IDS Bulletin,* 24(4): 27–31.

Shaw, J. (1967) 'Resettlement from the Nile in Sudan', *Middle East Journal,* 21(4): 462–87.

Sidahmed, A.S. (2011) 'Islamism and the State', in Ryle, J., Willis, J., Baldo, S. and Madut Jok, J. (eds), *The Sudan Handbook,* Oxford: James Curry.

Stoddard, A., Harmer, A. and DiDomenico, V. (2009) *Providing Aid in Insecure Environments: 2009 Update,* HPG Policy Brief 34, London: Overseas Development Institute, Humanitarian Policy Group.

Stoddard, A., Harmer, A. and Haver, K. (2006) *Providing Aid in Insecure Environments: Trends in Policy and Operations,* HPG Report 23, London: Overseas Development Institute, Humanitarian Policy Group.

Sudan Council of Ministers (2008) *Agriculture and Articulate a Future Vision and Action Plan for Agricultural Revival, Executive Programme for Agricultural Revival,* Khartoum: Council of Ministers General Secretariat.

Tanner, V. (2002) *Save the Children (UK)'s Response to Drought in North Darfur 2000–2001, Evaluation,* Save the Children UK, Unpublished.

Tanner, V. (2005) *Rule of Lawlessness. Roots and Repercussions of the Darfur Crisis,* Interagency Paper, Khartoum: Sudan Advocacy Coalition.

Traub, J. (2010) *Unwilling and Unable: The Failed Response to the Atrocities in Darfur,* Occasional Paper Series, New York: Global Centre for the Responsibility to Protect.

Tvedt, T. (1994) 'The collapse of the state in southern Sudan after the Addis Ababa agreement: a study of internal causes and the role of NGOs', in Harir, S. and Tvedt, T. (eds), *Short Cut to Decay: The Case of the Sudan,* Uppsala: Nordiska Afrikainstitutet.

UN OCHA (2013) *Humanitarian Bulletin, Sudan,* Issue 3, Khartoum: UN OCHA.

UN Sudan (2010) *Beyond Emergency Relief: Longer-term Trends and Priorities for UN Agencies in Darfur,* Khartoum: United Nations.

UN Sudan (2013) *Sudan: United Nations and Partners Work Plan 2013,* Geneva: OCHA.

UNICEF (2012) *Darfur Surveys 2010–2012,* UNICEF, Unpublished.

Verhoeven, H. (2011) 'Climate change, conflict and development in Sudan: global neo-Malthusian narratives and local power struggles', *Development and Change,* 42(3): 679–707.

WFP (1969) *Interim Evaluation, WFP Assistance to Three Land Settlement Projects at Khasm el Girba in the Sudan,* WFP Intergovernmental Committee 16th session, Rome: World Food Programme.

WFP (1970) *Khartoum Green Belt, Terminal Report,* WFP Intergovernmental Committee 17th session, Rome: World Food Programme.

WFP (1972) *Interim Evaluation Report, Afforestation and Wood Processing Operations in the Blue Nile and Kassala Provinces,* WFP Intergovernmental Committee 22nd session, Rome: World Food Programme.

WFP (1975) *Interim Evaluation Report, Scheme for Sawmilling and Other Forestry Operations in Bahr El Ghazal Province,* WFP Intergovernmental Committee 28th session, Rome: World Food Programme.

WFP (1983) *Project Summary, Restocking of the Gum Belt,* WFP Committee on Food Aid Policies and Programmes 15th session, Rome: World Food Programme.

WFP (1988) *Progress Report, Dairy Development in Khartoum Area,* WFP Committee on Food Aid Policies and Programmes 25th session, Rome: World Food Programme.

WFP (1994) *Review of WFP Policies, Objectives and Strategies,* Committee on Food Aid Policies and Programmes 27th session, Rome: World Food Programme.

WFP (2006) *A Report from the Office of Evaluation, Full Report of the Evaluation of EMOP 10339.0/1: Assistance to Populations Affected by Conflict in Greater Darfur, West Sudan,* Rome: World Food Programme.

WFP (2009a) *International Food Aid Information System.* Online. Available <www.wfp.org/fais> (accessed November 2011).

WFP (2009b) *Humanitarian Assistance in Conflict and Complex Emergencies, June 2009 Conference Report and Background Papers,* Rome: World Food Programme.

WFP, Government of Sudan and UNICEF (2008) *Food Security and Nutrition Assessment of the Conflict-affected Population in Darfur, Sudan, 2007, Final Report,* Khartoum: World Food Programme.

WFP, Government of Sudan (HAC and MoA), UNICEF and FAO (2009) *2008 Darfur Food Security and Livelihood Assessment, Final Report,* Khartoum: World Food Programme.

WFP Sudan (2008) *Emergency Operation (EMOP) Sudan 10760.0 (for 2009),* Rome: World Food Programme.

WFP Sudan (2010a) *Emergency Operation Sudan: EMOP 200151 (for 2011),* Rome: World Food Programme.

WFP Sudan (2010b) *WFP Operational Strategy: Darfur 2010–2011,* World Food Programme, Unpublished.

WHO, UNHCR, IFRC and WFP (2000) *The Management of Nutrition in Major Emergencies,* Geneva: World Health Organization.

Williams, P. and Bellamy, A. (2005) 'The responsibility to protect and the crisis in Darfur', *Security Dialogue,* 36: 27–47.

World Bank (2009) *Sudan: The Road Toward Sustainable and Broad-Based Growth,* Washington, DC: World Bank.

Young, H. (2007) 'Looking beyond food aid to livelihoods, protection and partnerships: strategies for WFP in the Darfur states', *Disasters,* 31(S1): S40–56.

Young, H. and Maxwell, D. (2009) *Targeting in Complex Emergencies: Darfur Case Study,* Boston: Feinstein International Center, Tufts University.

Young, H., Osman, A., Aklilu, Y., Dale, R., Badri, B. and Fuddle, A. (2005) *Darfur – Livelihoods under Siege,* Boston: Feinstein International Center, Tufts University.

9

FOOD INSECURITY AND AGRICULTURAL REHABILITATION IN POST-CONFLICT NORTHERN UGANDA

Winnie Wangari Wairimu[1]

Abstract

This chapter analyses efforts to address food security and recovery in post-conflict northern Uganda. This is achieved by exploring how local government actors, humanitarians and local communities try to build on recovery interventions and agricultural rehabilitation programming, and the consequences of their efforts for the food security of the local community. Its major focus is the shift from temporary relief to recovery (i.e., production, infrastructure and restarting basic services) amidst several challenges. The chapter shows that a convergence of multiple humanitarian aims, national development policies, and donor priorities presents a muddle that does not reflect the farmers' own recovery processes and the constraints they face in bringing land back into cultivation in general. The role of sitting allowances is analysed as a way to create a modestly enabling environment combining and adapting these various norms to the local context.

Introduction

Since 1986, Uganda has been in the spotlight as a potential 'model' for achieving food security despite its conflict history. There was a rise in consumption levels and household income in the 1990s following major economic reforms.[2] Balihuta and Sen (2001) noted relative price movement in favour of food crops compared to cash crops since 1995 that explained the increased number of farmers who seemed to prefer growing food crops over cash crops starting in 1995. This is important because self-production constitutes a significant portion of the consumption basket in rural Uganda (Opolot and Kuteesa 2006). Thus a rise in production of staples has a higher impact on household food security because higher proportions of the harvest (between 59 and 95 per cent depending on the crop) are usually retained for home consumption (Ellis and Bahiigwa 2003).

Partly because of the relatively good conditions for agriculture, Uganda is generally seen to be a country with good prospects for maintaining national and household food security, with the conflict-affected north constituting a notable exception. Conflict contexts like northern Uganda are recognized to pose specific challenges for the eradication of extreme hunger and poverty (Maxwell et al. 2012) given the high levels of food insecurity that can occur in such areas (Alinovi et al. 2007; FAO 2010). In such contexts state capacity to respond is weak (Macrae and Zwi 1992) while social, economic and political processes (or even conflict itself) generate high levels of vulnerability to food insecurity. Humanitarian assistance thus commonly becomes a major component of people's everyday lives.

In northern Uganda,[3] a long-term conflict dates back to the 1980s when the Lord's Resistance Army rose against the governing National Resistance Army/ Movement (NRA/M). The consequence was forced displacement of over 1.8 million people. Efforts to address food insecurity and facilitate recovery were largely implemented by humanitarian organizations,[4] both during the displacement period and in early post-conflict recovery. In order to understand the nature and impact of efforts to maintain and enhance food security in northern Uganda during the conflict and post-conflict years, it is important to investigate how humanitarian efforts addressed the severe constraints to household food security and how agricultural rehabilitation has been conceptualized and addressed with the return of peace. An important part of the restoration of food security has entailed (re)building public sector agricultural institutions at the local level.

This chapter looks at how national food security policies have been implemented in the north of the country and the role of building household food security in the broader processes of post-conflict recovery. Specific attention is paid to trends during the early recovery period after people started moving back home in 2006 after years of displacement. The analysis builds on the literature addressing food security and recovery in post-conflict contexts by exploring how local government actors, humanitarians and local communities try to build a bridge between relief and development, and the effects and consequences of this in relation to prospects for maintaining food security.

Specifically, this chapter unpacks the underlying theories of change in Ugandan agricultural rehabilitation. This relates to the way investment in public works linked to vouchers for seeds and seed fairs has been expected (by humanitarian actors) to contribute to household food security, and how other actors, including returning displaced people, manoeuvre within these programmes. Conclusions are drawn as to how such programming has worked in practice and the implications for the food security of formerly displaced persons.

In the wake of calls for 'resilience', expectations for humanitarian assistance to contribute to food security are becoming even more ambitious in comparison to past reliance on food aid as the de facto response. In northern Uganda, responses to food insecurity and rebuilding agricultural services connect two dynamic processes. First are the struggles of individuals, households and communities to survive and rebuild their livelihoods after long-term conflict, displacement and marginalization. Second are the efforts by local district government and humanitarian actors to make sense of the shifting institutional relationships that they are entangled in.

This chapter looks at the changing nature of food security–related interventions in northern Uganda amidst these processes with specific attention to the interactions between households and humanitarian organizations in production, infrastructure and restarting basic services. Here vouchers for labour undertaken on public works, agricultural extension services and seed fairs constitute interlinked modalities that are used to provide access to seeds and services assumed to be essential to enhancing household food security. Inter-agency processes and relationships shape these interventions. Sitting allowances are given particular attention here because of their role as a mechanism that is used to manage the relations between the local government and humanitarian actors in agricultural rehabilitation efforts. This analysis is based on research on household and community interactions with international humanitarian organizations, and analysis of the concurrent engagement between the humanitarian community and local government agricultural departments.

This chapter draws on two years of ethnographic work. This started with a short visit in 2009 and later a long extended stay in Pader district between 2010 and 2012, as part of research on food security and agricultural services in fragile contexts.[5] A survey was undertaken in 2011 covering two villages from which 30 households were chosen for recurrent household visits. The author 'followed' (Marcus 1995) these households and case study programmes by humanitarian organizations. These methods were complemented with focus group discussions, participant observation in seed fairs, meetings and training events and in-depth interviews with United Nations, humanitarian organizations and the district local government within the production, administration and planning departments. Decentralization and service delivery concerns benefitted from a programme review in nine districts in the north for the Netherlands Embassy in Kampala.

Country context

Because of its relatively favourable agro-ecological conditions, Uganda is generally seen as a country with good potential for maintaining both national and household food security. At independence in 1962 and in the years following, Uganda was marked by successes and growth. By the early 1970s, the country had plunged into political instability and macro-economic mismanagement. The effect of strong state intervention in most sectors of agriculture resulted in 'poor service delivery; shortage of agricultural inputs; infrastructural decay and deterioration of markets; and delayed payment of farmers, which in effect was a tax on farmers' income' (Opolot and Kuteesa 2006).

Starting in 1987 these policies were reversed with World Bank–influenced reforms. Liberalization and privatization proved effective in reversing many aspects of the earlier decline. In the 1990s, there was a noted rise in incomes for farmers, with productivity increasing for major export crops such as coffee, cotton and tobacco. However, the impacts of trade liberalization on food security, farmer incomes and poverty are contested. On the one hand Balihuta and Sen (2001) argued that the reforms led to fall in rural poverty, with households surveys supporting this by showing rises in consumption. The number of food insecure households declined. The poor benefitted as small producers were able to enter the coffee industry (Deininger and Okidi 2003).

Others however have not been so enthusiastic. Belshaw et al. (1999) argued that claims for success were overstated and questioned the sustainability of the heavily aid-dependent economy. The (slight) rise in incomes fell short of the minimum income needed to meet basic food requirements and failed to guarantee access to food and basic needs for a considerable section of the population (Opolot et al. 2006). Rural smallholders reported exploitation by private traders in the absence of farmer organizations that could contribute to relative price stability (Wiegratz 2010; Opolot et al. 2003 cited in Opolot and Kuteesa 2006). Quoting Ssewanyana et al. (2006), Ssewanyana and Kasirye (2010) support these criticisms of the much lauded reforms of this period by showing that between 1992–3 and 1999–2000 Uganda only managed to register marginal improvements in household food security status as measured in caloric terms – improvements which would start reversing by 2000–3. Thus the economic and production results in the agricultural sector, though promising at the macro level, were contested at the smallholder level.[6]

Starting in 2000, the sectoral policy framework shifted emphasis towards modernization and commercialization to accelerate and transform agricultural growth, improve farmers' income and achieve food security.[7] Those predominantly reliant on subsistence rural agriculture (over 90 per cent of the rural population) were encouraged to commercialize. Support to the poor rural population was reoriented to increase the marketed share of agricultural production, which was also expected to create on-farm and off-farm employment. It was assumed that such developments would increase rural incomes and enhance food security (MAAIF and MFPED 2000; NAADS Secretariat 2000). The impacts of the modernization and commercialization approach are mixed and contested. Generally, the literature suggests that there was an improvement in farmers' ability to take advantage of agricultural extension services. Participating farmers demonstrated high adoption rates of new and improved agricultural practices, technologies and new crop enterprises (Benin et al. 2007; OPM 2005). However, the impact on incomes and food security is disputed (Benin et al. 2012). The 2008–9 Uganda Census of Agriculture showed that 57 per cent of rural households were still periodically unable to maintain normal food consumption over a 12-month period.

With Uganda's economy still heavily dependent on the agricultural sector, a new sectoral policy came into force in 2010. The mission remains to transform subsistence farming into commercial agriculture with the stated objectives of increasing rural incomes and livelihoods and improving household food and nutrition security (MAAIF 2010: 51). In this new policy, the strategy has shifted away from direct poverty reduction to wealth creation, growth and transformation. It emphasizes helping farmers move 'up' the value chain.

Developments in the north

There is little literature that empirically analyses food security in the north prior to 1986. Interviews with local government agricultural workers and humanitarian organizations and literature such as those by the World Food Programme (WFP 2009; 1999) indicate that northern Uganda was considered the 'bread-basket' and 'granary' of Uganda before the conflict and 'consistently produced grain surpluses for domestic and sometimes international markets' (WFP 2009: 5). The Acholi (the main ethnic group) relied on a mixed farming system for their livelihoods and food security. The main food crops were sesame, finger millet, cow peas, sorghum, various vegetables, beans, groundnuts, maize, sweet potato and cassava. Livestock keeping and an integrated small cash crop growing economy of mainly cotton (Martiniello 2013) (and to a lesser extent tobacco) provided the cash for basic household needs and if necessary, to smooth consumption (Stites et al. 2006).

Civil unrest began in 1986 when President Museveni came to power. The most prominent of the rebel groups was the Lord's Resistance Army. Starting in 1996, the government forcefully moved over 1.8 million people into 'protected villages' established near trading centres and military bases as part of a counter-insurgency operation. In interviews, people described displacement as leading to loss of access to land, loss of assets and disruption of rural livelihoods, and a major cause of food insecurity. In a household food security analysis of Uganda, Bahiigwa (1999) showed that 51 per cent of the population in the north was food insecure in 1997, which rose to 74 per cent in the following year. As another example of the challenges to food security during this period, United Nations Department of Humanitarian Affairs (UNDHA) (1997) noted that despite a bumper crop resulting from good rains, most people were likely to experience food insecurity since they were unable to harvest, and that displacement and insecurity prevented them from preparing for the 1997 rainy season.

It is in this setting that the WFP became the leading front line humanitarian agency after the initial displacement in 1996–7 and throughout the mass displacements over the following years (Tusiime et al. 2013). This is exemplified by their domination of the 2002 consolidated appeal at the height of the displacement. WFP's requirements amounted to 52 per cent of the total appeal (68,114,892 USD), while support to general agriculture stood at 11 per cent, with the Food and Agriculture Organization (FAO) in the lead for provision of agricultural inputs (UNOCHA 2002). At this time, many of the other humanitarian organizations worked in collaboration with WFP,[8] and were initially restricted to traditional relief activities and later on food assistance, i.e. supplementary and institutional feeding and food for work (WFP 1999).

While states in principle have a primary role to play in humanitarian action and development, in this case there was largely a withdrawal of local government services, with minimal or no activity by local district agricultural production departments.[9] The humanitarian sector rushed to fill the gap with an upsurge in activities in 2004. Humanitarian response systems by the United Nations and international non-governmental organizations (NGOs) ignored the role of local government. However, reviewed documents and interviews with district officials and extension workers indicate that the government nonetheless implemented several agricultural programmes, albeit at a very small scale, providing a minimal level of agricultural extension services. For instance, using contact farmers, farmer-to-farmer seed multiplication was set up for drought tolerant crops like cassava to respond to reported cycles of drought in the district.

In contrast to the more stable parts of the country, the north largely missed out on most of the structural reform processes underway elsewhere in the country. The commercialization and modernization programme was introduced in one district in the north in 2003–4 at the height of forced displacement, and when most of the population had limited access to their land.[10] In other northern districts, some efforts were made during the return process in 2007 – at a time when this national programme was already largely paralysed by uncertainty, several suspensions, shifting reforms and heavy political interference (Kjær and Joughin 2012). The National Agricultural Advisory Services (NAADS) district coordinator in one of the districts summarized the results of their work in those contexts as being extremely limited.[11]

Northern Uganda: Addressing food insecurity for a returning population

According to the Inter Agency Standing Committee in Uganda, by August 2008, 64 per cent of the population had moved out of the original 'mother' camps to 'decongestion sites'[12] or back to their villages of origin. In the decongestion camps, livelihoods shifted from relief to more sustainable and diversified ventures. Individuals were able to capture new economic opportunities, and petty trading, casual labour and market exchange became important. Return also restored access to land – a key resource for many – and increased agricultural production. Humanitarian organizations shifted much of their attention to agricultural rehabilitation. The government, mostly through district local governments and the Office of the Prime Minister, started to reintroduce some services to the area.

Parallel realities shaped the recovery context. Unpredictability and uncertainty characterized people's everyday lives in a still very fragile recovery context where household food security remained highly vulnerable to small shocks. In interviews, people mentioned unpredictable climatic conditions and vagaries of weather, highlighting 2007 and 2010 as years when they lost a substantial proportion of their harvests. While some events seemed large and visible enough to generate some attention,[13] smaller and recurrent events of waterlogging or 'too much rainfall' led to little response from local government and development agencies.

The return and resettlement process involved a shift from temporary relief to recovery in terms of restarting production and basic service provision and rebuilding infrastructure. This was achieved largely through seeds and tools provision via vouchers for labour undertaken in public works.[14] These vouchers were redeemed in 'seed fairs' (discussed further below) and there was some provision of agricultural extension services. At the time the research was conducted, return and recovery efforts were still underway. Poverty levels in the region were twice the national average.

Agricultural rehabilitation as a response to food insecurity

In the initial stages of recovery, efforts shifted largely from food aid to agricultural rehabilitation programming. The assumption was that food security would best be achieved through improvement in agriculture, specifically improvements in crop production. Livestock farming did not feature strongly in interventions. At the same time, WFP phased out food aid partly because of reduced funding.[15]

Humanitarian policy narratives had framed household food insecurity as being related to seed/input insecurity, i.e. that there was claimed to be a lack of, inadequate or poor-quality basic agricultural inputs as a result of periods of displacements and reduction in agricultural activities. This was candidly captured in interviews and reviewed documents such as an excerpt from the mid-term review of the 2005 consolidated appeal which stated:

> the overall objective in this sector remains sustained food security with increased access of vulnerable farmers to agricultural inputs and technical advice for the second planting season. Most of the actions implemented so far targeted primary production potentials, which are in some cases constrained by access to land and/or by lack of inputs. However, if security and access to land improve, inputs in this sector must increase proportionally.
>
> *(UNOCHA 2005: 7)*

Seeds and tools emerged as the standard *modus operandi* of humanitarian programmes, justified by the objective to first of all 'kick start' production by providing material inputs. These were accompanied by demonstration and multiplication plots of seed varieties, with cassava featuring prominently. In interviews, people highlighted an appreciation for improved access to seeds at such an early return phase. At this early recovery stage humanitarian programmes linked relief and development through activities such as the formation of farmer groups who 'contributed' to rebuilding 'community assets and infrastructure' through public works. In return, people received vouchers (instead of food or money) redeemable with selected input providers (stockists) working with an agro-input dealers' association. This was accompanied by training in improved agricultural practices, also intended to create demand for the new inputs (CEM 2010).[16] The ultimate intended outcome was expected to be the revival of production using improved planting materials combined with creation of a sustainable market for agricultural services led by the private sector. The ultimate impact was expected to be enhanced household food security. The overall intent coincided with attempts to move towards more pluralistic agricultural service provision elsewhere in Uganda, but despite apparently similar intentions to engage private service providers, the modalities used were completely different.

Public works and labour contestations

The recovery projects were largely designed to combine employment with construction of what is referred to by humanitarian agencies as 'community assets'. The Ugandan modalities were similar to efforts undertaken in many countries where disaster- or conflict-affected people receive employment through public works programmes. This modality has come to be seen as a solution to address diverse concerns simultaneously, among governments, the international development community (McCord 2012) and the humanitarian arena.[17]

The efficacy of public works as a tool to address food and livelihood insecurity assumes the availability of household labour among the target population. Furthermore, readiness to commit household labour to projects is widely used as an indication that people are willing to contribute to these projects, and that they therefore value them, as they recognize that the 'asset' created will benefit the 'community'.

Given competing demands on household labour, these assumptions are questionable. The Acholi are reliant on family and community labour for production (Atkinson 1989; Martiniello 2013) – a factor that is key in re-establishing self-sufficiency (Martin et al. 2009). Public works create competing demands for household labour. The survey showed that 76 per cent of the households in the two villages were involved in communal (traditional) labour participation arrangements. At the same time the cropping season creates opportunities for short-term employment in the form of casual labour where at least 41 per cent of the households work on other people's land to earn income to supplement their food and other household related needs. Interviews and my observations revealed that the timing of public works coincided with the peak labour markets. For participating households, labour-based programmes did not generate additional income, but rather displaced income from other activities such as casual labour or happened at the expense of land clearance. The labour programmes thus made demands on an already scarce resource, and were tied to vouchers only redeemable in seed fairs and not cash. The informants mentioned that it was better to engage in parallel livelihood activities such as casual labour.

There are thus differences between how the humanitarian organizations and the local people perceived the public works modality. Some agencies justified their public works programmes as a way of empowering the local community. This discourse was fed by narratives of displacement and disempowerment. Particularly, several interviewees said that people were in camps for 20 years – conflating the conflict period with the displacement time. A project manager of one agency added, '[W]e need to reduce the dependency on being given free things. They stayed in camps for a long time receiving free food.' People were thus understood as having been displaced for years, during which they were dependent on relief handouts. Public works were seen in contrast to symbolize 'empowerment', i.e. that people earn their own income and take control over their own development. One programme manager stated, '[W]e don't give free things. . . . You cannot ask a cow to give you eggs. This is how we do our programmes.' A consistent picture arose in interviews and in observed meetings that the agencies which had previously provided human-itarian assistance were very eager to use public works programmes to 're-brand' themselves as development actors. However, households interviewed again held a different view. First, many were highly ambivalent about the empowerment logic. Those interviewed did not consider that short periods of displacement meant they were in need of 'empowering'. Also, they expressed little ownership of the 'com-munity assets' as shown by the lack of maintenance of this infrastructure once completed.

In addition to 'empowerment', public works were introduced as a means to transfer agricultural inputs – mainly seeds and tools. Humanitarian organizations claimed that, although they had a community benefit, this was only meant as sec-ondary to the main objective of improving household food security (defined in terms of household self-sufficiency) directly through access to seeds and tools and accompanying agricultural extension services. Seeds and tools were considered to have a direct and primary effect on household food security.

Most importantly, households interviewed described serious labour shortages as limiting their own efforts to achieve food security. Lack of labour limited their ability to open new land, which in turn resulted in fallowing for shorter periods or continuous cultivation – factors which when compounded by increased land conflicts could lead to declining soil fertility. Prevailing shifting cultivation practices in north Uganda require considerable periods of fallow to maintain fertility (Marti-niello 2013). These existing land management practices are under intense pressure, which has additional implications for the relevance of new seed varieties (discussed in the next section), as they require sufficient levels of soil fertility.

Some informants thus referred to the public works as employment generation schemes and not empowerment projects. This was in turn related to their dissatis-faction at being paid in vouchers tied to seed fairs (discussed in the next section), which restricted their 'empowerment', as they could not choose how to invest what they perceived as their own earnings. They thus referred to 'earning' and 'working' for the vouchers, in contrast to the aid agency narratives about 'contributing' to 'community assets'.

Seed fairs and contesting the rebuilding of markets?

Those participating in public works earned vouchers that were primarily exchanged for seeds. Initially this involved cashing in the vouchers for seeds sold by selected input providers (stockists) working with an agro-input dealers' association. This modality proved rather complex in such a quickly changing context. Outside of aid programmes, there was insufficient demand for the inputs and the stockists failed to establish themselves in these hurriedly created 'markets'. As an alternative, seed fairs were piloted and they soon became the main de facto interface between the aid agencies and public sector agricultural service providers.

Longley et al. (2006) highlight the role of seed fairs as part of innovative tempo-rary schemes aimed at rebuilding local markets. Farmers exchange vouchers issued by an agency for inputs at these events. An additional logic is that by restricting the farmers to buying seeds and tools, this investment can directly contribute to increased household production, and ultimately food security. A manager in an international NGO explained, '[W]e use vouchers and seed fairs because our projects are agricultural so we saw this as the main way to ensure the farmers get the seeds. So the aim is to get seeds for planting for their own production' (Field interview, 2011). Since the fairs promote choice and variety of agricultural inputs, they are regarded as a means to 'put farmers first' (Remington et al. 2002), or to highlight farmers' demands and needs (Birner et al. 2011) as part of a more demand-driven alternative to direct seed distributions. In addition, they can contribute to service provision and institutional development by creating conditions for normal markets to eventually function through promotion of the development of private input providers.

By restricting farmers' choices if they should wish to use their earnings in other ways, e.g. by purchasing cattle or engaging in non-agricultural pursuits, this modality could hardly be seen as being fully empowering. Seed fairs thus became contested in the research arena. Many returning households reported that they ultimately wanted to use the money they 'earned' on different needs such as health or education. During project meetings, the participants often suggested that they preferred cash instead of vouchers. In the survey the results were similar, with a preference for cash or a mixture of cash and vouchers to enable people to meet these other needs.

It was within these conditions that private traders from outside specific localities established themselves, but the market was apparently still not sufficiently developed to provide comprehensive services. Seed fairs are based on assumptions of creating a competitive market that would lead to increased accountability of service providers to clients (Longley et al. 2006). It is questionable whether a one-day encounter provides the necessary conditions for accountability between the vendors and buyers. From our interviews, service providers rarely indicated any measure of accountability to the participating farmers (who were supposed to be their 'customers'), partly because of a lack of alternative options for the farmers. This led to the development of a cadre of service providers who positioned themselves with seed fair financiers as their main clients, with little trust between them and the local communities they served. In interviews, local people commonly used the words 'stealing' and 'cheating' in reference to the transactions that took place at seed fairs. This culminated in a case in March 2011 where local farmers joined together to buy items from their own relatives, sidelining the input dealers who were considered outsiders and whom they accused of 'stealing' from them.

Contestations and accusations of 'cheating' were fuelled by differences in perceptions between the agencies and the local people – a reflection of the different ways the two groups perceived these practices (Kibreab 2004). Particularly, the differences concerned livelihood priorities, types of seeds promoted, quality concerns and input prices. At several seed fairs attended, the supply of vouchers exceeded the inputs on sale. As could be expected in such a 'market', prices of inputs were twice or thrice the average market price. Later, farmer choice was further restricted as a considerable proportion of the vouchers was tied to 'improved seeds'. While 'improved' and 'certified seeds' were promoted in some seed fairs, these were generally more expensive. Recurrent household visits showed that such seeds were rarely used, the households relying instead on seed from previous harvest or that bought from local markets.

Some farmers negotiated the exchange of vouchers for cash from the vendors – a practice considered as 'cheating' (this time from the aid organization's viewpoint rather than the farmer's). One NGO project manager stated, 'Exchange of vouchers for cash undermines the logic of vouchers and seed fairs of ensuring farmers got seeds and not cash.'[18] It also carried consequences for the vendor if found in possession of vouchers whose worth considerably exceeded the produce brought to the market. This was regarded as an indication of having 'bought' vouchers from the market. Elaborate practices like weigh-in (at start of the fair) and weigh-out of produce (at the end) were introduced to give agencies an idea of quantity transacted and thus ascertain whether traders were involved in malpractice. But by 2012, some participants could save enough seed for sale and bought seed only from their own relatives. In one seed fair attended, local farmers refused to buy seeds from traders who came from far away. As such, the seed fairs have come to be part of a larger political economy (of service provision and public sector reforms) that shapes provision of agricultural services in the district, as I will show in the next section.

Seed fairs, quality control and sitting allowances

The success of seed fairs or any input-related intervention critically depends on effective and independent quality control mechanisms for inputs (Christoplos 2008). Such mechanisms, however, were not well established in north Uganda. Poor quality seed is a broad concern within Uganda and attributed to a history of breakdown of state regulation (CEM 2010) in part a consequence of the two decades of national neoliberal reforms (Wiegratz 2010).

Seed fairs require the physical presence of an officer from the production department to provide quality control. In an assessment of governance challenges and the level of influence of different actors in agricultural livelihood programmes in north Uganda, Birner et al. (2011) note that outside of the army (which provided security when seed fairs were conducted in open conflict) and the programme coordinator, the district agricultural officer was the most influential person affecting the outcome of a seed and tools programme. Generally, issues of quality control became even more pertinent when the Ministry of Agriculture demanded better control on the movement of planting material such as cassava cuttings (which are largely the typical seed material in the early recovery phase) because of the appearance of a very invasive weed (Lubangakene 2010).

The re-establishment and provision of local government extension services (which included aspects of quality control) have not been without problems. Pader district faced low staffing rates, which meant that individual technical staff often had to cover responsibilities intended to be managed by two to three staff members.

With heavy reliance on central government financial transfers, which were generally inadequate and often late, the local government in the north faced huge challenges in fulfilling its responsibilities for service delivery.

In addition, there was a great push from the central government for local governments to be more proactive in recovery efforts. By 2010, the gradual withdrawal of the United Nations Office for the Coordination of Humanitarian Affairs and a highly publicized return process signalled that the humanitarian phase was over. In this situation, donor pressure for humanitarians[19] to work more with the government connected with the tendency of the district government to look to aid agencies to offset shortfalls in departmental funds. In such contexts, the humanitarians generally constitute the only real 'resource' (Olivier de Sardan 2011). At the time of this research, the many international and national NGOs, United Nations agencies, community- and church-based organizations implementing projects with a focus on household food security looked to the local district government departments for services such as quality control and for some extension services. In return for these services they provided operational funds like transport. In addition, they paid sitting allowances – monetary incentives through a cash transfer – to the extension workers engaging in these activities. In this way, the sitting allowances became a core component of day-to-day interactions in the post-conflict context and to a high degree shaped service delivery.

Sitting allowances became a central element in the process of negotiating responsibilities for quality control, service provision and development, as they were a mechanism to bridge the boundaries between the local government and aid agencies. It is possible to look at them as initially intended during the late 1980s within development circles – to encourage civil servants to concentrate on service provision, instead of looking for alternative means of income. In north Uganda they still supplement civil servants' salaries, which in most cases are inadequate or delayed.

One of the processes through which boundaries are bridged is through labelling. The term 'sitting allowances' is a colloquial expression that invokes the image of paying someone to literally 'sit through' a meeting or, as indicated above, pay them for their services. Other terms I heard were considered more appropriate. Both local government and the agencies preferred to talk about 'transport' and 'training' allowances; 'facilitation' was another commonly used term. These were considered more aligned with NGO discourse (and donor requirements) on capacity building. District staff relied on a privatization narrative because, despite their status as civil servants, they also described themselves as 'private service providers', since they argued they were contracted by the NGO on behalf of the community. This was much in line with government efforts to privatize extension services (which existed from 2000 to 2009). This was especially because the NGO 'beneficiaries' received training – at times outside and/or in addition to the agricultural extension service provided to groups under government programmes such as NAADS.

This arrangement has had significant influence on how local governments and aid actors legitimize shifting service provision norms. NGO staff use allowances to establish constructive relationships that are needed to influence the working of the local government and the quality of extension provided. Those interviewed claimed it enabled them to hold the district departments accountable for service provision and thus on such grounds, allowances have become something that can always be 'organized'. Several respondents added that in these cases sitting allowances represented a modest proportion of project costs, but could make a major impact on effectiveness.

When facilitation was 'organized', quality control was still a complex process. In seed fairs attended by the author, the district government representative sampled inputs and enquired on their source and storage. Both seeds and tools entailed physical examination. This senior agricultural officer explained that certified and packaged seeds must come with a certification card. Germination tests by the NGO or the production directorate were supposed to be required. However, this was rarely done in practice, because it would require a process of pre-qualifying vendors in order to obtain samples for germination tests. Instead, a physical examination was undertaken.

Even though the input control may not always have been realized as foreseen, the presence of the production department staff at seed fairs served different purposes in practice for different actors. The production department staff used the gathering as an opportunity to communicate with farmers on relevant issues, such as the outbreak of diseases, commodity price fluctuations and weather forecasts. In a number of seed fairs the district agricultural officers visited local input dealers, checking on their stores; they also collected food prices from local markets or even visited some gardens, for example, to assess the prevalence of the mosaic disease in cassava. Local people also took advantage of these events to seek specific advice on crops and diseases, or even just to clarify information provided earlier by the production department. They would, for example, request advice on whether or not to plant cotton in view of the price-fall in the previous year. The NGOs insisted on a report written by local government staff after the event to show their concern for quality assurance and to enhance their legitimacy in the eyes of their donors.[20]

Conclusion

In this chapter we have looked at two key aspects of agricultural rehabilitation programming in the early recovery phase after displacement in north Uganda. We first focused on the interaction between local communities and the aid organizations, highlighting the difficulty in a complex context of matching people's needs with interventions. Second, focusing on the ways that the aid agencies and local agricultural departments come together, we have seen that interventions and people's everyday lives are conditioned (partly) by the ways that local government deals with the seemingly disparate assumptions of national government and aid agencies regarding how to achieve household food security. This draws us to three major conclusions on the developments in the northern region, which also carry implications for supporting household food security in northern Uganda.

First, food security narratives in the north have shifted over time from being driven by food aid to re-establishing production. In addition, humanitarian service provision has gradually been re-conceptualized so as to reflect at least partially the current agricultural modernization policy. The idea is that 'returnees' need to be given material aid to attain food security, through a modernization-imbued agenda to be achieved through improved planting materials. Recovery and sustainability have come to be associated with quick-fixes for institutions and markets for service provision after the conflict. This is with an assumption that it leads to both greater household food security and 'empowerment' by increasing people's capacity to choose their own paths to recovery, despite the paradox that it is based on mechanisms that in many respects constrain people's choices. The public works/voucher/seed fair modality constitutes a convergence of the multiple humanitarian and development aims and national development policies that characterize the re-merging of northern Uganda into national processes.

Second, the result of this convergence is a muddle of humanitarian and development paradigms that does not seem to reflect the farmers' reality and constraints faced by locals in relation to bringing land back into cultivation in general, much less in the transformation to 'modern' intensive production that is envisaged. Failures to factor in conflicts between labour on 'community assets' and the need to clear sufficient land to use the resulting earnings/seeds in a sustainable and productive manner exemplify the problematic nature of this conceptual muddle.

Third, the modality of sitting allowances, while perhaps seeming particularly out of place in relation to the rhetoric of returning to sustainability after conflict, institutional development and modernization, may actually be one of the few mechanisms that creates space to sort out this mess. The arrangement creates a modestly enabling environment for negotiated relations wherein these policy goals can be combined and adapted to the demands of the population and the efforts of local service providers to find a new social contract with the local population. Currently, rebuilding agricultural services and addressing household food insecurity in north Uganda is much more than a transition from exceptional crisis to the normality of governance. In some ways, it coincides with major reorganizations of the Ugandan state, where semi-fulfilled policy shifts and the privatization of services considerably alter the conditions of (agricultural) services. While there seems to be little appreciation of this and the effect that it has for the north re-emerging from conflict, it highlights the complexity in recovery and addressing food security when the institutions intended to provide services are involved in negotiating and bridging at the same time as providing those services. Sitting allowances have become one of the main mechanisms of social ordering of working relations between the local government and aid actors in the domains of food security and agricultural services. In north Uganda, they play a role in the establishment of smooth collaboration between agencies and civil servants, and they have become a principal channel through which the mutual accountabilities between local government and aid agencies are negotiated. It was to a certain extent through sitting allowances that agencies could realize the expectations they had from civil servants, and civil servants could undertake a modicum of service provision.

Notes

1 PhD researcher, Humanitarian Aid and Reconstruction, Wageningen University.
2 Most studies examining the 1990s rely on national household surveys which are analysed in relation to poverty trends and not food security per se.
3 The phrase 'northern Uganda' is used here in a restricted sense, to refer to the seven districts referred to as the Acholi region that bore the brunt of the Lord's Resistance Army war between 1986 and 2006.
4 Humanitarian organizations increasingly seek to broaden their mandate and take on more development-oriented tasks. I use the term to capture even this broadening role of agencies with dual humanitarian and development mandates. The primary emphasis is on NGOs.
5 The study is undertaken within the framework of the IS academy on Human Security in Fragile States of the special chair Humanitarian Aid and Reconstruction at Wageningen University, the Netherlands. This particular PhD project is hosted and financed by the Dutch aid agency – ZOA (Dutch translation of South East Asia) under an interactive research arrangement. I am most grateful for this contribution. All usual disclaimers apply.
6 Opolot (2006) actually argues that per capita production in 2000 was half that of 1975 meaning that national food production could not keep up with what was required to feed the nation.
7 This is generally recognized as the Plan for Modernization of Agriculture (PMA) under which the National Agricultural Advisory Services (NAADS) programme would become the most successful and most widely known. Ironically, around the same time a food and nutrition policy that pushed for food security through self-sufficiency as opposed to the market came into effect. The latter has become the proverbial 'child that is meant to be seen but not heard' while the market bias has shaped the government's approach towards food security since then.
8 These included World Vision, the Church of Uganda, Feed the Hungry Children, Gulu Support the Children Organization (GUSCO), Catholic Relief Services (CRS) and the Italian Development Cooperation Associazione Volontari per il Servizio Internazionale (AVSI). Later the Norwegian Refugee Council (NRC) joined also.
9 This consists of crop, veterinary, commercial services, entomology, fisheries, trade, forestry and cooperative societies sub-departments.
10 People were only allowed access to a 'safe radius' (Adoko and Levine 2004) of 2 kilometres around the camps to cultivate.
11 Interview, 24 March 2011, Pader district.
12 Also known as transition or satellite camps, these were set up from 2006 and located about 2–5 kilometres from the original villages of displacement.
13 The Office of the Prime Minister distributed seeds after the 2007 floods. The 2010 events resulted in an assessment by the Pader local government in collaboration with agencies active in the sector.
14 Other variant interventions focused on direct seeds and tools transfer.
15 Interviews, head of WFP Pader sub-program, 2011.
16 The poster example of this linking modality was the Agricultural Livelihoods Recovery Project (ALREP). Targeting 43,466 participants, ALREP was an FAO project implemented through ZOA, Arbeiter-Samariter-Bund Deutschland (ASB), Agency for Technical Cooperation and Development (ACTED), and Caritas among others.
17 While McCord refers specifically to the development field programming where public works are based around social funds, these are similar in many ways to the humanitarian sector. In the development arena people receive money, and not vouchers or food.

18 Interview, project manager, INGO September 2010.
19 Increasingly, there has been pressure on humanitarian agencies (from both local and central government) to work more closely with the local government.
20 Other performance mechanisms included pictures for inclusion in reports to donors.

References

Adoko, J. and Levine, S. (2004) *Land matters in displacement: the importance of land rights in Acholiland and what threatens them,* Kampala: Civil Society Organizations for Peace in Northern Uganda.

Alinovi, L., Hemrich, G. and Russo, L. (2007) *Addressing food insecurity in fragile states: case studies from the Democratic Republic of the Congo, Somalia and Sudan,* Rome: Agricultural and Development Economics Division, Food and Agriculture Organization of the United Nations.

Atkinson, R. (1989) 'The evolution of ethnicity among the Acholi of Uganda: the precolonial phase', *Ethnohistory,* 36: 19–43.

Bahiigwa, G. (1999) *Household food security in Uganda: an empirical analysis,* Kampala: Economic Policy Research Centre.

Balihuta, A.M. and Sen, K. (2001) *Macroeconomic policies and rural livelihood diversification: a Ugandan case study,* Livelihoods and Diversification Directions Explored by Research (LADDER) Working Paper No. 3, Norwich: University of East Anglia.

Belshaw, D., Lawrence, P. and Hubbard, M. (1999) 'Agricultural tradables and economic recovery in Uganda: the limitations of structural adjustment in practice', *World Development,* 27: 673–90.

Benin, S., Nkonya, E., Okecho, G., Pender, J., Nahdy, S., Mugarura, S. and Kayobyo, G. (2007) *Assessing the impact of the National Agricultural Advisory Services (NAADS) in the Uganda rural livelihoods,* IFPRI Discussion Paper 724, Washington, DC: International Food Policy Research Institute.

Benin, S., Nkonya, E., Okecho, G., Randriamamonjy, J., Kato, E., Lubade, G. and Kyotalimye, M. (2012) 'Impact of the National Agricultural Advisory Services (NAADS) program of Uganda: considering different levels of likely contamination with the treatment', *American Journal of Agricultural Economics,* 94: 386–92.

Birner, R., Cohen, M. and Ilukor, J. (2011) *Rebuilding agricultural livelihoods in post-conflict situations: what are the governance challenges? The case of northern Uganda,* Uganda Strategy Support Program Working Paper 07, Kampala and Washington, DC: International Food Policy Research Institute.

CEM (2010) *Final evaluation of the Agricultural Livelihoods Recovery Project (ALREP),* UK: Cardno Emerging Markets.

Christoplos, I. (2008) 'Narratives of rehabilitation in Afghan agricultural interventions', in Pain, A. and Sutton, J. (eds), *Reconstructing agriculture in Afghanistan,* Rugby: Practical Action.

Deininger, K. and Okidi, J. (2003) 'Growth and poverty reduction in Uganda, 1999–2000: panel data evidence', *Development Policy Review,* 21: 481–509.

Ellis, F. and Bahiigwa, G. (2003) 'Livelihoods and rural poverty reduction in Uganda', *World Development,* 31: 997–1013.

FAO (2010) *The state of food insecurity in the world: addressing food insecurity in protracted crises,* Rome: Food and Agriculture Organization of the United Nations.

Kibreab, G. (2004) 'Pulling the wool over the eyes of the strangers: refugee deceit and trickery in institutionalized settings', *Journal of Refugee Studies,* 17: 1–26.

Kjær, A.M. and Joughin, J. (2012) 'The reversal of agricultural reform in Uganda: ownership and values', *Policy and Society,* 31: 319–30.

Longley, C., Christoplos, I. and Slaymaker, T. (2006) *Agricultural rehabilitation: mapping the link-ages between humanitarian relief, social protection and development,* Humanitarian Policy Group Report 22, London: Overseas Development Institute.

Lubangakene, C. (2010) 'Killer weed invades Pader district', *New Vision,* 11 March.

MAAIF (2010) *Agriculture for food and income security: agriculture sector development strategy and investment plan 2010/11–2014/15,* Kampala: Ministry of Agriculture, Animal Industry and Fisheries.

MAAIF and MFPED (2000) *Plan for the modernization of agriculture: eradicating poverty in Uganda,* Entebbe and Kampala: Ministry of Agriculture, Animal Industry and Fisheries, and Ministry of Finance, Planning and Economic Development.

Macrae, J. and Zwi, A. B. (1992) 'Food as an instrument of war in contemporary African famines: a review of the evidence', *Disasters,* 16: 299–321.

Marcus, G. (1995) 'Ethnography in/of the world system: the emergence of multi-sited eth-nography', *Annual Review of Anthropology,* 24: 95–117.

Martin, E., Petty, C. and Acidri, J. (2009) *Livelihoods in crisis: a longitudinal study in Pader, Uganda,* Humanitarian Policy Group Working Paper, London: Overseas Development Institute.

Martiniello, G. (2013) *Accumulation by dispossession, agrarian change and resistance in northern Uganda,* MISR Working Paper No. 12, Kampala: Makerere Institute of Social Research.

Maxwell, D., Russo, L. and Alinovi, L. (2012) 'Constraints to addressing food insecurity in protracted crises', *Proceedings of the National Academy of Sciences,* 109: 12321–5.

McCord, A. (2012) *The politics of social protection: why are public works programmes so popular with governments and donors?* Background Note, London: Overseas Development Institute.

NAADS Secretariat (2000) *National Agricultural Advisory Services (NAADS) master document of the NAADS Task Force and Joint Donor Group,* Entebbe: Ministry of Agriculture, Animal Industry and Fisheries (MAAIF).

Olivier de Sardan, J.-P. (2011) 'Local powers and the co-delivery of public goods in Niger', *IDS Bulletin,* 42: 32–42.

OPM (2005) *Evaluation report: the plan for modernization of agriculture,* Oxford: Oxford Policy Management.

Opolot, J. and Kuteesa, R. (2006) *Impact of policy reforms on agriculture and poverty in Uganda,* Discussion Paper No. 158, Dublin: Institute for International Integration Studies.

Opolot, J., Wandera, A. and Atiku, Y. A. (2003) *Trade reforms and food security in Uganda,* Final report submitted to the Food and Agriculture Organization of the United Nations.

Opolot, J., Wandera, A. and Atiku, Y.A. (2006) 'Uganda', in Thomas, H.C. (ed.), *Trade reforms and food security: country case studies and synthesis,* Rome: Food and Agriculture Organiza-tion of the United Nations.

Remington, T., Maroko, R., Walsh, S., Omanga, P. and Charles, E. (2002) 'Getting off the seeds and tools treadmill with CRS seed vouchers and fairs', *Disasters,* 26: 316–28.

Ssewanyana, S. and Kasirye, I. (2010) *Food insecurity in Uganda: a dilemma to achieving the hunger millennium development goal,* Research Series No. 70, Kampala: Economic Policy Research Centre.

Ssewanyana, S., Obwona, M. and Kasirye, I. (2006) *Understanding food insecurity in Uganda: a special study,* Report prepared for USAID Uganda Country Office.

Stites, E., Mazurana, D. and Carlson, K. (2006) *Movement on the margins: livelihoods and security in Kitgum district, northern Uganda,* Massachusetts: Feinstein International Centre.

Tusiime, H.A., Renard, R. and Smets, L. (2013) 'Food aid and household food security in a conflict situation: empirical evidence from northern Uganda', *Food Policy,* 43: 14–22.

UNDHA (1997) *Uganda: humanitarian situation report, 15 March 1997,* Geneva: United Nations Department of Humanitarian Affairs.

UNOCHA (2002) *Consolidated appeal for Uganda 2002* (Revision), Kampala: UN Office for the Coordination of Humanitarian Affairs.

UNOCHA (2005) *Uganda 2005 mid year review consolidated appeal,* Geneva: UN Office for the Coordination of Humanitarian Affairs.

WFP (1999) *WFP Assistance to internally displaced persons: country case study on internal displacement: Uganda, displacement in the northern and western districts* (final draft), Rome: World Food Programme.

WFP (2009) *Protracted relief and recovery operation 10121.3,* Rome: World Food Programme.

Wiegratz, J. (2010) 'Fake capitalism? The dynamics of neoliberal moral restructuring and pseudo-development: the case of Uganda', *Review of African Political Economy,* 37: 123–37.

10

POVERTY, FOOD SECURITY AND LOCAL WATER CONFLICTS IN SOUTHERN ZAMBIA

Mikkel Funder, Carol Mweemba and Imasiku Nyambe[1]

Introduction

Food security and water access

Access to water is increasingly recognized as a pivotal issue in household food security. Calow et al. (2010) identify three key links between food security and access to water, namely (i) the fundamental productive aspect of water for irrigation and livestock watering, (ii) water for drinking and hygiene as a prerequisite for subsistence and labour and (iii) the time, funds and other resources spent obtaining water which influences the means available for other purposes.

Access to water for these purposes is partly influenced by physical water availability (which may be susceptible to climate change effects), and the technological and economic constraints to exploiting water resources (e.g. groundwater in Zambia). However, relative water scarcity in terms of unequal distribution of water resources is equally critical to food security. Access to water is therefore contested and may be subject to conflict.

Much scholarly work on water conflicts has been directed at international water politics and transboundary water governance. Climate change has furthermore prompted the notion of possible future 'water wars' and a 'securitization' of the climate change discourse (Nordåsa and Gleditscha 2007). This has sometimes led to overly simplistic portrayals of major food security crises (such as those on the Horn of Africa) as vicious circles of drought, conflict and hunger (and see Willenbockel, Chapter 3).

Most conflicts over access to water are however local and typically play out in an everyday, non-violent fashion. They are rarely noted beyond the community level but are nevertheless of critical importance to local livelihoods and food security. This chapter seeks to provide insights into such everyday water conflicts and their implications for water access and food security among the rural poor in Zambia.

Despite being well endowed with both land and water resources, food insecurity in rural Zambia remains widespread, especially in rural areas (GoZ 2013). Although there are numerous reasons for this, lack of access to water for small-scale farmers and pastoralists is a critical factor, and local water conflicts are a significant fact of life in many communities.

Methodology

The chapter draws on the Zambian findings of the four-year 'Competing for Water' research programme, which during the period 2007–9 mapped and studied local water conflict and cooperation in Bolivia, Mali, Nicaragua, Viet Nam and Zambia from 2007 to 2011. During our research within this period, we collected three sets of data:

(i)   Quantitative inventory of water conflict and cooperation events in the study area. The inventory mapped all formally reported events in the Namwala District during the period 1995–2007, as well as unreported events from ten villages sampled randomly from a total of 427 communities in the district.

(ii) A household survey that explored the relationship between livelihoods and water access from a poverty perspective. The survey was conducted with a total of 200 households in 20 randomly selected villages in Namwala District. Informants were selected using stratified sampling of household well-being (i.e. divided into lowest, middle and highest well-being groups), to allow for analysis of differences of water access between different socio-economic strata.

(iii) Three qualitative case studies of how conflicts and cooperation unfolded as a social process. The case studies had a particular focus on the position of the poorest households in water conflict and cooperation, and traced the actions, strategies and roles of those involved through ethnographic interviews with them and with key informants.

This chapter draws on all of this information, but focuses particularly on two case studies that illustrate how poor households – who are the most food insecure households in Namwala – act during and are affected by local water conflicts. Although the study did not focus specifically on food security as such, relevant data was acquired, and it is our hope that the issues discussed here are of use for interventions and studies related to enhancing food security through improved water access.

Livelihoods, food security and water access in Namwala

Livelihoods and food security in Namwala

Namwala is a rural district located in the Kafue River Basin in southern Zambia, within the wider Zambezi River Basin. The district covers some 10,000 km^2 and has a population of 82,700, with an annual range of rainfall between 800 and

1100 mm. Although infrastructure in the area is now gradually developing, many parts of the district remain relatively remote and experience high poverty rates. The northernmost part of the district is dominated by the Kafue Flats, an extensive floodplain which reaches beyond the district and covers a total of about 6,500 km².

Traditionally, the population in the area has consisted of various ethnic minorities as well as Ila pastoralists, who migrate with their cattle to the Kafue Flats during the dry season. The district furthermore has a population of Tonga farmers, who combine livestock husbandry with cultivation of maize, cotton and vegetables as cash crops for sale locally or to markets in neighbouring districts. Namwala District sees seasonal in-migration of fishermen near the Kafue River, as well as increasing numbers of land-seeking farmers from surrounding districts (Haller 2010).

In a good year with average rainfall and timely access to inputs, Namwala has a net surplus in maize production (RoZ 2006). The district is however located in a region that is historically vulnerable to droughts, and which has experienced increasing temperatures since at least the 1960s (Jain 2007; Thurlow et al. 2009). In addition, rainfall patterns appear to be changing towards a later and less predictable onset of rains, as well as more heavy downpours when rains do come (Thurlow et al. 2009). Although the Kafue River floods annually, the 2000s saw a series of extreme floods which affected local livelihoods and crop production. During the 1990s and early 2000s the area was also subjected to serious epidemics of livestock diseases, which in some areas literally decimated livestock populations among both Ila pastoralists and mixed-farming Tonga households.

These factors suggest that food security in the area is precarious, even in the best of times. It is also unequally distributed among households: our 2008 household survey (conducted in 20 villages and comprising both pastoralist and mixed-farming communities) showed that while 28 per cent of households were food self-sufficient during that season, 51 per cent had experienced shorter periods of food insufficiency (below two months) during which they had to reduce the amount of food eaten or buy food. The remaining 21 per cent – typically the poorest segment – had experienced food insufficiency for more than two months that season. The latter group coped by reducing the number of meals, borrowing food or money to buy food, asking for food aid or engaging wives and/or children in day-labour to generate income for food.

Improving food insecurity through increased yields and livestock production are key development goals for Namwala District's local government and technical line agencies. Various bilateral donors and NGOs are working in the area to address food security. In this respect there has been a tendency to focus on sedentary mixed farming rather than pastoralism, with a focus on crop diversification, conservation farming and alternative incomes from e.g. gardening, fishponds, poultry and communal seed banks.

Food security and access to water in Namwala

Food security in Namwala is closely tied to water access, especially during the dry season. As in much of southern Zambia, water is in principle plentiful in the area, but is geographically unequally distributed and difficult to access outside the rainy season or when rains fail. The Kafue River is located in the extreme north of Namwala District, whereas the central and southern parts of the district have few streams in the dry season.

While there are good groundwater resources in Namwala, the district is characterized by a relatively low level of water infrastructure development. Both shallow and deep wells are widespread in communities, but during the dry season they frequently dry out or silt up, and are insufficient to meet demands in many parts of the district. The pressure on remaining water points is thus high during the dry season. Until recently only a limited number of boreholes existed in the district, but in recent years various borehole construction projects have been/are being implemented. Breakdown of boreholes and difficulty in obtaining parts is however a major problem. Approximately 46 per cent of the population has access to safe drinking water, typically those located in major settlements. The average distance between water sources is 2.8 km for most areas, but effectively this is greater during the dry season when water points dry out or boreholes break down from heavy use. Our survey thus showed that 13 per cent of households walked 20–40 minutes to the nearest domestic water source during the dry season, while another 13 per cent walked 40 or more minutes. Many households – typically women and children – thus spend a good deal of productive time collecting water.

Traditional livelihood strategies in the area have been adapted to the prevailing geographical conditions as far as possible. The Ila pastoralists thus continue to follow a migration pattern whereby they reside in the southern uplands of the district during the rainy season, and then move the cattle to the plains alongside the Kafue River in the northern part of the district. These plains – the Kafue Flats – are inundated on an annual basis and consist of rich grazing areas and plentiful surface water for watering livestock. The plains thus form a critical backbone of the pastoralist livestock economy and food security in the area. Apart from being a source of protein for the Ila pastoralists, cattle are sold both at local markets and for export to the urban centres in neighbouring districts, as well as to Lusaka and the Copperbelt.

Access to the water resources in the Kafue Flats is however not a natural given: the plains are controlled by different pastoralist clans and associated elites, who demand implicit support and sometimes in-kind payments to grant access for less well-off households. Pastoral elites furthermore control a system whereby poorer households are employed as cattle herders in return for e.g. milk, meat and water. The pastoral economy in the area is thus highly differentiated, and access to water is an important element in this. Because they control access to the plains, pastoralist elites have a more secure access to their water resources and can regulate the influx of less well-off households if required. Likewise, during the wet season when cattle are mainly in the southern uplands of the district, better-off households are better placed to withstand drought: many own private wells from which cattle can be watered, whereas less well-off households must resort to surface water such as small wetlands and streams until they dry up.

Turning to the area's Tonga farmers, maize is the main staple and is the pivotal crop for local subsistence. Cash crops including cotton are being promoted by extension officers and sold through middlemen outside the district. Most households rely on a single annual crop of rainfed maize (and cash crops), as large-scale irrigation does not currently exist in the district. Recently small-scale irrigation has developed in some sites, where small weirs provide water for through, for example, hand-pumps. Communal boreholes are also used for hand-irrigation of vegetable gardens. Such gardens are a sought after means for women to provide additional nutrition and extra incomes. Cattle husbandry among the Tonga relies on water from local streams and *dambos* (wetland) during the rainy season, and wells or – where available – boreholes during the dry season. Drinking water is obtained from the same sources.

Both Ila and Tonga production systems in the area are vulnerable to drought and erratic rainfall patterns. Studies have shown a close relationship between rainfall and maize production in southern Zambia (Jain 2007; Thurlow et al. 2009). Likewise, a direct correlation between rainfall trends and livestock populations has been documented for livestock economies of the region (MTENR 2007). During drought periods, cattle are sold in order to generate cash and food commodities, leaving reduced herds for the coming years.

These factors – and the increasingly harsh climatic conditions mentioned earlier – mean that both Ila and Tonga households in Namwala increasingly seek to supplement their dependence on rainfall with more reliable artificial sources of water. Water points such as boreholes, hand-dug wells and small-scale irrigation are thus becoming important for food security in the area.

But here again, socio-economic differentiation has significant implications: better-off households tend to have shorter distances to travel to water points and many own private wells located near their household/on their land which can provide water when surface water dries out (Mweemba et al. 2011). They are furthermore better disposed towards paying user fees at boreholes, and have the means to acquire pumps and other features for small-scale irrigation. Poor households, by contrast, typically rely on digging shallow wells in dry river beds, or use collectively owned water sources such as communal boreholes. The extent to which a household has access to water for human consumption, livestock and small-scale irrigation is thus to a large extent determined by socio-economic status.

The differentiated nature of water access in Namwala is illustrated in Figure 10.1 through the example of access to drinking water (see also Mweemba et al. 2011). The figure shows how the water sources on which the poorest households depend are either community or publicly owned, or are owned by others or are open access. Only a small number of poor households (7 per cent) own their own source of drinking water. By contrast, 41 per cent of wealthy households (the highest well-being group) have private access to drinking water, mainly in the form of deep wells near their houses.

The implication of this is partly that poor households are more dependent on water points that require collaboration, whereas wealthy households tend to have a

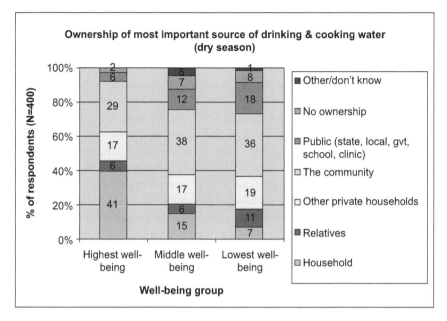

FIGURE 10.1 Ownership of the most important source of drinking and cooking water in Namwala

Source: data from a survey of 200 households in 10 Ila and Tonga villages in Namwala District under the 'Competing for Water' programme (see Mweemba et al. 2011).

greater degree of private control over water access. This in turn provides them with an important 'fall-back' position if for example a community-owned borehole dries out. However, as we shall see below the fact that many wealthy households have private water access does not deter them from seeking to improve their access to and control of communally owned water resources.

The nature of local water conflicts

A significant number of water conflicts take place in Namwala District. Our study identified 183 public water conflict and cooperation events over a 12-year period from 1995 to 2007[2]. Of these, 121 were conflictive events. In addition to these public events, there were numerous small private events between, for example, two neighbouring households[3].

Water conflicts in Namwala are very local. The large majority of conflicts (88 per cent) take place *within* individual communities rather between different communities or across district boundaries. Most of these intra-community conflicts are also unreported, i.e. they are not known to district staff or other public individuals and authorities outside the individual community. This very local nature of many water conflicts is often overlooked by water planners, who tend to have a river basin/watershed approach and therefore often focus on large-scale basin-wide conflicts. Yet taken together, the many small-scale water conflicts in Namwala District actually impact on a larger number of people than conflicts which cut across communities (Funder et al. 2010, Mweemba et al. 2010). These everyday 'invisible' water conflicts are thus highly significant for household water access and thereby food security.

Most water conflicts in Namwala originate in the dry season from May to October. Indeed, 82 per cent of all conflict and cooperation events identified began during this period. The majority of conflicts are a result of competition between different water uses aimed at securing household food production and security. The conflicts revolve around questions such as to what extent should a particular water source be used for cattle watering or farming or domestic use or gardening. Conflicts thus reflect different livelihood strategies among households in communities, but also different gendered interests within households. Two further issues are notable in this respect: first, the majority of conflicts (67 per cent of those identified in our study) concern communal/publicly owned water sources, on which the poorest households are particularly dependent. Second, most conflicts (almost three quarters of those identified) relate to the introduction of new water infrastructure, such as boreholes, irrigation or a small dam. New water points are often 'unsettled fields' where existing orders and rules can be challenged (Cold-Ravnkilde and Funder 2011).

It should also be pointed out that although there are many water conflicts in Namwala, there is also cooperation. Of the 183 water conflict and cooperation events identified in our study, 62 primarily involved cooperative measures – i.e. almost exactly one third of events. It is also significant that while conflicts increased in number during the dry season, cooperation increased by almost the same proportion. In other words, while increased water scarcity led to more conflict, it also led to more cooperation[4]. Although the current chapter deals mainly with conflicts, this is important to keep in mind.

In the following section the nature of water conflicts in Namwala and how they relate to food security and poverty are discussed in more detail. Two illustrative cases are used: one that describes the dynamics of access to and exclusion from new water infrastructure – in this case a borehole – and one that illustrates the local politics of water conflicts and food security, i.e. a planned irrigation scheme that never materialized.

A borehole in Muchila: The dynamics of gender, class and water access

From 2003 to 2009 a conflict took place between community members over access to a newly established borehole in the village of Muchila in the southern part of Namwala District. Although built with the aim of providing water for both domestic and productive purposes, the borehole soon became controlled by wealthy Ila and Tonga households, who used it to water their cattle during the dry season. This developed to such an extent that other community members were either directly excluded from using the borehole, or had to wait for long periods before being allowed to draw water. Complaints to the community committee who managed the borehole were of little use, as the committee was dominated by the same households that owned the cattle in question.

The conflict had both a gender and a class dimension. The gender aspect stemmed from the fact that most women in the village – including some from the wealthiest households – were opposed to the idea of only using the borehole for cattle. They also wanted to use it for domestic purposes (water for drinking, cooking, etc.), and for establishing vegetable gardens that would provide both a source of food and a source of extra income for *ad hoc* needs, including additional food supplies. The class aspect of the conflict related to the fact that very few poor households in the village owned cattle. Their primary concern (both women and men) was therefore to use the borehole for domestic purposes, for establishing gardens and for smaller livestock (chickens, etc.).

The poor households in the community responded to the situation in three different ways: some found themselves in the paradoxical situation that their main opponents in the conflict – the wealthy households – were also their patrons, for whom they worked as cattle herders and who lent them money or provided food in times of need. Not wanting to risk this valuable relationship, these poor households typically refrained from opposing their patrons in the conflict, and instead used their position as cattle herders to collect water for their own purposes. Other poor households simply chose not to engage with the borehole at all, or only to a limited extent. They preferred instead to seek water elsewhere, even if this meant walking much longer distances and poorer water quality. When asked why, they responded that this option allowed more individual decision making and flexibility on their part. The borehole, by contrast, required risky interactions with a dominating elite with little promise of ever obtaining equal access.

Lastly, some poor households did in fact decide to engage in the conflict. They did so indirectly, however, forming an unlikely alliance with dissatisfied women from the better-off and wealthy households. The latter used their social position and networks to engage various local authorities on the matter, including local government representatives and the customary chief of the Muchila chiefdom. As a result of pressure from these authorities, the wealthy cattle owners eventually agreed to a set of rules that provided for a more equitable access to the borehole. This included a timetable which allowed access for domestic water users, vegetable gardens and livestock at different times of the day. It furthermore provided for a more balanced gender ratio in the community committee that managed the borehole.

This outcome was however only temporary. As time wore on, women from the better-off and wealthy households – who had previously teamed up with the non-cattle-owning poor households – began to monopolize the time slots allocated to fetching domestic water from the borehole. Their argument for doing so was that they had been the main driving force in getting the timetable established, and that they therefore had priority of access. At the same time, they took over the land immediately next to the borehole for vegetable plots, and obstructed poor households from doing the same. Following this, the remaining poor households either gave up and sought alternative sources of water, or accepted having to wait several hours for their turn to collect drinking water.

What were the implications of this in food security terms? Boreholes in Namwala rarely dry out during the dry season, and are less at risk of doing so during droughts. They are therefore important strategic assets in minimizing risk in livestock husbandry, they allow diversification into vegetable production, and they ensure a reliable supply of drinking water for household consumption. This supports household food security in several respects: (i) it helps meet the fundamental everyday needs of the household, e.g. nutrition, good health and livestock for draught power, (ii) it provides a basis for enhancing livelihoods, e.g. increasing incomes through sale of surplus livestock and vegetables and (iii) it supports household coping strategies, e.g. building up a 'bank' of cattle that can be sold if crops fail.

In other words, the mechanisms of access and control that developed around the borehole in Muchila helped enhance the food security of better-off households. By contrast, they did little to support the needs of the most food insecure households in the area: they had no cattle to water in the first place, and were excluded from access to land near the borehole that would have allowed cultivation of vegetable gardens for nutrition and income. Those that were not engaged in direct patronage ties with the wealthy households had instead to spend valuable time and energy collecting water for drinking and cooking from more distant sources, including shallow wells dug in river beds, which are less reliable in times of drought when the water table sinks (for a similar case, see Mweemba and van Koppen 2010).

In broader terms, the case from Muchila thus illustrates how even a single individual water point may be subject to intense struggles over access between different community actors who are all seeking food security and livelihood improvement, but who do so from different vantage points (see also Peters 1984; Cleaver 2012). In this respect the case illustrates both gender and class differences, although it also highlights how economic differentiation may prevail even when gender barriers are overcome. The case further illustrates how poor households apply water access strategies that minimize risk *vis-à-vis* the better-off and more powerful households. Such agency should not be seen as passive behavior, but rather as an experience-based attempt to make the best of a bad situation by using whatever means are available. This directly corresponds to coping strategies applied in relation to other aspects of risk, such as drought. Nevertheless, the outcome of the process is a reproduction of existing dependency ties and power relations, and the new borehole in Muchila thus comes to solidify rather than change the structural distribution of food insecurity in the area.

The Mbeza irrigation scheme: The local politics of water for food

In the dry season of 2002, Namwala District experienced a prolonged drought along with the rest of Zambia's Southern Province. Food shortages were severe among many households, and the situation aggravated an already strained local economy, following the effects of a cattle disease that decimated livestock populations in the area.

Against this background, one of the area's influential chiefs – Chief Nalubamba of Mbeza – announced that he was collaborating with central government in Lusaka to secure Italian donor funding for a new irrigation project to be implemented within his chiefdom. The announcement was attended by the Minister of Agriculture, and described by the chief as an effort to address poverty alleviation and food security in the area by diversifying the economy away from the traditional cattle economy towards a more intensive and modern crop economy. The project was to cover some 3,600 hectares of the floodplains along the Kafue River, with a projected 3,000 beneficiaries in the area. The main emphasis of the project was on irrigation for household production of rice, sunflower, soy beans, maize and sorghum, as well as development of intensified grazing in some areas. The project also included small-scale feeder road development. The project was largely supported by government staff in the area, and by many Tonga farmers who saw an opportunity to enhance crop production.

The plan was however not welcomed by everyone. By proposing to irrigate a large section of the area's plains, the project threatened pastoralists' use of the plains for grazing and watering their cattle. At the same time, the project was seen as a direct affront to other chiefs and Ila clans in the area who customarily controlled access to the plains. Opponents to the project thus saw it both as a direct attack on pastoralist livelihoods and food security, and as an attempt by Chief Nalubamba to extend his control over land and water in the district with the project as the medium. Seeking to halt the project, members of a dominant Ila clan thus formed a powerful 'anti-irrigation' group. The group was backed by other chiefs opposing the scheme, and was led by an influential local MP, whose personal interests in developing range-land and cattle production in the area were also threatened by the irrigation project.

In the years that followed, the conflict fluctuated and gradually intensified. The opposing sides held a number of rallies, sought the backing of provincial and national authorities, and attempted to engage the media[5]. The conflict over the Mbeza irrigation scheme thereby became one of the few water conflicts in Namwala that was actually known beyond the individual locality. A key element in the conflict was its strong links to development discourses in the area: Chief Nalubamba and his supporters among government staff argued that irrigated crop farming would enhance food security and was an effective and modern development path. By contrast, the opposing chiefs and the MP argued that pastoralism was an indigenous and carefully adapted livelihood, and that the irrigation scheme would erode local livelihoods and adaptation strategies, such as migrating cattle seasonally or in response to drought.

Given the increased politicization of the issue, the government eventually withdrew its support and the project was cancelled. However, the idea of the irrigation scheme continues to be promoted by Chief Nalubamba when the opportunity arises, and during interviews district agricultural staff tend to back the idea of irrigation as the most viable road to food security in the Kafue Flats.

What were the implications of the conflict for food security in the area? At first sight, the conflict seems to represent a classic struggle between Ila pastoralism and Tonga mixed farming. On the face of it, pastoralist livelihoods thus seem to have prevailed, while the crop farming strategy lost out. However, closer investigation revealed a more complex picture. Among the Ila pastoralists of the area, many of the poorer households only own very few cattle or none at all, as poor households were hit particularly hard during the livestock disease epidemics of the 1990s and early 2000s. Some of the poor households did oppose the irrigation scheme, because they remained involved in the cattle economy as cattle herders for wealthy livestock owners. Other poor Ila households were however not averse to the irrigation scheme; they saw it as an opportunity to diversify their livelihoods and thereby improve food security, and were therefore disappointed when it did not materialize.

Whether or not the Mbeza irrigation scheme would in fact have helped support the food security strategies of poor households (and whether it would have been hydrologically and environmentally sound) will remain speculation. The point here, however, is that despite the rich discourse of food security and livelihood improvement forwarded by the leading figures of both sides, these issues were only one aspect of the conflict. Behind the scenes, the conflict was also to a large extent driven by the struggle over authority and legitimacy among chiefs and other 'big men' in the area, and the personal gain that was to be had from possessing such authority (see also Haller 2007; Funder and Mweemba 2010).

The case of the Mbeza irrigation scheme thus highlights the inherently political nature of food security and efforts to address it. There are several dimensions to this. First, what may increase food security in one type of production system (e.g. farming) may reduce food security in another type of production system (e.g. pastoralism). This is fairly obvious, even if it is sometimes forgotten in actual policy-making and project planning. Second, water conflicts that appear to be between different production systems may conceal internal differences and food security strategies within each production system (e.g. wealthy pastoralists vs. poor pastoralists). Finally, conflicts over access to and use of water for improved food security are frequently enmeshed in wider political and institutional conflicts over authority, legitimacy and personal gain – to an extent where the actual food insecurity issues become blurred or are not even the main cause of the conflict.

A final feature to note in the case of the Mbeza irrigation scheme is that poor households only engaged to a very limited extent in the conflict. Indeed, all the poor households interviewed said they had spent more time and given greater attention to seeking new and alternative sources of water than to engaging in the conflict over the irrigation project. This included helping to build a separate, small-scale local irrigation scheme, providing labour to the construction of a borehole, joining with other households to dig deep wells, or simply looking for good places to dig their own shallow well. In other words, the poor households of the area continuously sought alternative means of obtaining access to water both during and after the conflict.

Addressing local water access and water conflicts for food security

What do the local water conflicts in Namwala tell us more generally about how to address water access, and conflict in relation to food security?

First, they illustrate the ongoing effort by rural households to secure water access as a key element in achieving or consolidating food security and livelihoods. This applies to both poor and wealthy households. In the cases discussed above, it is notable that poor households continuously *act* to seek water access. Although they may often elect not to engage in conflict situations, they rarely resign in desperation. Rather, they continuously seek out both old and new options for water access, drawing on experience-based perceptions of what serves them best at a given time. Likewise, wealthy households do not step back from pursuing new water access opportunities that can complement those they already have. Despite already owning private wells and/or access to surface water for livestock, they assert power and influence where possible to gain control of new water sources such as boreholes, which can help consolidate food security and improve their economic basis. Meanwhile, water resources also serve as arenas for political struggles over control and authority, whether in relation to water governance specifically or political positioning more generally.

Access to water for food and livelihoods is thus contested and subject to ongoing agency and struggle by rural actors. Efforts to address food security through water access can build on the positive aspects of this – i.e. that water access is a high priority activity for all involved and that people are willing to engage – but the intensely political nature of the issue also needs to be considered. Interests and water access strategies differ widely among different actors and social strata, and universal solutions that seek to meet all actors' needs are rarely adequate.

Second, development of *new* water infrastructure such as boreholes, water harvesting arrangements and small- or larger scale irrigation is particularly prone to conflict. As mentioned earlier, almost three quarters of the conflicts recorded in Namwala were about new water infrastructure (see also Ravnborg et al. 2012). This is partly because such infrastructure ideally presents year-round water access and thus potentially helps to enhance food security and diversify/increase production. But it is also because such new water sources are 'unsettled' arenas in which actors seek to consolidate or change their control of and access to water, as well as the rules that guide this.

This has implications for efforts to support food security, because it highlights that it is not enough to merely enhance the supply of water by building more boreholes, water harvesting measures or irrigation schemes: 'more water' will not necessarily reduce conflicts over a scarce resource – it may in fact contribute to new conflicts. This does not of course mean that no new infrastructure should be developed, but rather that such developments need to take conflict prevention and resolution measures into account. This can include simple practical measures such as devising timetables for different types of water use, but should also recognize that conflicts will inevitably arise. Conflict resolution and mediation mechanisms are therefore a key aspect to consider when enhancing rural water access. These may include customary institutions or new institutions introduced specifically for the purpose, but it is critical that they are accessible for all actors. River basin committees, water user associations and other similar mechanisms are typically composed of representatives of different types of water *use,* but they tend not to take socio-economic differentiation *within* the various uses into account. As a result, they have often been dominated by elites and/or commercial interests, rather than by more subsistence-oriented users for whom food security is particularly uncertain (Funder and Ravnborg 2004). There is furthermore a need to ensure opportunities for neutral mediation and resolution of conflicts, and the existence of alternative spaces for voicing grievances if the conventional conflict resolution forum (e.g. a local chief or headman) is not impartial (Gómez and Ravnborg 2011).

Third, it is important to note that efforts to enhance rural water access through infrastructure development tend to involve collectively managed water resources. Boreholes, small dams and weirs, water harvesting, etc. are thus typically designated as community-owned facilities. In principle this is an advantage for poor households, as they do not normally have the capital available to invest individually in private water infrastructure such as wells. And clearly, cooperation does happen, as indicated by our mapping of conflictive and cooperative events discussed earlier. Moreover, even intense conflict situations usually include cooperative events. The case of the Muchila borehole thus included several cooperative events, such as the agreement to a timetable for water access. However, the problem for poor households is that cooperation requires assets. One obvious example of this is user fees, but cooperation also requires position, influence and voice to assert one's interests and ensure that they are incorporated and enforced in agreements. Poor households typically lack this, and therefore tend to end up being excluded from agreements, or are unable to effectively assert their rights when agreements are broken.

This situation is compounded when disagreements or power games over access to and control of water resources lead to conflict. As we have seen, poor households in Namwala are caught in dilemmas in such situations: their livelihoods and food security may be dependent on labour and/or clientilistic relations with the very same wealthy households that are in theory their 'opponents' in water conflicts. Or they are faced with the choice of (i) engaging in the conflict and opposing better-off households at the risk of exposing themselves to the domination and sanctions of the powerful, or (ii) seeking alternative water sources in e.g. streams or shallow wells and relying on natural rainfall – which provides a more independent means of accessing water, but also exposes them to natural fluctuations and hazards. In this predicament between social and natural risk, poor households often have little real choice but to employ the latter option (Funder et al. 2012).

Poor households are thus in an insecure position in both cooperative and conflictive situations, and efforts that seek to enhance the food security of poor households through better water access need to consider this. This involves ensuring that agreements include the interests of the poorest, and are monitored and enforced so that they cannot be ignored or overruled in everyday practice. This can entail community 'governance monitoring' measures (Child 2006), and/or can involve external actors such as local government water officers. The existence of multiple alternative spaces for voicing grievances, as mentioned above, is particularly critical for poor households. More fundamentally, measures to develop water resources that are reserved for poor or otherwise marginalized actors is a possible (though potentially controversial) means of avoiding the 'double jeopardy' situation of poor households *vis-à-vis* water cooperation and conflict.

Fourth and finally, it is critical to acknowledge that addressing food security through improved water access involves more than merely establishing the right institutional frameworks at local levels. It also involves addressing national inequalities and policies, and thereby some of the basic political and economic inequalities that underlie the dependency relations that constrain poor households from engaging in conflict. Measures to address food security through enhanced water access must therefore be complemented by efforts to support the production systems and livelihoods of poor households more generally. Enhancing access to water is a critical factor in ensuring food security, but it is important that this does not lead to a purely technical approach that focuses only on 'providing more water' and relying only on water management committees at various levels to deal with issues of conflict and exclusion.

Conclusion

Local everyday water conflicts jeopardize poor people's food security and livelihoods, but have drawn much less attention than the more dramatic, larger scale water conflicts. Policy-makers and analysts need to focus more on these small-scale non-violent conflicts. There is furthermore a need to move beyond simplistic notions of 'vicious circles of scarcity, conflict and hunger', towards a better understanding of the underlying dynamics and causes of such conflicts.

Efforts to address food security through provision of new water infrastructure such as boreholes, irrigation and small dams very often lead to conflict in themselves. This is partly because they potentially provide year-round water access and options for diversification for both poor and better-off households, but also because new water resources may serve as arenas for struggles over institutional authority and broader political power.

The food security of poor households is particularly dependent on water infrastructure that requires collaborative management. However, this also makes them vulnerable on two fronts: (i) they frequently do not possess the assets required to engage in the 'cooperation game', and (ii) when cooperation goes sour and turns to conflict, they are most at risk.

Poor households are far from passive in water conflicts, but are to a large extent constrained by dependency relations to the very same better-off households that are challenging their water access. The result is that they ultimately tend to lose out in the water conflicts – either in the sense of missing out on new opportunities for improving livelihoods and food security or losing existing access to water resources. Ironically, attempts to introduce new water infrastructure such as boreholes may reproduce inequality of access, as benefits are captured by the better-off.

Addressing water conflicts for the benefit of poor people's food security requires attention to water governance issues, such as enhancing conflict resolution mechanisms and providing alternative spaces for expressing grievances in order to overcome local monopolies of power. However, it is important also to address the underlying dependency patterns in production systems if impacts are to be sustained.

Notes

1 The authors would like to thank the Danish Ministry of Foreign Affairs for financial support, as well as colleagues in the Competing for Water programme, especially Barbara van Koppen, Helle Munk Ravnborg and Ingrid Mugamya.
2 For a more detailed discussion of our mapping of water conflict and cooperation in Zambia, see Funder et al. 2010.
3 In the study referred to, we defined 'public' water conflict events as events that (i) involved two or more parties, of which at least one party represented or consisted of a group of more than five individuals from different households; or (ii) involved at least three different types of parties, e.g. cattle herders, an industry and domestic water consumers. See Ravnborg et al. (2012) and Mweemba et al. (2010) for details.
4 In the dry season, the conflict to cooperation proportion was 66% and 34% of events, while in the wet season the proportion was 69% and 31% of events respectively. See Funder et al. 2010.
5 It is beyond the scope of this article to describe the conflict events in detail, but see Haller (2007) and Funder and Mweemba (2010).

References

Calow, R.C., Macdonald, A.M., Nicol, A.L. and Robins, N.S. (2010) 'Ground water security and drought in Africa: linking availability, access, and demand', *Ground Water*, 48(2): 246–56.

Child, B. (2006) 'Developing adaptive performance monitoring for economics and governance of community based natural resource management institutions', Unpublished report, University of Florida at Gainesville, USA.

Cleaver, F. (2012) *Development Through Bricolage: Rethinking Institutions for Natural Resource Management*, London: Routledge.

Cold-Ravnkilde, S. and Funder, M. (2011) 'Struggles over access and authority in the governance of new water resources: evidence from Mali and Zambia', Paper presented at the 13th Conference of the International Association for the Study of the Commons, 10–14 January 2011, Hyderabad, India.

Funder, M., Bustamante, R., Cossio Rojas, V., Huong, B.T.M., van Koppen, B., Mweemba, C., Nyambe, I., Phuong, L.T.T. and Skielboe T. (2012) 'Strategies of the poorest in local water conflict and cooperation: evidence from Zambia, Vietnam and Bolivia', *Water Alternatives*, 5(1): 20–36.

Funder, M. and Mweemba, C. (2010) *The case of the Mbeza irrigation scheme, Zambia*, Case Studies of Local Water Conflict and Cooperation Report 12, Copenhagen: Danish Institute for International Studies.

Funder, M., Mweemba, C., Nyambe, I., van Koppen, B. and Ravnborg, H.M. (2010) 'Understanding local water conflict and cooperation: evidence from Zambia', *Journal of Physics and Chemistry of the Earth,* 35(13–14): 758–64.

Funder, M. and Ravnborg, H. (2004) *Addressing water conflicts: governance, institutions and functions,* DIIS Working Paper, Copenhagen: Danish Institute of International Studies.

Gómez, L. and Ravnborg, H.M. (2011) *Power, inequality and water governance: the role of third party involvement in water-related conflict and cooperation,* CAPRi Working Paper 101, Washington, DC: International Food Policy Research Institute.

GoZ (2013) 2013 'In-depth vulnerability and needs assessment report', June 2013, Zambia Vulnerability Assessment Committee, Government of Zambia.

Haller, T. (2007) 'The contested floodplain: institutional change of common pool resource management and conflicts among the Ila, Tonga and Batwa, Kafue Flats, Zambia', PhD thesis, University of Zurich.

Haller, T. (ed.) (2010) *Disputing the Floodplains: Institutional Change and the Politics of Resource Management in African Wetlands,* Leiden: Brill.

Jain, S. (2007) *An empirical economic assessment of impacts of climate change on agriculture in Zambia,* Policy Research Working Paper 4291, Washington, DC: World Bank.

MTENR (2007) *Formulation of the National Adaptation Programme for Action (NAPA),* Lusaka, Zambia: Ministry of Tourism Environment and Natural Resources.

Mweemba, C. and van Koppen, B. (2010) *The case of the Iliza Borehole, Zambia,* Case Studies of Local Water Conflict and Cooperation Report 13, Copenhagen: Danish Institute for International Studies.

Mweemba, C.E., Nyambe, I., Funder, M. and van Koppen, B. (2010) *Conflict and cooperation in local water governance: inventory of water related events in Namwala district, Zambia,* DIIS Working Paper 15, Copenhagen: Danish Institute for International Studies.

Mweemba, C.E., Nyambe, I., Funder, M. and van Koppen, B. (2011) *Poverty and access to water in Namwala district, Zambia: report on the results from a household questionnaire survey,* DIIS Working Paper 19, Copenhagen: Danish Institute for International Studies.

Nordåsa, R. and Gleditscha, N.P. (2007) 'Climate change and conflict', *Political Geography,* 26: 627–38.

Peters, P. (1984) 'Struggles over water, struggles over meaning: cattle, water and the state in Botswana', *Africa,* 54(3): 29–50.

Ravnborg, H.M., Bustamante, R., Cissé, A., Cold-Ravnkilde, S.M., Cossio, V., Djiré, M., Funder, M., Gómez, L.I., Le, P., Mweemba, C., Nyambe, I., Paz, T., Pham, H., Rivas, R., Skielboe, T. and Yen, N.T.B. (2012) 'The challenges of local water governance: the extent, nature and intensity of water-related conflict and cooperation', *Water Policy,* 14(2): 336–57.

RoZ (2006) *Namwala District Development Plan 2006–2010,* Lukasa: Republic of Zambia.

Thurlow, J., Zhu, J. and Diao, X. (2009) *The impact of climate variability and change on economic growth and poverty in Zambia,* IFPRI Discussion Paper 890, Washington, DC: International Food Policy Research Institute.

11

STATE CAPACITY AND MALNUTRITION

A critical analysis of capacity support to Sierra Leone's nutrition sector

Lisa Denney and Richard Mallett

Introduction

Sierra Leone is one of the world's most malnourished countries. Despite considerable reductions in malnutrition rates since 2005, today's statistics paint a troubling picture: 22 per cent of children are underweight; 44 per cent are stunted or too short for their age; and 8 per cent are wasted or too thin for their height (Statistics Sierra Leone and UNICEF 2011; Koroma *et al.* 2012: 39). Malnutrition is one of the underlying causes of 46 per cent of deaths of children under five (Aguayo *et al.* 2003). This situation is both a result of, and compounded by, the country's high levels of food insecurity. In 2011, the World Food Programme (WFP) reported that 45 per cent of households experience food insecurity in the wet season (June through September), and in 2009, the country was ranked among the top six states most severely affected by and vulnerable to the effects of the global economic crisis by the International Food Policy Research Institute (ACDIVOCA 2011). Even in comparison with other sub-Saharan African countries with similar per capita GDP levels – such as the Gambia, Togo and Zimbabwe – Sierra Leone's malnutrition problem stands out as a particular concern (World Bank n.d.).

In the immediate aftermath of Sierra Leone's 11-year civil war, assistance to address malnutrition focused on delivering treatment. This approach endured into Sierra Leone's post-conflict period, and only recently have government and development partners attempted to refocus energies towards prevention. This is in recognition of the unsustainable nature of externally driven treatment approaches and the need for the government of Sierra Leone (GoSL) to take the lead in preventing malnutrition. It is also reflective of Sierra Leone's broader transition over the last 12 years – from a weak political system with limited resources to a relatively stable democracy (with three largely peaceful national elections, including transition to opposition rule in 2007) and a stronger economic base (since 2010 Sierra Leone has achieved growth rates of over 5 per cent) (World Bank 2012). The nutrition community in Sierra Leone is, therefore, at an important juncture, attempting to balance the ongoing need to save lives, with shifting attention to growing GoSL's capacity to prevent malnutrition.

This chapter seeks to capture the nature of capacity support provided by development partners to GoSL. In doing so, it provides a clearer picture of the nature of current forms of capacity development and how 'fit for purpose' they are in relation to the objectives. An entirely government-financed health sector in Sierra Leone is a long-term goal and the medium-term will be characterized by an ongoing aid relationship in which government and international development partners share roles and responsibilities. What is crucial during this time is the extent to which international assistance either builds or undermines state capabilities to prevent malnutrition.

The chapter proceeds as follows. The first section sets out the socio-political context of malnutrition in Sierra Leone. In the second section, the idea of capacity is introduced in greater depth through a review of key literature, shaping how we might think about this amorphous concept. The third section analyses whether the current model of capacity support being implemented in the nutrition sector is 'fit for purpose'. It does so by first examining dominant approaches to capacity building and what is being overlooked; and second, questioning whether the assumptions underpinning current approaches hold true in practice. The final section reflects on what this might mean for our understanding of 'capacity' and 'capacity building' as central organizing concepts of international development.

The findings presented here are based on qualitative primary research carried out in Sierra Leone between September and October 2013.[1] Sixty-two semi-structured interviews were carried out in Freetown and Kambia district, in the north of Sierra Leone, with providers and targets of capacity support. In Freetown, interviews focused on donors and NGOs providing capacity support in the nutrition sector, as well as government ministries across health, agriculture, education and social welfare. In Kambia, interviews were conducted with the District Health Management Team (DHMT), government ministries, the district council, paramount chief, NGOs and, in each of the seven chiefdoms, Peripheral Health Units (PHUs), Mother-to-Mother (M2M) Support Groups and Farmer Field Schools (FFSs).

The socio-political context of malnutrition in Sierra Leone

The political context is centrally important in attempts to reduce malnutrition, and capacity support must be understood through this lens (Gillespe *et al.* 2013). Sierra Leone is characterized by strong patriarchy and gerontocracy, meaning that women and children are generally at the bottom of inequitable social structures (Schroven 2006). In relation to nutrition, this translates into women and children receiving less protein-rich diets than older men. Moreover, as men tend to control household finances, women can struggle to assert their spending priorities (Denney and Ibrahim 2012: 8). Mothers are often blamed for unhealthy children, which can discourage them from seeking early treatment. There are also high numbers of teenage mothers who are often not aware of the importance of diet in child development (GoSL 2013). In rural areas in particular, customary authorities have traditionally been seen as repositories of knowledge regarding child birth, rearing and well-being, and can have an important influence (Denney and Ibrahim 2012).

Despite these challenges to addressing malnutrition at the local level, GoSL appears committed to improving the nutritional status of the population, reflected in the high priority accorded to issues of nutrition and food security in the country's National Development Programmes – including the 2008–10 and current Poverty Reduction Strategy Papers. GoSL recognizes the pivotal role of improved nutrition in curbing high maternal and child mortality rates, and enhancing the general health of the population.

In addition to the formal inclusion of malnutrition in national development plans, the placement of the Scaling Up Nutrition (SUN)[2] Secretariat in the Office of the Vice President has been important in strengthening the political weight of efforts to prevent malnutrition. There are some concerns about potential politiciza-tion because of the secretariat being located within a political office; however these concerns are outweighed by the imperative to ensure the secretariat has a high profile, good political access and influence. Alongside strengthened political sup-port, technocrats working on nutrition issues also gained more clout in 2013. The 'Nutrition Unit' within the Ministry of Health and Sanitation (MoHS) was elevated to a Food and Nutrition Directorate, with greater access and influence. The Min-istry of Agriculture, Forestry and Food Security (MAFFS) approved a nutrition budget line, allowing it to plan for nutrition-specific interventions.

Yet despite significant progress, key GoSL institutions remain susceptible to cor-ruption. Criminal investigations into MoHS are ongoing following accusations that staff misappropriated USD 1 million from the Global Alliance for Vaccine Immunisation. In June 2013, the Anti-Corruption Commission singled out MoHS and MAFFS as particularly vulnerable to corruption because of weak management procedures (Tommy 2013). Such susceptibility makes capacity development both critical (in terms of strengthening oversight to limit corruption) and risky for devel-opment partners to work through government systems.

Making sense of capacity: What the literature tells us

Drawing on four years of practitioners' experiences, James and Wrigley (2007) concluded that capacity development is confused, contested, contextual, counter-acted and complex. Although of profound importance, understanding how state capacities to deliver effective services can be strengthened and sustained is far from straightforward and debate rages about what it means and whether it is effective at all (Kühl 2009). In this section, we discuss the challenges of strengthening capacity, and set out some key features that characterize the process of capacity develop-ment. The discussion draws on an analytical framework developed by the Secure Livelihoods Research Consortium (SLRC forthcoming) that in turn builds upon work carried out by the European Centre for Development Policy Management (ECDPM) (Brinkerhoff 2007). The framework breaks down the larger concept of capacity into a set of constituent parts, assisting us in making sense of what capacity support is currently being provided, how it is working, and what is being left out.

Simister and Smith (2010) speak of capacity as 'the ability of people, organizations and society as a whole to manage their affairs successfully', and refer to capacity building and capacity development as the process of creating, strengthening and maintaining capacity (involving either external interventions and/or the society in which capacity is being developed). From a review of key contributions to the capacity development literature, we distil four main features that help explain, respectively, what capacity is; what shapes capacity; where capacity is 'located'; and how capacity can be externally influenced.

Capacity is made up of a set of constituent parts

One of the central messages to emerge from a five-year, multi-country study conducted by ECDPM was that capacity is formed of five specific capabilities (referred to as the '5Cs'). According to Morgan (2006: 8–16), these include the following:

- *The capability to self-organize and act.* Actors are able to mobilize resources (financial, human, organizational); create space and autonomy for independent action; motivate unwilling or unresponsive partners; and plan, decide and engage collectively to exercise their other capabilities.
- *The capability to generate development results.* Actors are able to produce substantive outputs and outcomes (for example, health services); sustain production over time; and add value for their clients, beneficiaries, citizens, etc.
- *The capability to establish supportive relationships.* Actors can establish and manage linkages, alliances and/or partnerships with others to leverage resources and actions; build legitimacy in the eyes of stakeholders; and deal effectively with competition, politics and power differentials.
- *The capability to adapt and self-renew.* Actors are able to adapt and modify plans and operations based on monitoring of progress and outcomes; proactively anticipate change and challenges; cope with shocks; and develop resilience.
- *The capability to achieve coherence.* Actors can develop shared short- and long-term strategies and visions; balance control, flexibility and consistency; integrate and harmonize plans and actions in complex, multi-actor settings; and cope with stability and change.

To the degree that the target of capacity development integrates these capabilities, capacity – being able to achieve a desired collective purpose – is generated and enhanced. Breaking down capacity through the 5Cs model is useful as it helps to unpick the different kinds of capabilities demanded of the state in dealing with cross-cutting problems like malnutrition. Thus, strong overall capacity to prevent malnutrition depends not just on the presence of adequate resources or sufficient levels of technical knowledge amongst frontline health staff, but also on a set of 'soft' skills for key personnel and the creation of institutional environments that support cross-sector collaboration and constructive negotiations.

Capacity is deeply shaped by context

Capacity is not merely a technical issue determined by an individual's knowledge or an organization's limitations, but is rather a function of socio-political and historical context. Understanding the role of context requires recognition of the depth of factors that shape capacity at different levels. Barma *et al.* (n.d.), for instance, conclude that three major categories of factors help explain why some organizations work and others do not: inner institutional/organizational workings; the external operational environment; and the broader socio-political and historical context. The central implication is that capacity is both a technical *and* political problem: staffing policies and skills enhancement courses which affect inner-organizational workings might be important, but so too is the way in which decisions to invest in state capacity are undertaken (the nature of power and politics). Unfortunately, the political dimensions of capacity development are harder to address and more resistant to external intervention than the relatively straightforward technical issues of boosting resources or improving knowledge.

A deep understanding of capacity that engages with the broader political context also highlights the importance of both formal and informal institutions. While formal characteristics may define the *form* of state structures, the actual way in which things are done may be heavily influenced by informal norms and processes (Barma *et al.* n.d.). This phenomenon can manifest in the form of hybrid service delivery – or plural health systems, as observed in Sierra Leone (Scott *et al.* 2013) – indicating that services are often not provided in a purely state-centric manner. Acknowledging the informal means seeing what is really there – in terms of service delivery modalities, state capacity and the ways in which governance plays out on an everyday basis – rather than assuming that things work (or ought to work) in a pre-determined way (Boege *et al.* 2008). Capacity development processes thus need to be attuned to the 'actually existing' nature of governance in particular contexts. While this may seem obvious, capacity development has not always been designed or implemented in ways that take context seriously (Pritchett *et al.* 2012; Andrews 2013).

Capacity is 'located' at multiple levels

Broadly speaking, capacity can exist at three different levels of what Pritchett *et al.* (2012) call an 'ecological space': agents (leaders, managers and frontline staff); organizations (firms, NGOs, line ministries); and systems (the broader administrative and political apparatus under whose jurisdiction the activity falls, as well as cross-organizational processes of governance such as procurement systems and delivery chains). For example, the District Health Management Teams (DHMTs) found in each district in Sierra Leone can themselves be understood as organizations. Through these organizations, decisions concerning health in the district are made and resources disbursed to lower level health structures. The DHMTs are, in turn, made up of sets of agents, such as the district medical officer and the district nutritionist – and the individual capabilities of these personnel influence, in part, the broader functional capacity of the DHMT. Finally, the DHMT not only forms part of the broader health system in Sierra Leone, but is also governed by rules and procedures (both formal and informal) that cut across organizations – such as the ways in which budgets and annual plans are designed and signed off, or the financial mechanisms that move resources between different organizations. Importantly, these three levels are related. The actions of agents, for example, are mediated by the norms, procedures and mandates of the organizations in which they work. Yet, despite these interrelations, capacity is not developed in a linear fashion (i.e. developing the capacity of agents does not necessarily translate into higher aggregate capacity at the organization or systems levels).

Capacity can be targeted in different ways

Given both the technical and political dimensions of capacity, support can be targeted in different ways. The extent to which a particular capability can be referred to as strong depends on the 'right' mix of factors or conditions being in place. Drawing on the ECDPM study, we can identify five broad dimensions which seem to matter for the strength of capabilities (Morgan 2006).

- *Resources (who has what)*. Interventions to increase resources might focus on budget support or provision of equipment.
- *Skills and Knowledge (who knows what)*. Interventions to enhance skills and knowledge might focus on training, technical assistance or technology transfer.
- *Organization (who can manage what)*. Interventions to strengthen organization might focus on restructuring, civil service reform or decentralization.
- *Politics and Power (who can get what)*. Interventions to address politics and power might focus on legislative reform, community empowerment or civil society advocacy.
- *Incentives (who wants to do what)*. Interventions to realign incentives might focus on sectoral policy reforms, improving the rule of law or strengthening accountability structures.

Capacity building programmes often focus on addressing the first two or three targets above – resources, skills and knowledge, and organization. This is partly because focusing on these more tangible targets is simpler and more in line with donor reporting requirements; it also has the benefit of being more easily measured to demonstrate results. However, without changes in the enabling environment encapsulated in the political and incentive categories, increased resources and better skills are unlikely to be enough to develop sustained capacity. What is needed, therefore, is for capacity development practitioners to identify which mix of targets needs to be addressed in a particular context.

These key features help to refine our understanding of what capacity is, what shapes it and where it can be located. A comprehensive approach to capacity recognizes that capacity is broad, deeply shaped by socio-political context, and is located at multiple levels that can be targeted in different ways. Failure to engage with this breadth of capacity development can mean that interventions are narrow (focusing only on some aspects of capacity), shallow (unconnected to the deeper political context that shapes capacity) and limited (engaging with only some of the repositories of capacity).

Capacity building in Sierra Leone's nutrition sector: Fit for purpose?

What then are development partners doing to build capacity in Sierra Leone's nutrition sector? Table 11.1 sets out the majority of international development partners working in the nutrition sector in Sierra Leone and the kinds of capacity support they are currently providing. The table does not offer an exhaustive inventory of all capacity support activities – nor does it provide a historical overview of this activity. What it aims to do is capture the essence of partners' current support, providing the basis for an analysis of dominant and missing modalities.

As can be seen from Table 11.1, it is apparent that capacity support primarily targets resources through provision of materials and funding; and skills and knowledge through training. The focus is on the individual and organization levels, primarily the MoHS (which houses the Food and Nutrition Directorate), the District Health Management Teams (government structures that oversee health policy and programmes at the district level), PHUs (the local-level government health units that carry out frontline delivery of services) and Mother-to-Mother Support Groups (local-level, community-based groups tasked with promoting preventative behaviour). Drawing on the features of capacity discussed above, Figure 11.1 highlights the dominant characteristics of current capacity support interventions and illustrates what is being overlooked.

Analysing the dominant modes of capacity support

It is clear from Table 11.1 that much is being done to address the problem of malnutrition. What is less clear is whether what is being done is sufficient to achieve the ends of improved state capacity. Below, we first examine what is lacking in current forms of capacity support, before questioning the assumptions embedded within the dominant modes of engagement.

Drawing on Table 11.1, we identify three key characteristics of the nature of capacity support.

First, treatment over prevention. Interviews with both development partners and GoSL revealed that a shift is underway within the nutrition community in Sierra

TABLE 11.1 Capacity support provided to GoSL to prevent malnutrition

Organization*	Type of capacity support	Details	Recipient of support	Target level of support
Action Contre la Faim	• Training • Technical expertise/advice • Resources/supplies	• On-the-job training on CMAM for DHMT • Supporting screening by PHUs and Community Health Workers (CHWs) • Training on database management for DHMTs • Supporting M2M groups • Supporting micronutrient survey and review of surveillance system.	• DHMT • M2M Support Groups • CHWs • PHU staff • Food and Nutrition Directorate	• Individual • Organization
CARE	• Training • Resources/supplies	• 'Window of Opportunity' programme to promote IYCF finished 2012. • At PHU level, provided equipment and promotional materials, training and counselling • Established M2M groups and train DHMT to supervise	• M2M Support Groups • Traditional Birth Attendants (TBAs) • PHU staff • District council • Koinadugu and Tonkolili	• Individual • Organization • System
Catholic Relief Services	• Training • Resources/supplies • Supportive supervision	• Training of PHUs in screening, sensitization and CMAM • Some supportive supervision • Support to M2M groups through training and provision of materials for backyard gardens	• M2M Support Groups • CHWs • PHU staff • Kailahun (9 of 15 chiefdoms)	• Individual • Organization

(Continued)

TABLE 11.1 (Continued)

Organization*	Type of capacity support	Details	Recipient of support	Target level of support
Food and Agriculture Organization	• Resources/supplies • Paying government staff salary • Technical expertise/advice • Training	• Mainstreaming nutrition into smallholder agriculture (advertising for nutrition expert who will be placed in MAFFS; also involves training); national early warning system • Development of food-based dietary guidelines and nutrition modules for Farmer Field Schools (FFSs)	• National level staff in MAFFS • District level staff in district councils (helping to integrate nutrition into planning) • Civil Society Organizations (CSOs)	• Individual • Organization • System
GOAL	• Training • Resources/supplies	• CMAM and IYCF support to 19 PHUs in western area • Training of CHWs through cascade model in IYCF • Supporting government campaigns on basis of demand • Supporting micronutrient survey • Provision of vehicle and modem to Food and Nutrition Directorate	• PHU staff • CHWs • Nutrition Directorate	• Individual • Organization
Helen Keller International	• Supportive supervision • Technical expertise/advice • Research	• Technical support to integrate vitamin A supplementation into routine health services • Production and provision of training materials and research • Assistance with nutrition planning at national and district levels; development of protocols and fortification standards • Supporting monitoring through development of M-Health to allow health staff to record and transmit data by smart phones.	• National and district level staff within the MoHS	• Individual • Organization • System

Organization				
Plan	• Resources/supplies • Supportive supervision	• Food aid in partnership with WFP • Supplementary and school feeding programmes • Monitoring of PHUs • Financial support to MoHS and SUN to hold meetings • Supporting micronutrient survey	• Lactating mothers, people living with TB and HIV, and children in Port Loko, Kambia, Kenema and Moyamba	• Individual
SNAP	• Training • Resources/supplies • Supporting supervision	• Promote behaviour change through training and M2M groups • Distribution of food supplements (blanket rations in lean season) • Food/cooking demonstrations • Promotion of family planning • CMAM training for PHU staff	• M2M Support Groups and PHUs in 18 chiefdoms across 4 districts: Bonthe, Tonkolili, Koinadugu, Kailahun	• Individual • Organization
Trocaire	• Training • Resources/supplies	• Funding recently secured for initial nutrition work • Will involve training farmers and raising awareness about food and diet diversity, and reviewing government policies	• Makeni and Kambia	• Individual • Organization
UNICEF	• Training • Resources/supplies • Paying government staff salaries	• Support to CMAM on nationwide basis • Prevention support at national level (provision of promotional materials; assist with revision of curriculum in nursing schools; funding event costs) and district level supportive supervision of PHUs in 4 districts) • Pays salaries of 8 district nutritionists and some staff in Food and Nutrition Directorate	• At national level: MoHS, SUN Secretariat and key stakeholders • At district level: (Moyamba, Kenema, Pujehun, Kambia): DHMT and PHU staff and MCH aide schools	• Individual • Organization • System
Welt Hunger Hilfe	• Training • Technical expertise/advice	• Training of MAFFS and agriculture extension workers • Focus on building capacity of CSOs • Review government policies and provide advice	• MAFFS • CSOs	• Individual • Organization

(Continued)

TABLE 11.1 (Continued)

Organization*	Type of capacity support	Details	Recipient of support	Target level of support
World Food Programme	• Secondment of staff member to Food and Nutrition Directorate • Training • Resources/supplies	• Secondment of WFP staff member to MoHS for two years to oversee pilot prevention project • Provision of fortified micronutrient food through PHUs • Training in sensitization and clinical practices • School feeding and under-2 feeding programmes • Training of M2M Support Groups and MCH aides	• Fortified micronutrient food provision in 9 districts • School feeding in 12 districts • Under-2 feeding in Moyamba • M2M Support Groups • MCH aides	• Individual • Organization
World Health Organization	• Training • Technical expertise/advice • Resources/supplies	• Training on nutrition surveillance and data analysis • Provision of logistical equipment and materials • Technical inputs into nutrition planning and reviews • Review nutrition curriculum	• National and district level MoHS staff	• Individual • Organization
World Vision	• Training • Resources/supplies	• Training on screening and sensitization at community level • Provision of materials for outreach, seeds for backyard gardens, drugs to PHUs for CMAM, malaria, deworming and iron folate • Provision of logistical support and equipment for national campaigns and processes	• Food and Nutrition Directorate and MAFFS • Bo, Pujehun, Bonthe, Kono: M2M Support Groups; CHWs; DHMTs • SUN Secretariat	• Individual • Organization

*This table includes information about capacity development activities related specifically to nutrition as provided by representatives of the development partners listed. These organizations may provide other forms of capacity development not captured here and some organizations providing capacity development might have been unintentionally excluded.

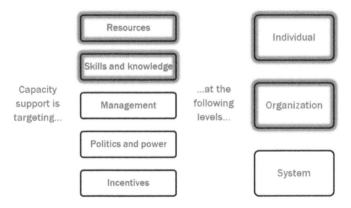

FIGURE 11.1 Analysing the dominant modes of capacity support

Leone from focusing primarily on treatment of malnutrition to prevention. In part, this reflects the transition that Sierra Leone is undergoing more broadly from a post-conflict country to a more stable low-income country, and the concomitant shift from emergency response to sustainable development on the part of development partners. In the interests of sustainability, development partners indicate that they are refocusing operations on the prevention of malnutrition. However, the majority of assistance from development partners in fact still focuses primarily on treatment. This continued focus on malnutrition treatment initiatives becomes more accentuated the closer one gets to communities, with DHMTs and PHUs almost solely focused on treatment. When PHU staff in Kambia were asked about their roles, they routinely replied that they were responsible for community-based management of acute malnutrition (CMAM) treatment. Only on further questioning did they mention infant and young child feeding (IYCF) practices and, rarer still, community awareness about nutrition issues. In part, this is because IYCF has only recently become an official part of the role of PHUs. The continued focus on treatment also reflects a division of labour at the local level, where PHUs focus on treatment, while prevention activities are essentially outsourced to Mother-to-Mother Support Groups and Community Health Workers. In any case, the discursive focus on prevention so tangible and widespread at the national level has not yet altered the focus at the PHU level.

The difficulty of operationalizing commitments to shift from a treatment to prevention focus may be explained by the fact that it is not possible to stop treatment work because this would result in loss of life. When cases of malnutrition are diagnosed, they must be treated. 'Shifting' to prevention, therefore, is not about reallocating resources away from treatment to prevention, but about allocating *more* resources to nutrition work so that treatment can continue while prevention capacity is built. In the longer term, increased prevention capacity should mean reductions in spending on treatment of malnutrition are possible, but in the interim a shift to prevention is only possible with increased nutrition budgets. Until such funding is available, it will be difficult for development partners to step up their support to prevention of malnutrition, as they will remain committed to the priority of saving lives through treatment.

Second, tangible outputs over deeper change. In general, capacity support activities are designed to increase 'resources' and 'skills and knowledge'. Activities focusing on these targets tend to be the most straightforward forms of capacity support, generally proving to be cheaper and less time-consuming than those associated with addressing incentives or politics and power – to say nothing of providing more tangible outputs in the short term. There is an obvious demand for support which targets 'resources' and 'skills and knowledge', with the importance of building basic technical capacity, particularly among PHU staff and Community Health Workers, as regularly cited in interviews.[3] However, it is not clear whether these kinds of investments on their own will lead to sustained, deep improvements in state capacity.

Indeed, what emerges quite clearly from Table 11.1 is the frequency with which training appears as a form of capacity support. This suggests not only the dominance of this particular modality amongst development partners but also the tendency for capacity development to target 'skills and knowledge' almost exclusively through training. The interviews were illustrative in this regard: respondents, when asked about the forms of capacity support provided to government, immediately cited training, indicating that this is how capacity development is often framed and understood. The words of Potter and Brough (2004: 336) seem fitting: 'Too often [capacity building] becomes merely a euphemism referring to little more than training.' This point was not lost on many development partners and MoHS staff interviewed, but it did not appear to be transformed into programming. As Leppo (2001: 9) points out, 'The need for training and skill development are normally accepted as necessary in any change for improving systems, but there is much less recognition of the fact that trained personnel will be effectively utilized only in organizational settings with certain characteristics.' In other words, the knowledge and skill-sets of individuals matter, but so too does the institutional environment in which they operate.

Third, 'individual' and 'organization' over 'system'. Capacity development is focused overwhelmingly at the individual and organizational levels, with relatively little engagement apparent at the level of systems (both in the sense of processes, such as procurement and delivery chains and the broader enabling environment in which organizations and individuals operate). In part, such a focus can be explained by the tangibility of individuals and organizations that make for more straightforward logical frameworks, measurability and demonstrable results. Moreover, strengthening the capacity of individuals and organizations has a more immediate timeframe in which results can be expected. Engaging at the 'systems' level is less straightforward, as it potentially entails organizational restructuring and fundamental changes in the way organizations work within a system. It is a long-term process, the results of which cannot easily be anticipated. Our sense from interviews was less that stakeholders failed to recognize the importance of building capacity in these more strategic ways, but more that doing so was simply much harder and therefore less common.

Focusing on strengthening the capacities of individuals and organizations is also less overtly political than working at the systems level to alter incentives and power relations. Given current donor discourse emphasizing the centrality of country ownership and ensuring that government is 'in the driving seat', engaging in the kinds of political reforms that capacity building of systems can entail can appear more interventionist than good practice principles advocate (Kühl 2009: 570). There is a danger that being seen to conform with good practice principles could undermine the ability of development partners to engage in 'deep' (that is, political or systemic) capacity building that is likely to achieve more sustained change.

These three trends highlight that the nature of capacity support currently being provided by development partners represents just a fraction of possible targets, levels and approaches. Opportunities to extend capacity support that take into account other layers and components of the multiple factors that shape capacity are largely being overlooked.

How is capacity support working in practice? Assumptions and realities of programming

The above discussion of the dominant trends in capacity support highlights the kinds of activities that are *not* being implemented. However, as the following examples demonstrate, there are also challenges regarding the forms of capacity support that development partners *do* provide. A number of interventions are relatively common across development partners, including training, and engagement with Mother-to-Mother (M2M) Support Groups and Farmer Field Schools. Each of these interventions incorporates some kind of 'theory of change' – how a particular intervention is intended to work[4] – whether explicit or implicit, which is in turn based on a number of assumptions. Here, we tease out the theories of change within these interventions, demonstrating that the assumptions they are based upon do not always hold.

Training. Much capacity support is provided by way of training to improve the skills and knowledge of individuals at various levels of the nutrition system. In addition, training almost uniformly shares two similarities: a focus on technical skills; and the transmission of knowledge through a 'cascade' model. These features are based on certain assumptions about skills needed and how to transfer knowledge that do not necessarily hold in practice.

The logic underpinning training programmes runs crudely as follows: identify capacity gaps; identify personnel whose low capacities are the cause of these gaps; provide them with knowledge, information and skills necessary to address their capacity deficits; and reinsert them back into the system whereupon the gaps will be filled. However, whether staff are trained or not risks missing the point if more basic requirements for doing the job (such as being paid on time, having access to supplies such as therapeutic foods, as well as vehicles and fuel) are not in place. Increased knowledge can only be put to use if it is deployed in an appropriately equipped and incentivized environment. Moreover, technical know-how is not the only set of skills required for good nutrition work. The cross-cutting, multi-sectoral nature of nutrition, demanding cross-government action, means that a high degree of coordination is required. The challenge of getting nutrition-related issues onto the agenda (and budget) in a conflict-affected, low-income country where there are multiple competing priorities means that significant influencing and negotiation skills are required. The oversight of approximately 100 PHUs per district means that a high degree of management expertise is required. These capacity needs make a focus on just technical practices involved in nutrition work particularly problematic.

The 5Cs model introduced above argues that improving the capabilities of key personnel to, for example, 'establish supportive relationships' and 'achieve coherence', is of similar necessity to building technical skills (Morgan 2006: 8–16). Drawing on systematic review findings, Gillespie *et al.* (2013) conclude that:

> Human and organizational capacity need to encompass not only nutrition know-how, but also a set of soft-power skills to operate effectively across boundaries and disciplines, such as leadership for alliance building and networking, communication in the case of collaboration, leveraging of resources, and being able to convey evidence clearly to those in power.
>
> *(Gillespie et al. 2013)*

This is consistent with comments from the Food and Nutrition Directorate that training in managerial, leadership and communication skills is often deprioritized in favour of technical skills so that staff may have the requisite technical knowledge, but not the social or managerial skills to share it and influence others.

In order to reach the 1,228 PHUs across Sierra Leone, development partners use a training system known as 'cascade training', allowing them to reach the target beneficiaries in a cost- and time-effective manner. A relatively small group of individuals are trained initially, who are then responsible for passing on the skills and knowledge acquired in the original training to a larger group. Members of this group may then also train others, and so on. Cascade training can be cost-effective in disseminating key information among large groups, but it operates on the assumption that information (both amount and quality) survives throughout the various levels to which it is cascaded, which does not always hold in practice. The quality of training diminishes at each level as it cascades down, meaning that the last to receive it (frontline healthcare staff) receive significantly diluted or flawed information. This was apparent, for instance, in the Mother-to-Mother Support Groups, who receive training from a combination of NGOs, PHU staff and Community Health Workers. The training reportedly includes information about ante-natal visits, exclusive breastfeeding, complementary feeding, nutritious diet and food diversification, as well as basic hygiene. However, the lead mothers we interviewed spoke almost solely about exclusive breastfeeding and it was not clear whether any information beyond this is getting through.

Mother-to-Mother (M2M) Support Groups. The M2M Support Groups are described by development partners and the MoHS as groups of up to 15 women where members learn about a broad range of child nutrition-related issues (including exclusive breastfeeding, complementary feeding, diversified and nutritious diet, home gardening, cooking practices and hygiene) via training, activities and demonstrations. The objectives of the groups are to facilitate optimal infant and young child feeding practices within communities, and to promote uptake of routine preventive services at PHUs through community mobilization and sensitization. In theory, each group is supposed to be headed by a lead mother,[5] who has responsibility for convening the group and passing on knowledge to the other members, which is initially acquired through a training course run by NGO or MoHS staff.

Our research found that M2M Support Groups are considered by GoSL and development partners to be one of the primary structures for sensitization on nutrition at the community level. Currently, M2M Support Groups are present in only 18 per cent of the country, although there are plans to increase this to 50 per cent in 2014. They are seen to constitute a key mechanism through which malnutrition can be prevented, and one of the main ways hard-to-reach households are targeted. Their role is particularly important given the constraints that PHU staff face in conducting outreach. Although PHU staff are meant to undertake phases of outreach activity, in reality this is made difficult by a lack of transport to travel the often considerable distances and, even when motorbikes are available, a lack of fuel. Outreach also becomes more problematic during the rainy season when roads can become impassable and the M2M Support Groups can thus fill an important gap in service delivery.

Yet their successful operation and impact is dependent on a range of assumptions. Given the centrality of the M2M Support Groups in prevention of malnutrition activities, we sketch out in Figure 11.2 a rudimentary theory of change that seems to underpin them. The boxes running across the top represent the steps leading to a successful outcome, while the dashed boxes beneath describe the assumptions that must hold for each to step to be realized.

Our research suggests that not all of these assumptions hold true. Interviews in Kambia's seven chiefdoms suggest that M2M Support Groups often do not adhere with the model explained by the Freetown-based nutrition community. In practice, we found groups operating in different ways in different communities. For instance, there were variations in how closely linked the M2M Support Groups were to PHUs. In some cases, relationships were strong, with regular interactions and use of the PHU facility for meetings. In other instances, we found the groups were essentially disconnected from the PHU. In one chiefdom, neither the PHU staff nor the lead mother knew the other, and it was clear that very little in the way of coordination or information sharing was taking place. There also appeared to be variations between the different M2M Support Groups in terms of structure and remit. A number of groups had more than one lead mother, while some lead mothers did not have an M2M 'group', instead conducting outreach at the PHU and in the community. In one chiefdom, the M2M Support Group and the local Farmer Field School had the same membership, with no clear division of labour between the two. In contrast to other M2M Support Groups, this group met on an ad hoc

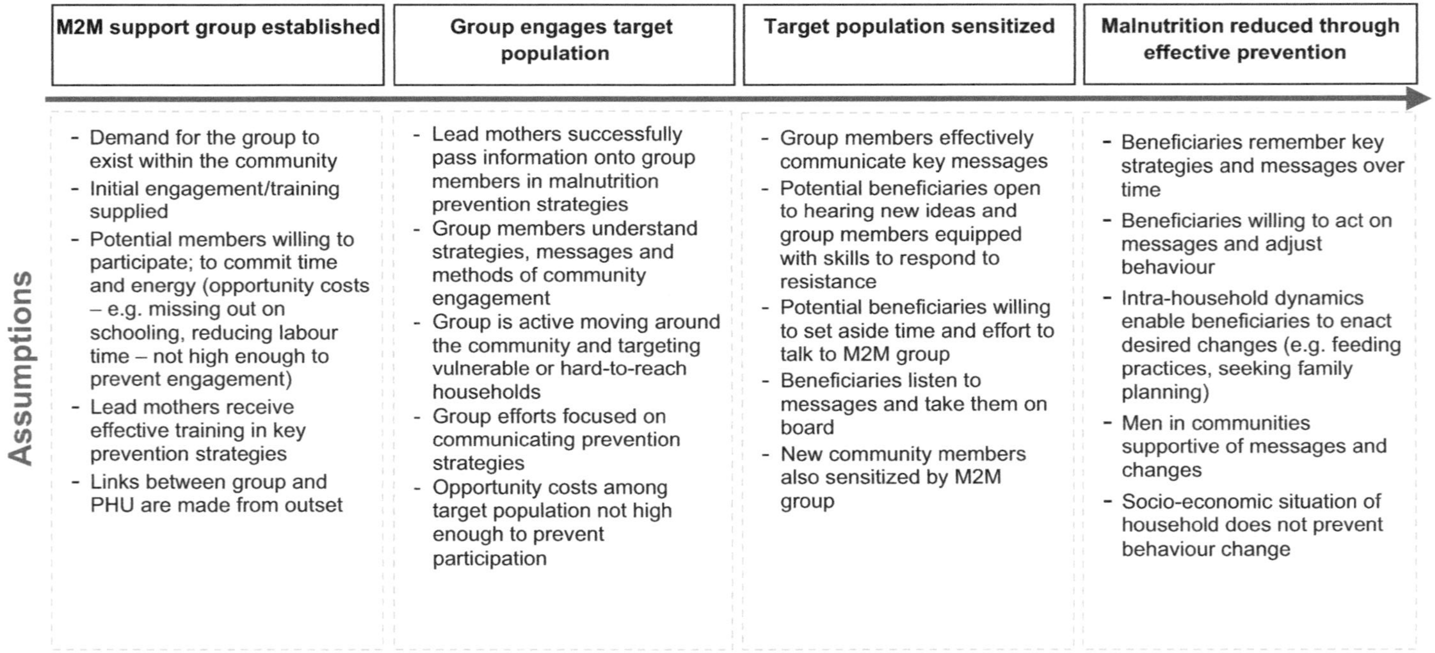

M2M support group established	Group engages target population	Target population sensitized	Malnutrition reduced through effective prevention
- Demand for the group to exist within the community - Initial engagement/training supplied - Potential members willing to participate; to commit time and energy (opportunity costs – e.g. missing out on schooling, reducing labour time – not high enough to prevent engagement) - Lead mothers receive effective training in key prevention strategies - Links between group and PHU are made from outset	- Lead mothers successfully pass information onto group members in malnutrition prevention strategies - Group members understand strategies, messages and methods of community engagement - Group is active moving around the community and targeting vulnerable or hard-to-reach households - Group efforts focused on communicating prevention strategies - Opportunity costs among target population not high enough to prevent participation	- Group members effectively communicate key messages - Potential beneficiaries open to hearing new ideas and group members equipped with skills to respond to resistance - Potential beneficiaries willing to set aside time and effort to talk to M2M group - Beneficiaries listen to messages and take them on board - New community members also sensitized by M2M group	- Beneficiaries remember key strategies and messages over time - Beneficiaries willing to act on messages and adjust behaviour - Intra-household dynamics enable beneficiaries to enact desired changes (e.g. feeding practices, seeking family planning) - Men in communities supportive of messages and changes - Socio-economic situation of household does not prevent behaviour change

Assumptions

FIGURE 11.2 Rudimentary theory of change underpinning effectiveness of M2M Support Groups

basis. These examples illustrate that, in practice, the M2M Support Groups operate less uniformly than is understood in Freetown or even the district capital.

Farmer Field Schools. An important intervention attempting to link nutrition and food security are the Farmer Field Schools. These schools were initially set up to improve farmer productivity in growing staple crops. More recently, their potential for improving nutrition has been recognized and an emphasis on training farmers in growing nutritious and diversified crops has emerged. In addition, efforts have been made to link farmers up with markets in order that they can sell their produce.

The Farmer Field Schools proved problematic to research as it was difficult to find operational schools in Kambia. On a number of occasions, Farmer Field Schools indicated by MAFFS were not operational or could not be located, with community members not having heard of them. In several cases, farmers involved in Farmer Field Schools reported that they had been operational in the past but that funding had dried up, and MAFFS and development partners had not visited for up to 10 months. There was genuine interest amongst the farmers in either revitalizing, or simply starting, Farmer Field Schools, but the schools were far from institutionalized and, in general, were not clearly recognized or understood. This is surprising given the impression from development partners and GoSL is that Farmer Field Schools are active throughout the country. It may well be that some Farmer Field Schools are active, particularly in other districts where they have received more sustained support, but it is certainly not the case that they are operating nationwide.

Implications for capacity building

Our research reveals a committed community of practice around nutrition in Sierra Leone and genuine momentum to prevent malnutrition. Yet the methods employed by development partners to build capacity appear fairly homogeneous. They focus overwhelmingly on the transfer of resources, skills and knowledge at the individual and organizational levels. What we see, therefore, is a major emphasis on training of health staff; provision of equipment and materials to clinics; and creation of new local-level prevention structures, such as M2M Support Groups. This is perhaps unsurprising given that it is relatively straightforward to deliver these forms of support, as they are more easily measurable and less politically difficult. But it is not enough to assume that capacities have been built as long as workshops were attended, staff were trained and equipment was provided (Johnston and Stout 1999). This chapter thus points to two main problems with the present state of capacity support. First, what is being done represents just a fraction of the possible modalities that could be used to develop capacities. Second, it is not clear whether current capacity building approaches are working in the way they are expected to.

More broadly, this raises questions regarding the constructability of 'capacity'. While the development community appears to overwhelmingly equate capacity building with giving training or resources in order to 'fill a deficit', as Clarke and Oswald (2010: 4) put it, our research suggests that the relationships within a system – between individuals and organizations (both formal and informal) – matter just as much, if not more. While skills and resources may be the 'building blocks' of capacity, the broader system is the 'glue' of capacity, holding those blocks together, to the extent that if external actors attempt to build capacity by focusing solely on the blocks, it is unlikely that such efforts will be sustained. In other words, seeing capacity as tangible and constructable, rather than relational and power-laden, can divert attention from what really matters (Clarke and Oswald 2010).

Ultimately, this kind of constructable, materialist approach mistakes capacity building for a technical, rather than political, exercise. Of course, developing the capacities of a state emerging from civil war in a context where much of the population experiences food insecurity for one third of the year was never going to be a purely technical task. The challenge for development partners is whether they are able to undertake the more political aspects of capacity building that are necessary for achieving deeper and more sustained improvements.

Notes

1 This chapter is based on a report authored by the Overseas Development Institute and Focus 1000 as part of the Secure Livelihoods Research Consortium's Sierra Leone country programme, funded by Irish Aid. The report can be found at <www.securelivelihoods. org/publications_details.aspx?ResourceID = 290>. See also Denney et al. 2014.
2 Scaling Up Nutrition (SUN) is a global movement focused on achieving the right to food and good nutrition for all by uniting governments, civil society, donors, businesses and researchers in a collective effort to improve nutrition. SUN Secretariats have been established in 46 countries.
3 It may also be the case that demand for training is high because of the financial benefits associated with participation.
4 A 'theory of change' is a framework used in development programming that sets out the various components of an intervention, providing the causal logic of how it will achieve intended outcomes. This causal logic is in turn based on a number of assumptions.
5 We spoke with eight M2M Support Groups; at least one in each chiefdom. However, as some groups are headed by more than one 'lead mother', in some interviews we spoke with two lead mothers simultaneously.

References

ACDIVOCA (2011) 'Sierra Leone Sustainable Nutrition and Agriculture Promotion (SNAP) Program'. Available <www.acdivoca.org/site/Lookup/SierraLeone-SNAP/$file/Sierra Leone-SNAP-USAID-branding.pdf> (accessed 30 May 2013).

Aguayo, V.M., Scott, S. and Ross, J. (2003) 'Sierra Leone – investing in nutrition to reduce poverty: a call for action', *Public Health Nutrition,* 6(7): 653–657.

Andrews, M. (2013) *The Limits of Institutional Reform in Development,* Cambridge: Cambridge University Press.

Barma, N.H., Huybens, E. and Vinuela, L. (n.d.) 'Institutions Taking Root: Building State Capacity in Challenging Contexts', Draft. Available <http://siteresources.worldbank. org/EXTGOVANTICORR/Resources/3035863–1289428746337/ITR_Rome_paper. pdf> (accessed 3 September 2013).

Boege, V., Brown, A., Clements, K. and Nolan, A. (2008) 'On Hybrid Political Orders and Emerging States: State Formation in the Context of "Fragility"', Berghof Handbook Dialogue No. 8, Berghof Research Center for Constructive Conflict Management.

Brinkerhoff, D.W. (2007) 'Capacity Development in Fragile States', Discussion Paper 58D, Maastricht: European Centre for Development Policy Management (ECDPM).

Clarke, P. and Oswald, K. (2010) 'Introduction: why reflect collectively on capacities for change?' *IDS Bulletin,* 41(3): 1–12.

Denney, L. and Ibrahim, A. (2012) 'Violence against Women in Sierra Leone: How Women Seek Redress', Overseas Development Institute Country Evidence. Available <www.odi. org.uk/sites/odi.org.uk/files/odi-assets/publications-opinion-files/8175.pdf> (accessed 2 November 2013).

Denney, L., Jalloh, M., Mallett, R., Pratt, S. and Tucker, M. (2014) 'Developing State Capacity to Prevent Malnutrition in Sierra Leone: An Analysis of Development Partner Support', Research Report, Secure Livelihoods Research Consortium. Available <www.securelivelihoods.org/ publications_details.aspx?ResourceID=290> (accessed 18 March 2014).

Gillespie, S., Haddad, L., Mannar, V., Menon, P., Nisbett, N. and the Maternal and Child Nutrition Study Group (2013) 'The politics of reducing malnutrition building commitment and accelerating progress', *The Lancet,* 382(9891): 552–69.

GoSL (2013) 'Let Girls be Girls, Not Mothers! National Strategy for the Reduction of Teenage Pregnancy, 2013–2015', Government of Sierra Leone. Available <www.k4health. org/sites/default/files/National%20Strategy%20for%20the%20Reduction%20of% 20Teenage%20Pregnancy.pdf> (accessed 17 January 2014).

James, R. and Wrigley, R. (2007) 'Investigating the Mystery of Capacity Building: Learning from the Praxis Programme', Praxis Papers 18, International NGO Training and Research Centre (INTRAC).

Johnston, T. and Stout, S. (1999) *Investing in Health Development Effectiveness in the Health Nutrition, and Population Sector,* Washington, DC: World Bank.

Koroma, A.S., Chiwile, F., Bangura, M., Yankson, H. and Njoro, J. (2012) 'Capacity development of the national health system for CMAM scale up in Sierra Leone', *Field Exchange,* 43: 39–45. Available <www.cmamforum.org/Pool/Resources/fx-43-govt-experiences-with-CMAM-scale-up-part-1–2012(3).pdf> (accessed 3 June 2013).

Kühl, S. (2009) 'Capacity development as the model for aid organizations', *Development and Change,* 40(3): 551–77.

Leppo, K. (2001) 'Strengthening Capacities for Policy Development and Strategic Management in National Health Systems', Background paper for the Forum of Senior Policy Makers and Managers of Health Systems, 16–18 July, WHO, Geneva.

Morgan, P. (2006) *The Concept of Capacity,* Maastricht: European Centre for Development Policy Management (ECDPM).

Potter, C. and Brough, R. (2004) 'Systemic capacity building: a hierarchy of needs', *Health Policy and Planning,* 19(5): 336–45.

Pritchett, L., Woolcock, M. and Andrews, M. (2012) 'Looking Like a State: Techniques of Persistent Failure in State Capability for Implementation', Working Paper 2012/63, Helsinki: UNU-WIDER.

Schroven, A. (2006) *Women after War: Gender Mainstreaming and the Social Construction of Identify in Contemporary Sierra Leone,* Berlin: Lit Verlag.

Scott, K., McMahon, S., Yumkella, F., Diaz, T. and George, A. (2013) 'Navigating multiple options and social relationships in plural health systems: a qualitative study exploring healthcare seeking for sick children in Sierra Leone', Original Paper, *Health Policy and Planning,* first published online 27 March 2013, doi:10.1093/heapol/czt016

Simister, N. with Smith, R. (2010) 'Monitoring and Evaluating Capacity Building: Is It Really That Difficult?', Praxis Papers 23, International NGO Training and Research Centre (INTRAC).

SLRC (forthcoming) 'Towards an Analytical Framework for RQ2 (State Capacity)', Secure Livelihoods Research Consortium.

Statistics Sierra Leone and UNICEF (2011) *Sierra Leone Multiple Indicator Cluster Survey 2010*, Freetown, Sierra Leone: Statistics Sierra Leone and UNICEF.

Tommy, E. (2013) 'In Sierra Leone anti corruption indicts ministries of health, agriculture and local government', *Awareness Times*, 7 June. Available <http://news.sl/drwebsite/publish/article_200522928.shtml> (accessed 20 October 2013).

World Bank (2012) 'Sierra Leone Overview'. Available <www.worldbank.org/en/country/sierraleone/overview> (accessed 30 May 2013).

World Bank (n.d.) 'Nutrition at a Glance: Sierra Leone'. Available <http://siteresources.worldbank.org/NUTRITION/Resources/281846–1271963823772/SierraLeone41211web.pdf> (accessed 30 May 2013).

12

SOCIAL INEQUALITY AND FOOD INSECURITY IN NEPAL

Risks and responses

Adam Pain, Hemant R. Ojha and Jagannath Adhikari

Introduction

In 2009 the World Food Programme (WFP) in Nepal (Hobbs 2009) published a report entitled 'The cost of coping: a collision of crises and the impact of sustained food security deterioration in Nepal'. The report argued that Nepal had experienced 'a sharp and sustained decline in food security' (Hobbs 2009: 2) leading to levels of food insecurity and nutritional deprivation that were the worst in Asia and were more comparable to the levels to be found in certain sub-Saharan African countries.

In the view of the report, while the proximate cause was rising food prices, underlying the levels of food insecurity to be found in Nepal was a complex of five interrelated and mutually reinforcing enduring crises that had contributed to a relative decline of domestic food production which, with high prices, had led to a reduction in food availability. The crises that they referred to concerned first 'a domestic food production crisis' reflecting the neglect of agriculture by the Nepalese government since 1990 and stagnation of production; second a crisis induced by climate change effects that had led to severe weather events, both droughts and floods in 2006–9; third a crisis due to conflict and ongoing instability that not least had severely disrupted markets and food supply chains; fourth a global food price crisis which had affected international food prices and led India to enforce a trade ban on exports to Nepal of several food commodities; and fifth a domestic economic crisis that had contributed to Nepal having the lowest GDP per capita in the region.

While there is evidence to support these claims made by the WFP, it should be noted that poverty rates in Nepal have declined over the past decade, and some social development indicators such as maternal health have in fact improved. The aim of this chapter is to examine the genesis of these emerging crises that came together in 2009 and which are still evident, but it will also assess household and government policy responses to this food insecurity. A core argument that will run through this chapter is that the root causes of the vulnerability (Wisner et al. 2004) of much of Nepal's population to food insecurity can be traced to deeper patterns of inequality that have limited people's access to resources and power, driven by an exclusionary political and economic system. The crises that WFP referred to have been played out over these underlying factors, giving rise to unsafe conditions for many of the Nepalese poor and making them even more susceptible to disasters. Thus while WFP (Hobbs 2009) acknowledges some of the social and geographical patterns of vulnerability to food insecurity in Nepal, the report stops short of a deeper analysis of them.

This chapter investigates in more detail these root causes and patterns of exclusion before examining the evidence in relation to the congruence of crises that have led to the current levels of food insecurity in Nepal. It starts with a discussion of the country context. As will become clear there are three distinct factors that are linked to patterns of inequality – location, caste and land ownership. However, tracing the food security consequences of the quintuple crises down to impacts on specific caste groups in particular locations with given land resources is all but impossible. In part this reflects a lack of suitably disaggregated data but it is also a reflection of the complexity of Nepal's physical and social landscape. The chapter concludes by examining responses to this food insecurity by government, donors, NGOs and households.

Country context and underlying patterns of social inequality

Country context

Landlocked Nepal, located in the central Himalayas has three distinct ecological zones – the mountains, hills and *terai* or plains. Historically the bulk of Nepal's population has lived in the middle mountains but since the 1960s there has been significant migration of hill and mountain people into the more productive *terai* where over 50 per cent of Nepal's population now lives on just over 17 per cent of the land area.

Nepal remains one of the poorest countries in the world, ranking at 195th out of 210 countries (World Bank 2010) in terms of Gross National Income per capita. While the poverty incidence has fallen from about 42 per cent in 1995–6 to 25.4 per cent in 2009 (National Planning Commission 2010), the levels of poverty remain high and inequality has been rising over the same period. Out of a population of 26 million, WFP estimated (Hobbs 2009: 2) that 3.4 million were highly to severely food insecure during the period 2007–9 because of price rises and winter droughts. An additional 5 million people were at risk of falling below the poverty line in Nepal. Two years later in August 2011, estimates of the population suffering from acute food insecurity were, at 3.48 million, largely unchanged (GoN and WFP 2011).

The country has recently emerged from ten years of an armed conflict (1996–2006) between the government and the Communist Party of Nepal (Maoist) which seriously challenged the state and caused significant death, destruction and displacement. Since 2006, when the comprehensive peace agreement (CPA) was signed, there has been a long, drawn-out process of developing a new constitution. In May 2012 the constituent assembly charged with developing this reached a fourth extended deadline without resolution and was dissolved. New elections were called for and were finally held in November 2013. Nepal still faces difficult processes of transition from war to peace, from monarchy to republican state and in social and economic relations. Underlying these difficulties and central to a view of Nepal as a state with limited capabilities is the ongoing challenge to its legitimacy and the failure of the state to deliver basic public goods. A feudalistic political culture persists based on old social hierarchies and resource control practices, leading to enduring patterns of social exclusion.

The roots of Nepal's current predicament date back to the establishment of an absolute monarchy when Nepal was unified in the eighteenth century, and the capture of that monarchy by the Ranas by the middle of nineteenth century. Even after the overthrow of the Rana regime in the early 1950s, there has been a process where windows of opportunity for change have emerged but been resisted by the ruling elite. Six decades of development aid has had limited impact, and has been seen at times to fuel the conflict (Bonino and Donini 2009). The effect has been an increasing awareness of, and agitation for, democratic rights that has fuelled increasing political contention over political representation at multiple levels.

The Ranas, who remained in control from 1847–1951, created a repressive regime based on social hierarchy, reinforcement of the caste system and patriarchy with heavy taxation of the rural population (Ganguly and Shoup 2005). In the 1950s, when an alliance of the Congress Party of Nepal and the monarchy toppled that rule, there was a brief period of democratic possibility and the creation, in 1959, of a multi-party system. However, this was thwarted by the king, who in 1962 banned political parties and instituted a multi-level *panchayat* system from village to national level. This arrangement remained contested and during the 1980s there was a rise of a pro-democracy movement; eventually in 1990 a multi-party democratic system was enacted as a compromise between the left, centrist parties and the monarchy. With the establishment of a constitutional monarchy, democratic elections were held, giving way to a series of unstable governments. In part this has resulted from a lack of commitment to a democratic culture by the elite, combined with a liberal view of addressing the issues of social exclusion and inequality that exist in the country.

Frustrated by the lack of progress on reform, the Communist Party (Maoist) declared war in 1994 against the government and for the next ten years a major insurgency took place in rural areas, although there were periodic phases of truce and negotiation. In 2005 the king dismissed the government and took control but the combined opposition of all political parties led ultimately to the transfer of power to parliamentary parties, the comprehensive peace agreement (CPA) with the Maoists and the stripping of the king's powers. With the Maoists brought into government, the newly elected constituent assembly abolished the monarchy and in 2008 Nepal was declared a republic. However, the constituent assembly was dissolved without a new constitution, pushing the country into an uncertain process of transitional politics.

Patterns of social inequality

In Nepal, social inequality is multifaceted. Although a caste-based discrimination system was abolished in 1963 and democracy established in Nepal in April 1990, the old social order that underlay the previous political system has remained entrenched. That social order was reflected in a caste pyramid which according to the Country Code or *Muluki Ain* of 1854 gave differential privileges and obligations to each caste and sub-caste within it. The Rana rule (1847–1951) reinforced both the caste system and a patriarchal gender system. Four strata were defined within this system; the top stratum was seen to be ritually pure and high caste and was occupied predominantly by *Brahman* and *Chhetris*. The second stratum was also 'pure' but made up of non-caste indigenous ethnic or *Janajati* people. Beneath these two upper strata were two 'impure' groups – the third stratum consisting of Muslims and foreigners and therefore outside the caste system. At the bottom were the ritually impure 'untouchable' or *Dalits* of the Hindu caste system. Within each of these strata there are finer gradations of difference between the higher status hill people and the lower status *terai* or plains people. These cultural distinctions serve as severe structural constraints to the ability of many households and individuals to exercise agency in relation to their lives.

Two examples illustrate the daily patterns of exclusion experienced by social groups lower down the pyramid experience. First, Bhattachan et al. (2004) reported over 200 forms of common and continuing practices of caste-based discrimination in Nepal including literal practices (refusing to share water sources, avoiding direct body contact) of untouchability. Second, although land ownership is not a requirement for citizenship, officials often demand it. The effect is that the many landless *terai Dalits* and *Janajatis* are denied citizenship. Even when they are officially provided land this does not give them land security (see Box 12.1). But there are also considerable vertical inequalities within each social/cultural or caste group. There are 'elites' in most of the high caste and *Janajati* groups that have been closely linked with the ruling class. In the case reported below (Box 12.1), the conflict was between a *Janajati* group (Tharu) – some of whom are landed class – and *Dalits* (who occupy the bottom rung in terms of ownership of land and access to political power).

**BOX 12.1   FOOD INSECURITY FOR A LANDLESS *DALIT*
COMMUNITY IN THE *TERAI*

(ADAPTED FROM ASIAN HUMAN RIGHTS COMMISSION 2011)

In 1993 Gandharva, a nomadic *Dalit* community, was provided with a piece
of land from the Commission for Resolving Problems with Landless People
(Sukumbasi Samasya Samadhan Aayog) without legal title. In October 2010,
members of the neighbouring Tharu community who own the land in the village came and destroyed the crops, demanding that the *Dalits* leave the land,
as the Tharu wanted to apply for the community forest. Despite the Gandharva calling the police for help, no assistance was received. The police came
to the village two days later but did not make any intervention to address the
violence committed by the Tharu. The Gandharva neither cultivated crops
last year nor planted crops this year, fearing that the Tharu would come again
to destroy the crop. Facing serious food insecurity, many male Gandharvas
migrated to India while several women migrated to the Middle East seeking
employment. The children were left behind, deprived of their education and
adequate food security and often discriminated against because of their caste
and poor living conditions. A complaint was submitted to the local administration twice in December 2010 and April 2011, but no actual response has
been received. Thirty-five persons including some from Gandharva filed a
public interest litigation on the case in January 2011.

There is also a body of evidence that shows sharp differences in terms of poverty
levels based on social identity, as well as marked variations within each social group.
Data on caste and ethnicity was first collected in 1991 when 60 caste and *Janajati*
groups were listed. The census of 2001, which is the most recent, identified 103
social groups based on caste, ethnicity, religion and language. However the definition of who is and is not a *Dalit* and the subdivisions (and hierarchies or sub-castes)
within the *Dalit* group remain contested (World Bank and DfID 2007). Similarly, a
schedule listing the *Janajati* groups also remains under negotiation; the 2001 census
only listed 43 *Janajati* groups but 59 are now officially recognized (World Bank and
DfID 2007: 63). These issues of classification are not trivial since aggregation tends
to mask micro-level variation in poverty between different social groups within
Nepal's extremely heterogeneous physical and social environment.

Table 12.1 provides a broad breakdown of the population by caste and census
groupings.

Differences based on identity is one variable but there are two more to take into
consideration. First is the variation in population density with the mountains and

TABLE 12.1 Nepal's population by caste/ethnic groupings

% Total Population	Hindu Caste Groups (57.5%)	Janajatis (37.2%)	Muslims (4.3%)	Others (1%)
	Brahman/Chhetri (Hill)	Newars	Muslims	Others
	Brahman/Chhetri (Terai)	Janajatis (Hill)		
	Terai Middle Castes	Janajatis (Terai)		
	Dalits (Hill)			
	Dalits (Terai)			

Source: adapted from World Bank and DfID (2007).

hills holding about half of Nepal's population, with a further half of the population now living in the *terai*. The second dimension is that of geographical distribution of different social groups. There has long been a historical distinction and hierarchy of the hill dwellers *(Parbatiya)* over the plain dwellers *(Madhesi)*. But there are other spatial dimensions as well. For example, the *Dalit* populations are spread throughout the country but there are particular points of concentration of the hill *Dalits* in the mid-western and western regions (which holds 50 per cent of their population), whereas 85 per cent of the *terai Dalits* live in the central and eastern regions (World Bank and DfID 2007: 57–8). In contrast the concentration of *Janajati* groups is greatest in the eastern part of the country and in the northern hill and mountain areas.

Drawing on the evidence of income and consumption poverty derived from the Nepal Living Standard Surveys (CBS 2011), an overall fall in poverty rates has taken place between 1995 and 2009 (Table 12.2). However the starting levels and the degree of that decline differs between social groups. In 1995–6 the highest poverty rates of 50 per cent or more were found in the *Dalits* and the *Janajatis*. By 2004 despite the overall decline in poverty rates both hill and *terai Dalits* recorded poverty rates of 48 and 46 per cent respectively and hill *Janajatis* 44 per cent. In contrast the *Brahman* and *Newars* had poverty rates of 20 per cent or less; for the *Brahman/Chhetri* group poverty rates had declined by 46 per cent in contrast to the 21 per cent decline of the *Dalits*. In sum, per capita consumption in 2003–4 of *Dalit, Janajati* and Muslim households was still between 15 and 13 per cent lower than in the *Brahman/Chhetri* households. As the World Bank (2007: 20) put it, '[T]hese differences in consumption levels can be called the "penalty" that certain groups pay because of their caste, ethnicity or religious identity.'

As with horizontal (between social groups) inequality in poverty outcomes, the vertical inequalities in poverty are also significant. For example, some *Janajati* groups are better off economically than some groups within high caste *Brahmans* and *Chhetris*. For example, *Newars (Janajati* group) have, on average, the lowest poverty rates (10.25 per cent) in the country (Table 12.2). Their education and health outcomes are also better. The *Thakalis*, another *Janajati* group also tend to be better off.

TABLE 12.2 Trends in the incidence of poverty by caste/ethnicity between 1995–6 and 2003–4 (figures are per cent of group population)

	1995–6	*2003–4*	*2010*
All *Brahman/Chhetri*	34	19	Hill *Brahman* – 10.34 Hill *Chhetri* – 23.40 *Terai Brahman* – 18.61
Terai Middle Castes★	29	21	28.69
All *Dalits*	59	47	Hill *Dalit* – 43.63 *Terai Dalit* – 38.16
Newar	19	14	10.25
Hill *Janajatis*	49	44	28.25
Terai Janajatis★	53	36	25.93
Muslims	44	41	20.18
Others			12.34
Nepal	42	31	25.16

★Note that the composition of these two groups differs between the three surveys.
Source: adapted from World Bank 2007: 18; CBS 2012: 10.

On the other hand, there are wide disparities within *Brahman* and *Chhetri* groups. The *Brahmans* and *Chhetris* from the Karnali region, for example – a region synonymous with hunger – tend to be relatively poor (CBS 2012).

Land poverty is both a cause and a correlate of poverty, but an accurate current assessment is extremely difficult to get because of problems of conflicting data, methods and concealment (Alden Wily et al. 2009). Thus while Table 12.3 reports the most recent set of figures on land distribution which reveals the degree to which there is a considerable degree of inequality in land holdings, trying to assess the long-term trends is problematic. Alden Wily and colleagues' (2009) critical assessment suggests that although there has been a gradual decline in land inequality, it is difficult to be precise about this. However, her analysis provides a number of relatively stark conclusions.

First, most rural households do not have enough land to subsist on and Alden Wily, et al. concludes that up to 58 per cent of farmers or 2.7 million rural households are functionally landless (with less than 0.5 ha). Second, landholdings correlate with caste and ethnicity in predictable ways with higher castes owning more land and renting less, particularly if they live in the hills. In contrast lower castes own less land and rent more, and the *terai Dalits* are worse off in terms of land than the hill *Dalits:* 80 per cent of the former are functionally landless in contrast with 75 per cent of the latter. The *terai Janajati* have the least land of all ethnic groups and more commonly rent land. Third, while land ownership is one dimension it is also clear that feudal land relations persist with extractive forms of tenancy, exploitation of labour and even bonded labour despite legislation designed to address this. There

TABLE 12.3 Distribution of land by farm size and households

Farm size (ha)	Percentage of farm households	Cumulative percentage of households	Percentage of farm area	Cumulative percentage of farm area
Less than 0.1 ha	9.1	9.1	0.6	0.6
0.1–0.2 ha	12.1	21.2	2.5	3.1
0.2–0.5 ha	31.5	52.7	15.4	18.5
0.5–1 ha	27.4	80.1	28.3	46.8
1–2 ha	15.5	95.6	31.1	77.9
2–3 ha	2.8	98.4	9.9	87.8
3–4 ha	0.7	99.1	3.7	91.5
4–5 ha	0.5	99.6	3.4	94.9
5–10 ha	0.3	99.9	3.3	98.2
10 ha or more	0.1	100	1.9	100
	100			

Source: CBS 2011: vol. 2, 12.

are two major land reforms on the statute book but Alden Wily, et. al. concludes that the main beneficiaries of these have been the landed elite. Redistributive measures and attempts to establish rights for the poor have effectively been evaded. It is against this setting of social inequality and uneven access to resources that the effects of the various crises have to be assessed.

Five crises

The domestic food production crisis and the effects of the rise in international food prices

Around 85 per cent of Nepal's population live in rural areas and draw their livelihood from agriculture and related activities, although the share of agriculture in GDP has been falling. The agriculture sector contributes around 40 per cent of GDP and has provided employment to more than two-thirds of the population (CBS 2011). Self-employment in agriculture and agricultural wage labour are the country's most important sources of household income. But agriculture remains largely subsistence or semi-commercial with only 15 per cent of gross outputs being sold in 2003–4. However, growth in agricultural production has not kept up with the population growth rate.

Until 1987 Nepal was a net exporter of rice although even during this period there are reports of chronic food shortages, especially in the hills and mountains (Adhikari 2008). But between 1974 and 1992 there was a sharp decline in per capita gross food production and since the late 1980s Nepal has been a net importer of food and only in surplus between 2000 and 2004–5 (Agriculture Project Services Center and John Mellor Associates 1995). In part this has been an outcome of a long-term decline in public expenditure allocated to agriculture. Adhikari (2010) reports that in the 1970s the agricultural sector received about 30 per cent of the total budget; this fell to 16 per cent in the 1980s, to 10–12 per cent in the 1990s, and to 2.79 per cent from 2000 to 2010.

The Nepal Agricultural Perspective Plan (APP), a 20-year plan (1995–2015) for the agriculture sector financed by the Asian Development Bank (ADB), was designed to address the prevailing low crop and livestock productivity. Investments were to be made in irrigation, fertilizers, research and roads with the aim of promoting the adoption of modern technologies, enhancing productivity and supporting agricultural growth. However the APP was only partially implemented because of the general paralysis of government during the long, drawn-out conflict that began in 1996. Problematic assumptions about the conditions for market-led agricultural growth (including, for example, that there was a basic functioning of public infrastructure such as roads), and insufficient budget combined with falling levels of investment in the agriculture and irrigation sectors, have also had their effects. In 1996 capital expenditure on the agricultural sector was just under 6 per cent of agricultural GDP, but since 2000 it has been under 3 per cent, falling as low as just over 1 per cent in 2005 (ADB 2012: 35). The agricultural sector remains characterized, according to ADB (2012), by low labour agricultural productivity, limited technical change and low levels of commercialization. There have been major supply constraints on inputs for crops and livestock, and rice yields are amongst the lowest in the region.

As a result, agricultural growth has been poor, with cereal production increasing by only 5 per cent between 2004 and 2009 while demand, driven by a rising population, has increased by more than 20 per cent (Hobbs 2009). Between 1995–6 and 2009–10 the agricultural trade deficit has risen from US $124 million to US $373 million (ADB 2012: 21).

The increasing contribution of food imports to national food availability has had a number of consequences. While the *terai* has remained a surplus producer of food, the levels of food deficit in certain hill and mountain districts, particularly those in the mid-west and far west of Nepal, have become acute, with local production only meeting 3–5 months of household needs in an average year. Given the poor infrastructure and lack of roads, food imports even at the best of times are costly and unreliable. With the rapid rise in international food prices during 2007–8 there was a very steep food price inflation, which lasted well beyond the decline in international prices in the latter part of 2008 and into 2009 (Hobbs 2009). Thus even in March 2009 food price inflation rates of 17.1 per cent were being reported, with levels of 10.5 per cent in 2012 (Awasthi 2013). Nepal's access to core food commodities from its main trading partner, India, which has provided up to 80 per cent of agricultural imports, was also stopped when India imposed a trade ban on selected food commodities. This lasted into 2009, leading to more expensive imports from the international markets.

More recently there have been initiatives to prepare a new 20-year Agricultural Development Strategy (with financial support mainly from core donors). However this has been subject to considerable protests by various civil society organizations that have objected to its prioritization of commercialization, private-sector-led development and trade over food security considerations.

The effects of climate change

A decline in domestic food availability and a greater exposure to the international food markets and the risks associated with this have been core drivers of increasing food insecurity. This has been compounded however by the effects of climate change. Nepal is a country that is seen, on account of its mountainous landscape, poor governance and fragmented and limited infrastructure, to be particularly vulnerable to hazards. These include the annual monsoonal rains, floods, landslides and forest fires. These natural occurrences and governance-related risks have been exacerbated by climate change (e.g. increasing frequency of intense rainfall events), leading to increased chances of major flash floods and landslides, while glacial melt associated with climate change has increased the risk of glacial lake outburst floods.

A total of 64 out of Nepal's 75 districts have been assessed as prone to some type of disaster. Nepal was ranked twenty-third in the world when it comes to total natural hazard-related deaths, from 1988 to 2007, with total deaths reaching above 7,000; it ranks seventh for the number of deaths resulting from all floods, landslides and avalanches and eighth for flood-related deaths alone (Ministry of Home Affairs 2009). In June 2013, for example, far western Nepal suffered from floods and landslides that killed 38 people and displaced over 3,500 households (UN TSS 2013). The disaster was triggered by an early high-intensity monsoon, an event consistent with predicted climate change effects. Globally, Nepal therefore ranks very high in terms of risks for earthquakes and water-related disasters respectively (Ministry of Home Affairs 2009: 17).

The country's climate is influenced by the Himalayan mountain range and the South Asian monsoon. There is evidence of rising temperatures (Ministry of Environment 2010: 8) and of Himalayan glacier melt although no clear evidence of systematic changes in rainfall patterns. Climate change projections for Nepal identify a rise in mean annual temperatures as likely and of particular significance in winter compared to the monsoon season and in western and central Nepal in comparison with the east of the country. Climate change models do not identify any changes in winter rainfall but summer rainfall is projected to increase by 15–20 per cent. These projections, which the Ministry of Environment (2010) reported, are consistent with observations suggesting a warming at higher altitudes leading to reduced snow and ice coverage (see for example McSweeney et al. 2008), increased climatic variability and more frequent extreme events (floods and droughts), particularly in the *terai*. However, as noted by Shrestha and Aryal (2011: 66–7), Nepal's physiographical and topographical variability creates enormous climatic and ecological diversity, making it extremely difficult to project the impact of climate change on ecosystems in the country.

Whatever the formal scenarios, the recent experience of unprecedented winter droughts in 2006–7 and 2008–9 significantly affected food production, and, in combination with high food prices, net food availability to households. WFP (Hobbs 2009) noted that compared to the 2008 harvest, the 2009 harvest decreased across all the major regions with the mountain and hill districts suffering, respectively, a 40 and 25 per cent decline in production. The report further estimated that as a result of the drought about 2.2 million people in Nepal, particularly in the far western and mid-western districts, faced problems of access to food.

Thus through a combination of landscape features characteristic of mountainous countries, a largely subsistence agrarian sector, high poverty levels and limited government capability, Nepal has been ranked as the fourth most at risk country according to the Climate Change Vulnerability Index (2011). Natural hazards – especially landslides and droughts in the mountains and hills, and floods in the *terai* – accentuated by extreme weather events are likely to have a significant impact on agricultural production especially in marginal locations farmed by the more food insecure households.

Effects of conflict and instability

The entrenched nature of inequality in Nepalese society has been an enduring source of conflict, but its nature and manifestation has changed over time and varied across contexts. While some conflicts around rights such as the bonded labour[1] movement (Nepal 2001) have contributed to improved land rights and hence to some extent food security, other more violent forms have at times disrupted food supply in remote areas (Seddon and Adhikari 2006). Conflicts therefore can be causally linked to food insecurity.

When conflict has not necessarily been visible, structural violence[2] has exacerbated food insecurity. The structural violence is seen in the violation of the rights of tenants, farm workers and women, low caste *(Dalits)* and marginal ethnic groups and, despite legislation, it continues in one form or another. From the unification of the country until the 1950s, the state imposed on peasants heavy taxation and unpaid labour demands *(corvee)*. Feudal landlords expropriated a high proportion of the production and exploited their tenants physically and economically. This caused continuous out-migration to different parts of India because of livelihood insecurity in Nepal (Shrestha 2001). *Dalits* were primarily affected by this violence, and they formed the major portion of out-migrants. Even now, the tendency to migrate to India is high among the *Dalits*.

However, the armed conflict that started in 1996, when the Communist Party of Nepal (Maoist) launched a 'People's War', mobilizing the grievances of excluded groups, led to open violence. Under serious threat from the Maoists the state used the army to control the armed struggle of the 'rebels'. The rural population was caught between the two armed forces – that of the rebels and that of the government. An estimated 13,300 people lost their lives during the conflict. There were frequent disruptions to transport, frequent strikes, *bandhs* (forced temporary closure of businesses and schools and the restriction of movement by vehicle) and blockades. This affected the movement of people and of goods and commodities including food, causing food supply problems to food deficit districts as well as affecting the livelihoods of those dependent on trade. Many internally displaced persons (IDPs) fled to escape conflict and significant out-migration from Nepal to India occurred,[3] as well as the destruction of property and infrastructure (Adhikari and Gurung 2010).

Because of the conflict, public food depots (discussed in more detail later) were closed in the most food insecure regions such as Karnali (Adhikari 2008), mainly because of the looting of food from them by the Maoist rebels. As a result, people who used to have a small amount of subsidized food from the state lost access to this source as the conflict undermined the limited state mechanisms for the delivery of food. Damage to infrastructure and displacement of people led to increased food insecurity – and the abandonment of productive agricultural land. Given the origins and location of rebel activities in the mid-western and far western regions of the country, it was the people in these already food insecure regions that suffered most from the conflict.

The decade long civil war had major negative effects on agricultural production, markets and food security. Continuing political unrest including *bandhs,* strikes and blockades has had major effects on food supply, markets and food prices in the years since. WFP (Hobbs 2009) noted for example how a 13-day strike in the *terai* in April 2009 led to major supply problems of basic food commodities to nearly 40 per cent of the district markets in the hills and mountains, and the shortages contributed to price rises.

Economic crisis

In comparison with its regional neighbours, Nepal has performed poorly with respect to economic development, and growth has been the lowest in the region since 1990, ranging from 4 to 7 per cent annually. While the economy has traditionally been agrarian in nature, its share of GDP has fallen from 51 per cent in 1985 to about 40 per cent in 2000 and to 33 per cent in 2007 (World Bank, 2010). However, this has not meant a rise in opportunities elsewhere. The contribution from industry has shrunk and while there has been growth from the service sector this has been largely urban based. The most significant change has been the rise in remittances from the estimated 2 million Nepalese working overseas; these now contribute approximately 25–30 per cent of GDP (NIDS 2012).

There are marked regional disparities in contributions to GDP, with the central region accounting for about 42 per cent of GDP in 2001, the mid-western and far western regions together accounting for 18 per cent, and the eastern and western regions contributing 21 per cent and 19 per cent respectively (ADB et al. 2009). Most of the market-based economic activity is found within the *terai*. In contrast markets have limited penetration in the hills, which remain a largely subsistence economy.

The private sector is poorly developed and tends to be urban based, focusing on services and construction and reflecting the absence of a market-oriented rural economy (World Bank, 2010). On paper, official policy since the mid-1980s (1985–90) and especially since the eighth plan (1992–7) has been to liberalize the economy and reduce state control of key economic sectors, but in practice this has not happened in a comprehensive way. Key sectors related to agriculture, such as input provision (e.g. fertilizer) were under state control and subsidized. In 1998 the government undertook to withdraw from the fertilizer market. However, in the face of falling supply through formal market channels because of the level of imported low quality fertilizers through informal channels, the government reintroduced in 2008 a price subsidy on chemical fertilizers to be managed by the Agricultural Input Company Limited (AICL). The private sector is not entitled to import and distribute the subsidized fertilizers and has proved unable to compete with the subsidized fertilizer and unofficial imports. However, the volume of fertilizer (100,000 metric ton) that AICL is able to import and distribute at a subsidized rate through farmer cooperatives falls far below demand, and there is limited uptake of unsubsidized fertilizer (Shrestha 2010).

Two sources point to the core issues of unequal access in Nepal's markets, a report by the Asian Development Bank (ADB) et al. (2009) and an anthropological study by Rankin (2004) which focuses on a specific *Newar* trading community living close to Kathmandu. The study points strongly to the social embeddedness of markets and their regulation by non-economic factors, supporting the more general conclusions of the ADB study, summarized below. It is also widely believed that much of the international trade is in the hands of a few traders, who benefit from connections with political leaders and government bureaucrats, although this is not well documented.

The ADB report points to a number of factors that have constrained growth including political instability and weak governance, inadequate infrastructure and labour unrest. Climate effects are not included. The report also identifies a number of constraints that limit inclusiveness of economic growth and contribute to unequal access to markets, the outcome of which has been indicated in the rising levels of inequality. These constraints include the following:

- Limited productive employment opportunities particularly in rural areas where most poor live;
- Lack of rural employment as a result of in particular the poor growth and performance of agriculture; this is evidenced by the high migration rates within and out of the country, with about 15 per cent of adult men commonly away from their usual place of residence;
- Unequal access to agricultural inputs and opportunities to commercialize and diversify agricultural activities, based in part on size of landholdings and unequal access to irrigation;
- The poor and other disadvantaged groups – based on identity/caste and gender – have fewer opportunities for non-agricultural employment;

- The poor and disadvantaged groups have fewer opportunities to migrate to countries other than India;
- Inequality in accessing infrastructure and productive assets such as credit and land.

ADB et al. (2009: 93) argue that the root causes underlying the constraints to inclusiveness in economic growth are linked to systematic patterns of exclusion and inadequate public service delivery.

Impacts of the crises on patterns of inequality and food security outcomes

There is therefore considerable evidence to support the argument made by WFP (Hobbs 2009) that a number of long-term processes came together in 2009 leading to a marked deterioration in food security in Nepal. Long-term trends associated with poor agricultural growth and a stagnant economy on which were superimposed shocks linked to climate change, conflict and a rise in global food prices all had effects on access to food. There is unfortunately no disaggregated data or research available to critically scrutinize the differentiated effects of these crises against other long-term trends on patterns of inequality and food security outcomes of marginal people (Shively et al. 2011). The effects these crises have had in reinforcing inequalities and responses by different social groups in contrasting locations are complex.

For example, populations who live in geographically marginal mountainous places such as Humla in the northwest of Nepal traditionally depended on long distance trade systems to Tibet and parts of southern Nepal to gain food security (Roy et al. 2009) given the limits of production in meeting household food needs. But cross-border restrictions and declining demand for trade goods within Nepal have removed this opportunity. More generally, long-term trends of economic decline have already driven household responses prior to 2009, evidenced both by the movement of hill populations down to the *terai* over the last few decades and patterns of international migration that have given rise to a remittance economy (discussed further below). Thus migration and remittance economies may have become more of a solution to food insecurity than evidence of distress, although the evidence of the landless *Dalit* community (see Box 12.1) being driven out of the *terai* also points to a level of compulsion driving migration. As one of the authors[4] of this chapter found during a recent visit to a Gurung village in Lamjung, 85 per cent of the households were in receipt of income from external sources, making agricultural cultivation a secondary occupation and reducing the dependence on farm production for food security.

There is also evidence of past landslides in Lamjung causing complete resettlement of whole villages to the *terai*. Ongoing research[5] in the district of Lamjung and Dolokha has found places with a long history of landslides, and where the richer households have moved out first to resettle in urban areas. Moreover, although more recent extreme rainfall events may trigger a significant landslide, the underlying determinants of that landslide may relate to a longer history of poor road construction undertaken by contractors with powerful political connections, giving rise to the unsafe conditions in the first place.

It is clear therefore that there is a complex patterning of food insecurity across Nepal's landscape determined by caste, gender, geography and history, but it is the geographical dimensions of food insecurity that are most easily identified. For example, WFP (Hobbs 2009) found that the areas most similar to sub-Saharan African countries in terms of food insecurity were the hills and mountains in the far west and mid-west of Nepal. These are also the regions which have been the priority areas of humanitarian food-aid programs like that of WFP. In these regions high caste *Brahmans* and *Chhetris* are a dominant part of the population, and there is evidence that the food security status of *Janajati* groups is better than that of high caste groups. This may be linked with a longer history of migration and remittance flows for this group.

In contrast Lovendal (2004), drawing on livelihood analysis, argued that the households most vulnerable to food insecurity were largely marginal farm households located in the hills and *terai* and low status rural castes. Within these the most vulnerable to food insecurity were sub-populations of women, children, *Janajatis* and low castes.

The evidence on geographical patterning of food insecurity is consistent with evidence on nutritional outcomes. The Nepal Demographic and Health Survey (New Era and USAID 2011) for example found that about 41 per cent of Nepalese children under five years have a low height for age or suffer from stunting due to chronic malnutrition and that the highest rates of hunger were to be found in the far and mid-western hill and mountain regions.

With respect to the food prices rises in 2009, WFP suggested (Hobbs 2009) that these had particular effects on two social groups. The first were the landless poor of the *terai,* who are mainly *Dalits* and are entirely dependent on the market for food. The second were food deficit households in the hills and mountains, who were faced with food prices that were almost three times those of the *terai* (Hobbs 2009), leading to households spend as much as 78 per cent of their income on food. One study showed that nine remote districts of Nepal experienced an increase in food prices of nearly 40 per cent, from July to December 2008, compared to the same period of study in 2007 (Ministry of Agriculture and Cooperatives and WFP 2008 cited in Ghale 2010).

Responses to food insecurity

In Nepal's planning circles, the concept of 'food security' emerged in government planning documents in 2008, although there has been a long history of government initiatives in relation to food policy (Pyakuryal et al. 2005). A food availability crisis in 1971–2 led to the establishment of a National Food Corporation (NFC), which initially had a major responsibility for distributing subsidized food to remote food deficit areas. However for much of the period up to the late 1990s it mostly sold subsidized food in the Kathmandu valley, primarily to people working for the government, including the army and the police (Pant 2012). Since 1998–9 the policy has been to reduce the role of the NFC to allow private sector development. In the late 1980s the NFC was handling about 80,000 tons but this was reduced to about 30,000 tons a year in 2008; by 2010–11 it was handling only about 0.4 per cent of total food grain production in the country. Even up to 2006–7 about 60 per cent of subsidized food sales took place in the Kathmandu area, although by 2010–11 this had changed and 74 per cent of sales now go to more remote districts (Pant 2012).

However even in more remote areas, the majority of subsidized food sales were to government employees and teachers, indicating that these subsidies primarily serve as a salary top-up rather than a food security measure. Pant (2012) noted that in Karnali, 60 per cent of the allocation was to government servants. Roy et al. (2009) observed that even in 2008, 50 per cent of the subsidized grains (rice) went to government servants and the amount available for rural households would at best provide for a week or so. Thus the NFC has played a limited role in addressing the grain deficits of food insecure households.

After the political change in April 2006 in Nepal, civil society activists focused on gaining rights to food and food sovereignty and these are now enshrined in the interim constitution 2007: clause 18 (3) states that 'every citizen shall have the right to food sovereignty as provided for in the law'. (GoN 2007a: 8). In line with this, Nepal's Three-year Interim Plan (GoN 2007b) has, for the first time, included a chapter on food security. This political change helped to maintain the public distribution system of the NFC although as noted its role is somewhat limited.

After 2006 in the immediate aftermath of political change there was an increase in humanitarian support and the distribution of food, with the WFP distributing even more food than the NFC. Donors supported food-for-work programmes to help conflict victims and repair damaged infrastructure. There have also been school feeding programmes and relief and emergency operations to address food insecurity. The WFP has provided support for the school feeding program in 12 food deficit districts, in which the government is also implementing the 'basic and primary school' program (WFP 2012). During emergencies created by natural disasters, food has also been provided to the victims under the 'relief and emergency operations' program but by design these programs do not address the fundamental causes and factors related to food and livelihood insecurity. The role that WFP has played has been scaled back since 2009, particularly in some of the most food insecure areas of Nepal, as a result of decreased funding (Shively et al. 2011). This may well mean that food security conditions in such areas have deteriorated further.

There has been action, supported by national and international NGOs, to promote community-based initiatives to enhance food security. Roy et al. (2009) for example describe one such programme in Humla that introduced improved water mills, the promotion of local foods and the cultivation of non-timber forest products, but evidence on improved food security as a result is limited. Such projects and approaches remain piecemeal and fragmented and there has been limited policy response at the national level.

Although multiple attempts have been made, especially after 1990, to reform land policies and institutions there has been little headway (Alden Wily et al. 2009); some good provisions have been crafted but have never been translated into practice, owing to the active resistance by the landed political elites. For example, after 2006, the government formed two 'land-reform commissions', and both of these commissions submitted reports. But these reports were never publicly disclosed or acted on. The landed groups, which have links with most political parties, have been able to stop any action.

If government and donor response to persistent food and livelihood insecurity has been limited, action by people has not. Labour migration has become a key strategy at the household level to ensure food security. Nearly 44 per cent of households have at least one absentee member currently living away from home (CBS 2011). The latest census reveals that as many as 1.92 million Nepalis (87 per cent of which are male and 13 per cent female) were working in foreign countries in 2010–11 for more than six months (CBS 2011). About 20 per cent of the population were recorded as 'absentee' (absent from their original homes), with just under half of these (43 per cent) living outside Nepal (CBS 2011).

As a result of foreign labour migration, remittances from abroad have become important for both household and national economy. In 2011–12 more than USD 5.0 billion and formal remittances contributed as much as one quarter of Nepal's GDP (NIDS 2012). Migration takes place for various reasons, but food insecurity has been one of the main causes of migration undertaken by poorer households. This is particularly so for those who go to India for work, especially from the far and mid-western hills and mountains (WFP and NDRI 2008).

Conclusions

There can be no doubt that a significant proportion of Nepal's rural population is acutely vulnerable to food insecurity. Who becomes food insecure and to what extent depends on at least three interrelated sets of factors: social inequality and the relative position of the households and individuals in the social and cultural hierarchy; the location of the household in relation to potential and actual climate change impact; and third, the access to non-agricultural employment opportunities both in Nepal and outside. An agricultural transformation, even if the competence to deliver and the necessary institutional support to drive it were present, would at best be limited to the more favourable areas with good market access and thus primarily in the *terai*. However, given the protracted political instability and largely unresponsive public policy system, significant reforms with respect to food security are unlikely.

Agricultural policy and planning have suffered from a lack of clear vision and strategies, continuing with the new Agricultural Development Strategy, which has failed to integrate food access issues with a classic production policy narrative.

Moreover, it is difficult to see what agricultural futures there can be in the hills and mountains for the many functionally landless households, and while marginal improvements in productivity may be possible, they may only serve to maintain these areas as a poverty trap. Indeed the evidence of the scale of out-migration and the rise of the remittance economy point to how the poor assess and respond to their current and future possibilities when the public policy system remains unresponsive. Structural reforms including those of land may offer some respite and these, at least at the level of rhetoric, are on the political agenda even if they are not matched by action. For the past many years, the 'land reform' agenda has emerged and subsided, with little impact. Alden Wily et al. (2009) see some possibilities for a more community-based approach to land reform but such an approach cannot take root without an enabling environment and political support. Land policy reform in itself will not be adequate, and broader policy framework is needed to address the diverse roots of food insecurity.

What is strikingly not on the agenda is any wider commitment, either by government or donors, to more robust and widespread social protection measures that would systematically address inadequate income and consumption at the household level in rural areas (Ellis et al. 2009). Without such measures it is difficult to be optimistic for the future of those who are currently food insecure. It is equally difficult to be hopeful that the political elite have either the interest or the competence to transform the current political and economic impasse to drive Nepal's economy forward and improve living standards.

Notes

1 Bonded labour happens when an individual has to provide labour or other services in repayment for a loan or debt. The nature and duration of the services to be provided are often unspecified and debt bondage can be passed from one generation to the next.
2 A form of violence where social structures or institutions harm people by preventing them from meeting their basic needs.
3 Most of the victims of the conflict flocked to the main urban centres, in particular to the capital, Kathmandu. The number of IDPs was estimated to be between 37,000 to 400,000, not including those who had gone to India.
4 Hemant Ojha, unpublished field notes, 2013.
5 Unpublished findings from the Climate Change and Rural Institutions research project.

References

ADB, Department for International Development, UK and International Labour Organization (2009) *Nepal: Critical Development Constraints,* Manila, Philippines: Asian Development Bank (ADB).
ADB (2012) *Assessment Report: Technical Assistance for the Preparation of the Agricultural Assessment Strategy,* Manila, Philippines: Asian Development Bank (ADB).
Adhikari, J. (2008) *Food Crisis in Karnali: A Historical and Political-economic Perspective,* Kathmandu: Martin Chautari.
Adhikari, J. (2010) 'Food insecurity, conflict and livelihood threats in Nepal', in B.R. Upreti and U. Müller-Böker (eds), *Livelihood Insecurity and Social Conflict in Nepal,* 73–130, Kathmandu: Swiss National Centre of Competence in Research.

Adhikari, J. and Gurung, G. (2010) *Livelihoods and Migration. A Case of Migration Between Nepal and India,* Kathmandu: NIDS (Nepal Institute of Development Studies).

Agriculture Project Services Center and John Mellor Associates (1995) *Nepal Agricultural Perspective Plan,* Kathmandu: Government of Nepal and Asian Development Bank.

Alden Wily, L. with Chapagain, D. and Sharma S. (2009) *Land Reform in Nepal: Where Is It Coming from and Where Is It Going? The Findings of a Scoping Study on Land Reform for DFID Nepal,* Kathmandu Department for International Development (DfID) Nepal.

Asian Human Rights Commission (2011) *Nepal: incomplete land distribution process causes serious food insecurity to the Gandharva Dalit community currently facing crop destruction and abuse by another community.* Online. Available <www.humanrights.asia/news/hunger-alerts/ AHRC-HAC-003–2011> (accessed 20 May 2013).

Awasthi, G. (2013) 'Disappointing economy of the country', *Kantipur,* February 11: 1.

Bhattachan, K.B., Tamrakar, T., Kisan, Y.B., Bagchand, R.B., Sunar, P., Paswan, S., Pathak, B. and Sonar, C. (2004) *Dalits' Empowerment and Inclusion for Integration in the National Mainstream: Strategy for Influencing Policy and Institutions,* Kathmandu: Dalit Empowerment and Inclusion Project.

Bonino, F. and Donini, A. (2009) *Aid and Violence: Development Policies and Conflict in Nepal. A Background Report,* Boston, MA: Tufts University.

CBS (2011) *Nepal Living Standard Survey 2010/2011,* Kathmandu: Central Bureau for Statistics.

CBS (2012) *Poverty in Nepal,* Kathmandu: Central Bureau for Statistics.

Climate Change Vulnerability Index (2011). Online. Available <http://maplecroft.com/ about/news/ccvi.html> (accessed 2 July 2012).

Ellis, F., Devereux, S. and White, P. (2009) *Social Protection in Africa,* Cheltenham: Edward Elgar.

Ganguly, S. and Shoup, B. (2005) 'Nepal: between dictatorship and anarchy', *Journal of Democracy,* 16: 129–43.

Ghale, Y. (2010) 'Corporate globalization: hunger and livelihood security in Nepal', in B.R. Upreti and U. Müller-Böker (eds), *Livelihood Insecurity and Social Conflict in Nepal,* 131–82, Kathmandu: Swiss National Centre of Competence in Research.

GoN (2007a) *Interim Constitution of Nepal,* Kathmandu: Government of Nepal (GoN).

GoN (2007b) *Three Year Interim Plan (2007/08–2009/10),* Kathmandu: National Planning Commission, Government of Nepal.

GoN and WFP (2011) *Nepal Food Security Bulletin,* Issue 32, August 2011, Kathmandu: Government of Nepal and World Food Programme.

Hobbs, S. (2009) *The Cost of Coping: A Collision of Crises and the Impact of Sustained Food Security Deterioration in Nepal,* Kathmandu: United Nations World Food Programme and Nepal Khadya Surakshya Anugaman Pranali (NeKSAP).

Lovendal, C.R. (2004) *Food Insecurity and Vulnerability in Nepal: Profiles of Seven Vulnerable Groups,* Food Security and Agricultural Projects Analysis Service Working Paper No. 04-10, Rome: The Food and Agriculture Organization of the United Nations.

McSweeney, C., New, M. and Lizcano, G. (2008) UNDP Climate Change Country Profiles: Nepal. Online. Available <http://countryprofiles.geog.ox.ac.uk> (accessed 26 December 2012).

Ministry of Environment, Nepal (2010) National Adaptation Programme of Action (NAPA), Government of Nepal. Online. Available <www.napanepal.gov.np> (accessed 26 December 2012).

Ministry of Home Affairs (2009) *Nepal Disaster Report: The Hazardscape and Vulnerability,* Kathmandu: Government of Nepal and Nepal Disaster Preparedness Network.

MoAC [Ministry of Agriculture and Cooperatives] and WFP [World Food Programme] (2008) *Report on Rapid Emergency Food Security Assessment; Far and Mid-West Hills and*

Mountains, Kathmandu: Government of Nepal and United Nations World Food Programme.

National Planning Commission (2010) *Poverty Measurement Practices in Nepal and Number of Poor,* Kathmandu: National Planning Commission.

Nepal, M.S. (2001) *Kamaiya: Slavery and Freedom,* Kathmandu: Mandala Book Point.

New Era and USAID (2011) *Nepal Demographic and Health Survey,* Kathmandu: New Era and USAID.

NIDS (2012) *Migration Year Book 2012,* Kathmandu: Nepal Institute of Development Studies and Swiss National Centre of Competence in Research.

Pant, P. K. (2012) *LDC Issues for Operationalization of the SAARC Food Bank, Nepal Case Study,* South Asia Watch on Trade, Economics and Environment (SAWTEE) Working Paper No. 01/12, Kathmandu: SAWTEE.

Pyakuryal, B., Thapa, Y.B. and Roy, D. (2005) *Trade Liberalization and Food Security in Nepal,* Market, Trade and Institutions Division Discussion Paper No. 88, Washington, DC: International Food Policy Research Institute.

Rankin, K.N. (2004) *The Cultural Politics of Markets: Economic Liberalization and Social Change in Nepal,* London: Pluto Press.

Roy, R., Schmidt-Vogt, D. and Myrholt, O. (2009) 'Humla development initiatives for better livelihoods in the face of isolation and conflict', *Mountain Research and Development,* 29(3): 211–19.

Seddon, D. and Adhikari, J. (2006) *Conflict and Food Security in Nepal: A Preliminary Analysis,* Kathmandu: RRN. Online. Available <www.internal-displacement.org/8025708F004 CE90B/%28httpDocuments%29/13C80D92223B06B5802570B700599A54/$file/eu-conflict.pdf> (accessed 02 January 2014).

Shively, G., Gars. J. and Sununtnasuk, C. (2011) *A Review of Food Security and Human Nutrition Issues in Nepal,* Agricultural Economics Staff Paper 11–05, West Lafayette: Purdue University.

Shrestha, A.B. and Aryal, R. (2011) 'Climate change in Nepal and its impact on Himalayan glaciers', *Regional Environmental Change,* 11(Suppl 1): S65–77.

Shrestha, N.R. (2001) *The Political Economy of Land, Landlessness and Migration in Nepal,* New Delhi: Nirala Publication.

Shrestha, R.K. (2010) 'The development of fertilizer policy in Nepal', *Journal of Agriculture and Environment,* 11: 126–37.

UN TSS (2013) *Nepal: Flood and Landslide Update,* Kathmandu: United Nations Transition Support Strategy (UN TSS).

WFP (2008) *Nepal Market Watch,* Kathmandu: World Food Programme (WFP).

WFP (2012) *Nepal: better education through school meals.* Online. Available <www.wfp.org/ photos/gallery/nepal-better-education-through-school-meals> (accessed 02 January 2014).

WFP and NDRI (2008) *Passage to India: Migration as a Coping Strategies during Crisis Times,* Kathmandu: World Food Programme (WFP) and Nepal Development Research Institute (NDRI).

Wisner, B., Blaikie, P., Cannon, T. and Davis, I. (2004) *At Risk: Natural Hazards, People's Vulnerability and Disasters,* Abingdon: Routledge.

World Bank (2010) *Nepal: Development Performance and Prospects: A World Bank Country Study,* Washington, DC: World Bank.

World Bank and DfID (2007) *Unequal Citizens: Gender, Caste and Ethnic Exclusion in Nepal,* Kathmandu: World Bank and Department for International Development, UK (DfID).

13

CHANGING APPROACHES TO FOOD SECURITY IN VIET NAM

Ian Christoplos, Le Thi Hoa Sen and Le Duc Ngoan

Abstract

Viet Nam is in a process of redefining its policies and practices related to national and household food security so as to respond to the multiple risks faced due to climate change and demographic pressures. A series of radical changes in food security policies have taken place since the end of the war in 1975. These changes also reflect intentions to take advantage of the opportunities that exist to enhance the livelihoods of the population and develop commercially within a dynamically changing region. At the same time, goal conflicts are appearing between a move towards a commercialized and urbanized society and efforts to maintain the security currently provided by smallholder farming and to protect highly productive agricultural areas from conversion to urban and industrial purposes. This chapter considers the range of factors that are influencing how food security is developing a new meaning in both national policies and in the practices of rural households.

Chapter overview

Viet Nam has undergone a profound transformation over the past 25 years in relation to the production of food, the sources of income through which people obtain entitlements to food and the patterns of food consumption. These changes have been sparked by reforms in state policies, but have then been taken up by farmers, market actors and consumers. This chapter looks at the context for these changes in terms of overall perspectives on the role of the state and rapid urbanization and commercialization. The evolution of policies is reviewed, from an early post-war focus on rice production to newer emphasis on diversification, export promotion and a growing concern for food safety, sometimes associated with 'modern' retail. The chapter also looks at how a former focus on food security has gradually come to be characterized by commitments to entitlements, where even in rural areas smallholder agriculture is not expected to be the primary or self-evident basis for future livelihoods.

The context of Vietnamese food security policies

Viet Nam's approach to food security reflects its status as a state with a clear developmental character. The country has often been characterized as a 'developmental state' because of its role in promoting a given development path and guiding markets (Gainsborough 2010). Even if the approaches to food security have varied over the years, the central role of the state in choosing national models for achieving food security, while guiding markets to support these aims, has never been in question. The historical relationship between the state and the citizenry, where trust is given in a developmental process, was forged most strongly during the wars against the French colonial forces, and after that the United States, and most recently with China in 1979. The state is widely perceived to be accountable for providing basic security for its citizens – not just political security but also food security and protection from natural hazards and other threats. The state is also proactive in investing considerable resources of its own in chosen development paths, at both national and local levels, and in both urban and rural areas. Some of the reforms in the agricultural sector that have been undertaken over the past two decades may resemble those proposed by the World Bank and other international institutions, but they have been driven and supported by the Vietnamese state. Policies today reflect lessons learnt from abroad and growing experience with markets, but the state has made its own decisions.

Viet Nam's economic performance and poverty reduction have been extraordinary since reforms began at the end of the 1980s. Recently, the positive trajectory in the Vietnamese economy has slowed considerably partly as a result of domestic policies. These difficulties have arisen at the same time as popular expectations have grown enormously. There are thus particular pressures on the government to maintain and reinforce its role in leading national development.

The Vietnamese economy remains largely rural, but has undergone a radical process of urbanization and diversification over the past two decades since liberalization of the economy began. Although rural incomes have improved, the gap between urban and rural incomes and between the rich and the poor has widened. Presently, annual average income per capita of urban dwellers is more than double that of the rural population (GSO 2010). In rural areas, the income gap between the rich and the poor gradually increased (MARD 2009: 15). Small urban centres and provincial capitals have grown rapidly, especially in coastal and other lowland areas with relatively good market access, where private and public investment have been high. The Red River Valley and the Mekong Delta, which have been the source of most of the national rice production, have also witnessed growth in other forms of agriculture, food processing and industry in the areas near urban centres. Household incomes have become diversified as youth obtain urban employment. Education levels are high and this has historically been highly valued as a way to expand livelihood opportunities beyond agriculture.

Urbanization has been very rapid in recent years, with industrial and residential zones spreading into areas that were previously used for rice production. Only in the period from 2001 to 2005 has the total land area in the country absorbed by urbanization reached over 366,000 hectares (3.9% of agricultural land) (MONRE 2007). From 2000 to 2008, agricultural land decreased annually by approximately 70,000 hectares and more than 100,000 agricultural jobs were lost (MARD 2009). Nguyen Mau Dung (2011) reported that for every hectare agricultural land transformed for other purposes, ten members of the agricultural workforce lose their jobs. Furthermore, land acquisition and evictions for the urbanization process have negatively impacted on agriculturally based livelihoods. During five years (2003–8) land transformation has affected the lives of 2.5 million people throughout the country (MARD 2009). Urbanization and industrialization in Viet Nam are mostly concentrated in the most fertile land (Hoang Thi Van Anh 2009). For example, in the lowland rice-producing Van Lam district of Hung Yen province a total of 928.52 hectares of land were converted, displacing 14,260 farm households, with losses of 736.50 hectares of rice paddy (Nguyen Thi Hong Hanh et al. 2013). Also during the period 2000–10, more than 20,000 households in Da Nang City lost over 1,900 hectares of agricultural land (CNPR 2008) for the enlargement of the city and for the development of industrial zones. These processes are set to continue, as large areas on the outskirts of many urban centres today are crisscrossed with new four-lane thoroughfares, with new residences and factories gradually displacing the rice paddies that remain in these suburban grids.

Partly in response to the displacement caused by urbanization and industrialization, but also because of increasing pressures on land in general, there has been a series of reforms to land laws. The classification and zoning of land is a highly contentious issue, as it is in many respects where objectives related to food security, expansion of plantation farming (especially perennial crops such as rubber), industrialization and environmental conservation converge and often come into conflict. The outcomes of these goal conflicts will have major implications for food security in the future.

Despite trends toward urbanization and industrialization, the Vietnamese economy is still based on agriculture. The growth rate of GDP from agriculture is increasing dramatically, from 3.69% in 2006 to 12.33% in 2010. However, agriculture as a proportion of GDP is declining (24.5% of total GDP in 2000 and 20.6% in 2007) because of the fast development of the industrial sector. Agriculture currently still provides employment for over half of the workforce.

Agriculture and those dependent on an agriculture-based livelihood have borne the brunt of the effects of extreme climate events and gradual climate change, which are already having considerable impact on the country. Despite increasing urbanization, rural forms of vulnerability are projected to endure (DARA and the Climate Vulnerable Forum 2012).

Rice production and diversification

After the end of the war in 1975 the top priority of the government was to ensure national and provincial level food security. This was done by setting and working toward achieving targets for rice production, as rice self-sufficiency has traditionally been viewed as synonymous with food security. These targets were enforced through directives and control, whereby farmers were told what to plant, and given little leeway in even determining how to plant.

These directives focused strongly on provincial rice self-sufficiency, as each provincial government struggled to ensure that it met targets. There were, for example, restrictions in the movement of rice across provincial borders. This led to local food insecurity when markets could not function in allowing surpluses to move to deficit areas, and when production of alternative crops better suited than rice to certain agroecological zones was discouraged. Without control over land or access to markets, there was neither ability nor motivation for farmers to improve production. Households were paid according to the number of days they had laboured and not according to their production or profitability.

Things began to change with the introduction of liberalization reforms referred to as *Doi Moi* starting in 1986–7. One of the first results of liberalization was that rice production grew at a phenomenal rate (see Figure 13.1). This process of moving away from rigid state policies began with some provinces practicing 'fence-breaking' in the late 1970s and early 1980s, i.e. testing whether they would be allowed to depart from central edicts. This process accelerated when reforms were formalized (starting by directions of Central Executive Committee No. 100/1981/ CT-BBT in 1981 and No. 10/1986/CT-BBT in 1986), most notably with the introduction of a new law on land ownership in 1993, which gave farmers control over their land. Once control over land was ceded, incentives appeared and people could adapt farming systems to local conditions. Resolution No. 10 in 1987 and Resolution No. 5 in 1992 allowed farmers to produce as they wished. Before 1988, the country imported at least 4.5 million tonnes of rice annually (Truong Huu

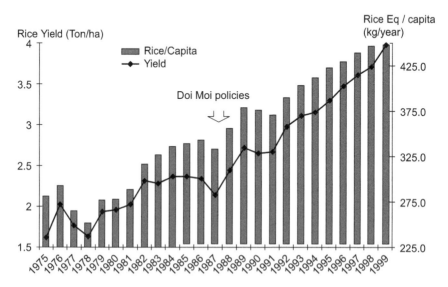

FIGURE 13.1 Effect of *Doi Moi* reform on rice productivity and produced rice equivalent amount/capita/year in Viet Nam (example in period 1975–99)

Quynh et al. 2003: 1129–34). Since 1989 Viet Nam has been a net rice exporter. The *Doi Moi* reforms have shifted the country from a food/rice insufficient country to the second biggest rice exporter in the world.

As these production increases eased concerns about availability of rice, the past 20 years have seen a cautious and gradually growing readiness within the government to encourage diversification beyond rice production. Farmers themselves have also gradually recognized the advantages of diversified production in terms of profitability and spreading of risk. Although rice yields and production have increased, economic return has been decreasing. Thus, there is a trend toward diversification of crops by converting rice land to other crops. In the Mekong Delta especially, farmers started by converting from rice to high value fruit and aquaculture in the early 1990s. In the North the early shift primarily occurred close to Hanoi in resource-rich areas that could take advantage of the urban market.

Alternative crops have resulted in six to ten times more economic return for farmers than rice (Anh Tuyet 2013). Climate change has to some extent encouraged this trend as in some areas affected by salinity and other climatic factors the former rice-based economy has experienced shrinking yields and uncertainty due to climatic variability.

In coastal areas in particular, aquaculture has grown rapidly, and in 2004 aquaculture was Viet Nam's third largest source of export income (Rasmussen and Tran 2005). In the Mekong Delta this has involved a spectacular expansion of *Pangasius* catfish production (Belton et al. 2011), whereas in other areas there has been more focus on shrimp (though shrimp has been of declining relative importance in recent years). Farmers have primarily shifted from rice to aquaculture in rainfed areas that are less productive for rice and/or where they do not need to follow an irrigation plan for a larger area. In the last few years, the total area of paddy land converted to aquaculture land has increased to around 350,000 hectares, mostly concentrated in the Mekong Delta (Nguyen Chu Hoi 2012).

Access to export markets for these new products, particularly for aquaculture, requires increasingly strict compliance with standards such as product certification. This has knock-on effects for food security in Viet Nam, as internationally the ability of farmers to retain access to export markets is reliant on such certification, and failure to attain these standards may exclude them from dynamic market opportunities. Smaller aquaculture producers in Viet Nam are facing increasing difficulty maintaining the foothold in the market that they had when they started production for export, though they may find a niche in international markets with lower quality standards, such as the Middle East and Latin America, or return to producing for the domestic or local markets (Belton et al. 2011). Rice producers are being supported to meet quality standards by planting new varieties, where quality is given precedence over quantity, and in promoting private investments in rice dryers to ensure better quality processing. These dryers have an added benefit in terms of climate change adaptation as they help farmers to ensure that their crop is not damaged when early rains after harvesting prevent the use of traditional sun drying methods.

BOX 13.1   FROM RICE TO SHRIMP TO POLYCULTURE

The Tam Giang-Cau Hai Lagoon in central Viet Nam is the largest lagoon in Southeast Asia. Prior to 1995 only state enterprises were allocated production areas in the lagoon for aquaculture, which at that time focused solely on seaweed production. Farmers struggled to produce rice on marginal lands, which were affected by salinity in the dry season and frequent flooding in the wet season. Lagoon communities were very poor. In 1994, the prime minister issued Decision No. 773-TTg, which launched a programme exploiting underutilized aquatic resources. Following this national policy, shrimp culture in the lagoon quickly increased almost 50-fold in the five years from 1990 to 1995 (Le Van Mien et al. 2000). In some areas, local households claimed their fishing grounds from the collectives that were then in rapid decline and occupied other fishing ground areas for aquaculture production. Some of the production is in ponds. In other areas nets have been set up to create corrals. These have expanded to the extent that capture fisheries have declined significantly and the flow of water in the lagoon is increasingly restricted. Production areas appear to be reaching or exceeding their physical limits, with little space available for additional corrals and increasing problems with spread of disease, presumably associated with declining water quality due to insufficient water circulation.

The focus of aquaculture has recently begun to shift from monoculture shrimp production to lower risk and less intensive polyculture, including seaweed, shrimp, crabs and fish. This shift has occurred as a reaction to two factors. First, it is because of repeated disease outbreaks, primarily affecting monoculture shrimp production systems. These are thought to result from the increasing density of production leading to disease transmission between production areas. Second, increasing occurrence of freshwater inflows due to heavy and unpredictable rains has affected all aquaculture and even capture fisheries, but here again, primarily monoculture shrimp production, as these systems are highly sensitive to changes in salinity. Together these factors are stimulating a shift from monoculture shrimp production to lower risk polyculture.

Discussion is now underway in central government circles on encouraging a shift from food to animal feed production because feed security for livestock (and reduction of feed import costs) may be just as important as food production in the livelihood and food security of a rural population looking for new ways to diversify. In 2011, imported feed amounted to 8.9 million tonnes, a three-fold increase since 2006 (Vu Duy Giang 2012). Therefore, consideration is also being given to encouraging the use of poor quality rice for feeding animals, as well as shifting less productive rice areas to maize production for fodder.

There are clearly different food security impacts and opportunities to diversify in the less fertile uplands than in the irrigated deltas and other coastal areas with good market access. The shift to diversification in the uplands has been slow, but now even in these areas there are new investments in cash crops such as rubber, cassava and coffee. New investments are starting to reach these areas, but households still rely heavily on upland rice and/or maize production to meet household food security needs (Nguyen Viet Khoa et al. 2008: 19), despite declining yields in some areas due to deteriorating soil fertility (Nguyen Viet Khoa et al. 2008: 37). Research into new upland varieties has not produced the same level of dramatic results as for irrigated rice and thus the research and extension systems have been less effective. Poverty and food insecurity levels are far higher in the uplands where, in many areas, coffee and rubber production have grown, but as monocultures, in contrast to the diverse production patterns in the lowlands. The gap between the determinants of food security in areas with a subsistence economy or reliance on a single cash crop, and the livelihood patterns in the dynamic and diverse lowlands, is growing. The lack of alternative employment opportunities for upland farmers has meant that they are particularly affected when there are price fluctuations in plantation crops (Pandey et al. 2006). In 2013 winds from Typhoon Wutip devastated rubber production in Quang Binh province, leading to widespread indebtedness among smallholders who had invested in rubber over the past decade.

Despite these moves towards diversification, the state is still very involved in guiding agriculture. There are still provincial targets for rice production. Resolution No. 63/NQ-CP in December 2009 decreed that the area for rice production shall not fall below 3.8 million hectares to ensure food security for the country (until 2020). Based on this, each province is expected to preserve specific areas for rice production, and areas producing two rice crops per year are to be particularly protected. In irrigated areas the freedom for individual decision-making is constrained by watershed level decisions on irrigation timing. Concerns about environmental degradation and new risks associated with climate change have also put limits on what farmers are allowed or encouraged to produce. But overall there has been a huge change in that farmers have been given leeway to focus on their own food security and well-being rather than national goals. Efforts to increase production of a given crop generally use incentives and subsidies, rather than directives and regulations.

The new role of the state in supporting food security and agricultural production

A central aspect of the shift from directives to provision of incentives has been that national and local governments have, since 1990, focused increasingly on the introduction and development of agricultural services and improved rice varieties. In the post-war years, the role of local agricultural offices was primarily one of directly managing the production process. In the early 1990s these offices shifted a large proportion of their efforts to advice rather than control. This coincided with research efforts (particularly driven by researchers in the Mekong Delta), where rice varieties more appropriate to Vietnamese conditions were being developed. In higher potential areas, first in the Mekong Delta and later elsewhere, farmers have been supported and encouraged to try a range of new types of production, as well as to increase their investment in rice.

Over the years, some of the government campaigns and provincial schemes to promote new products and varieties have proven effective, while others have flooded weak markets or otherwise failed. More recently the approaches to supporting production have become more refined. Subsidies (partly in conjunction with demonstrations of new varieties and production methods) are still provided for poorer smallholders, while technical advice, market linkages and market information are emphasized for more commercial farmers.

The development of extension services exemplifies the changing relationship between central government, local government and citizens in agriculture. Trends toward empowerment of local agricultural authorities to respond to farmer needs can be traced back to an overall process of creating and reforming extension services that began in 1993, when a decision was made to reconstruct a major part of provincial and district agricultural bureaucracies from being institutions mandated to instruct farmers in what they should plant, to advising them on a range of agricultural issues instead. In March 1993, the public extension system was founded and organized into five levels: Central (National), Provincial, District, Commune and Village/Hamlet. Extension functions were initially focused on organizing demonstrations and field-days; training; and organizing science-technology forums related to crops, livestock, veterinary care, forestry, water resource management, agro-forestry processing and engineering. Choice of technologies was based on recommendations from research institutions, universities and abroad. The extension system also provided farmers with information on new policies and market prices.

Since being established, the government extension system of Viet Nam has undergone many structural reforms and policy changes to make it more effective and respond better to farmer concerns. After Viet Nam joined the WTO in 2007, the extension system began to shift increasingly toward commercialization, with market information playing a central role. In 2008, market information systems were functioning in 23 provinces and 106 districts (Hoa et al. 2008).

Another aspect of reform in agricultural extension has been a gradual 'socialization', a term referring to a shift to provision of services by NGOs and private firms. These sometimes continue to be financed by public resources. One aspect of this 'socialization' is the new and varied forms of farmer organizations that are also emerging, some of which are operating as cooperatives while others are more informal, acting as intermediaries for information and brokering contacts between members and wholesalers. This includes supporting their members in understanding markets and addressing production quality factors in order to gain access to supermarket value chains (Moustier et al. 2010). The new types of farmer organizations are notably different from those of the past, when collective farming was compulsory. In some places new cooperatives are emerging as a major force in empowering farmers, and are being financially supported by provincial governments as a way of encouraging commercialization and access to markets. In other areas (particularly in the South), there is still a lingering distrust among farmers towards collective action mechanisms.

In addition to extension and farmer organizations, investments related to Vietnamese food security include public goods investments in infrastructure, drainage and dykes. The need for such investments is becoming increasingly apparent in conjunction with concerns regarding climate change. The effects of climate change are, in many of the most productive areas of the country, expected to be in the form of increased salinization and extreme and/or unseasonal floods. This is on top of hazards that have always existed related to storms. A major historical role of the state in Viet Nam has been to invest in irrigation and drainage systems to bring new land under production. For example, the history of the Mekong Delta is one of transformation of a formerly inhospitable swamp into an area of extremely high productivity, made possible largely through state investments in drainage to enable effective production, to manage flooding and to reduce acid sulphate levels in the soil.

Viet Nam also has a long tradition of constructing sea and river dykes to protect the population and the irrigation structures upon which they depend. Because of its location and geography, sea dykes are very important for Viet Nam to protect low-lying paddy areas from storm surges and other forms of flooding. As early as the Ly dynasty (1009–1225) sea dykes were built along the banks of the Red River, Ma River and Lam River in northern and central regions. Since that time the construction and repairing of dykes has been a major priority in disaster risk reduction efforts of the country. Over the centuries, the building of dyke systems has been so central to protecting the population as to become associated with the culture and economy of the nation (Nguyen Nguyen Hoai 2011). Some years ago it was suggested that the decline of central planning and collective institutions associated with the *Doi Moi* liberalization process was leading to declining investments in maintaining this system of dykes (Adger 1999), but this seems to have changed after a series of extreme floods (especially in 1999 and 2000) and with growing national awareness of climate change.

Urbanization, consumption patterns and the changing meaning of 'food security'

As noted above, Viet Nam is urbanizing at a very rapid rate and demographic pressures driving urbanization and land use change are having an impact on areas under production. They are also multiplying risks related to flooding. Urban and industrial sprawl is, in many places, reducing the areas of paddy land, and also the marginal swampy areas where run-off can be absorbed and dissipated, thereby increasing risks of flooding. Infrastructure to protect drinking water supplies for coastal cities from the influx of saline water in the dry season may also affect water flows, with impacts on agriculture. Other construction, especially new roads, further restricts and alters water flows, often causing flooding. As noted above, regulations exist to prevent transformation of land use away from agriculture but goal conflicts are inevitable.

It is not only the production systems that are changing. Consumption is also undergoing a transformation. Changes in livelihoods have increased purchasing power, allowing people to obtain more desirable foods. Vietnamese consumers are eating more meat and fruit, as well as new processed foods, especially in urban areas. Consumer demand is also growing for safe vegetables, free from pesticides, implying a shift of concerns from food security to food safety, again particularly in urban and near urban areas. Fears of food poisoning and disease due to unsafe food have increased. The government has increased efforts to manage food supply chains and control quality. However, the public is still concerned about food safety, particularly pesticide residues (Moustier et al. 2010). Many have doubts about the reliability of vegetables labelled as 'safe' (Mergenthaler et al. 2009).

Changing consumption patterns, together with the legalization of private advertising in 1991, have been described by Figuié and Moustier as leading to the 'birth of the Vietnamese consumer' (Figuié and Moustier 2009: 210); 'the "producer citizen", given pride of place in previous official policies, is giving way to the "consumer citizen" ' (Figuié and Moustier 2009: 215). There are also changing forms of retail and commodity chains, partly driven by concerns related to quality and food safety. Supermarkets are growing, particularly in urban areas (Maruyama and Trung 2007), and this shift has been promoted by government policy in relation to 'modernization' and 'civilization' (Moustier et al. 2010), a perception encouraged by the media (Maruyama and Trung 2007). In some urban areas local authorities are also active in supporting a shift to 'modern' retail outlets (Cadihon et al. 2006). It has even been suggested that consumers' preference for supermarkets is related to their desire to be associated with this modernity. 'Today the supermarket is replacing yesterday's "Honda Dream": in the mid-1990s the symbol of social success in Vietnam was to exchange a bicycle for a motorcycle' (Figuié and Moustier 2009: 215). Farmer organizations are playing an important role in ensuring that supplies to these new retail outlets meet quality requirements (Moustier et al. 2010).

However, the shift to supermarkets has not been so dramatic as in some other countries, since Vietnamese consumers continue to demand fresh produce and fish. This has meant that traditional markets still overwhelmingly dominate (Cadihon et al. 2006). Consumers demand extremely fresh food and tend to primarily use supermarkets to purchase non-food or dry staples and processed foods (Maruyama and Trung 2007). Consumers who do choose to shop for vegetables at supermarkets, do so because they have greater trust in the safety of the products on sale than those in traditional markets (Cadihon et al. 2006).

New risks related to climate change

Viet Nam is likely to be among the countries hardest hit by climate change, mainly through rising sea levels and changes in rainfall and temperatures. Studies for the

Southeast Asian region show that climate change and its related impacts could lower agricultural productivity by 2–15 per cent (Zhai and Zhuang 2009: 4–5). For Vietnamese agriculture, a 1°C increase in minimum temperature could lead to a 10 per cent decrease in rice yields. Fish and shrimp species develop more favourably in a water temperature in the range 20° to 25°C. The aquaculture yields may therefore decline if water temperatures rise above 30°C (Tuan 2010: 7).

Viet Nam's main rice growing areas, the Red River Delta in the North and the Mekong River Delta in the South, are particularly vulnerable to sea level rise (Dasgupta et al. 2007: 28; Boateng 2012: 26–28; Hanh and Furukawa 2007: 52). A large proportion of Viet Nam's population and economic activity is also located in these two river deltas. If the sea were to rise by one metre, almost 11 per cent of Viet Nam's population would be affected, and 35 per cent would be affected by a five-metre sea level rise (Dasgupta et al. 2007: 28). A one-metre sea level rise would impact 6–7 per cent of the agricultural sector (Dasgupta et al. 2007: 32–3). In one assessment, by 2100, 4.4 per cent of Viet Nam's land area could be permanently inundated (Carew-Reid 2008: 2). If the sea level rises by one metre, 39 per cent of the Mekong River Delta, more than 10 per cent of Red River Delta, more than 2.5 per cent of Central Coast and 20 per cent of Ho Chi Minh City will be inundated (MONRE 2012: 1–2). Coastal and estuary mangrove forest wetlands and other submerged aquatic vegetation, which provide habitats and nutrient sources for fish, may also be severely affected by sea level rise (Tuan 2010: 7).

Saline intrusion is a particularly serious problem for coastal agriculture (Hanh and Furukawa 2007: 56). According to a study by the International Food Policy Research Institute [IFPRI] (Yu et al. 2010), projected sea level rise will profoundly affect rice production around the Mekong River Delta. The report assesses that in the rainy season the areas inundated with rainfall more than 0.5 metre would increase by 276,000 hectares, and in the dry season areas affected by saline intrusion with a concentration greater than four grams of salt per litre would increase by 420,000 hectares. The loss of areas devoted to paddy rice production could lead to a rice production decline of about 2.7 million metric tonnes per year (based on 2007 rice yields), 0.9 million tonnes in the rainy season as a result of inundation and 1.8 million tonnes in the dry season because of saline intrusion. This is an equivalent of about 13 per cent of the 2007 total rice harvest in the Mekong River Delta (Yu et al. 2010: 8), where 50 per cent of the nation's rice is produced (Hanh and Furukawa 2007: 56).

One major response to salinization has been the investments in aquaculture described above. In some places this has been undertaken in areas that had formerly been used for crop production, but in some areas climatic uncertainty and increased salinization have made these production systems unviable. Places that had just one poor rice crop are now being converted to aquaculture. Elsewhere, the majority of expansion has been in coastal lagoons, which had previously been used for capture fisheries.

Viet Nam is in the process of taking a strong stance in managing the increasing risks associated with climate change. As mentioned above, construction of sea dykes has been a traditional role of the state. In addition to infrastructure, over the past 20 years the government has established strong flood and storm control institutions. There are explicit policies, regulations, roles and task assignments in place at all levels. Public sector actors are thus held accountable for preparing for and responding to potential disasters according to their place in the chain of command. Agricultural authorities are advising farmers how to adapt to the risks that they increasingly face, as described in the following example.

BOX 13.2 LOCAL AGRICULTURAL AUTHORITIES RETHINKING RECOMMENDATIONS TO MANAGE CLIMATE RISK

In the highly productive irrigated areas in the central Vietnamese province of Quang Binh, provincial policies promote double cropping of rice to meet production targets. In interviews by the authors it emerged that the district agricultural departments had questioned these goals as they encountered an emerging practice where farmers allow rice from the first season to regenerate as a low risk way of obtaining a small second crop instead of two main rice crops per year as has been traditional. The farmers were doing this because they had recognized that because of the increasingly unpredictable onset of the rainy season, they very often lost their second crop to floods (CCCSC 2013). The province had urged conventional replanting to maximize production, while the districts pointed out that the low risk methods were preferred and more profitable for farmers (who could then focus on other livelihood opportunities). The district agricultural officials advocated supporting the farmers in this approach to optimizing, rather than maximizing, production.

It appears that these local responses to climate-related food security risk focus largely on those risks that are already being experienced. By contrast, response to longer term climate change scenarios is not (yet) well anchored in approaches at provincial and district levels. National climate change strategies have been established, but they are not very explicit in relation to food security. Provincial governments have been tasked with reviewing scenarios and designing responses, but as yet most provinces have not begun implementing these new strategies in earnest. It is widely assumed that the majority of resources for these investments will come from international rather than domestic sources. Indeed, where large investments are underway (e.g. in some provinces in the Mekong Delta), these focus on infrastructure to protect high potential agricultural areas. Elsewhere however, there does not appear to be an equally strong 'social contract' regarding longer term adaptation as compared to addressing immediate, visible risks. The roles of different institutions for long-term adaptation are not equally clear as they are in immediate responses to floods and storms.

Shifts in rural development policy

A major shift in government perspectives in relation to food security is that of recognizing that, because of a rapidly diversifying, urbanizing and commercializing economy, and a growing population, food security is currently more related to access to employment and entitlements rather than just food production. Particularly in the densely populated rural lowlands, a shift is underway from seeing rural development as synonymous with agriculture, to supporting instead the creation of different on-farm and off-farm livelihoods. This is reflected in new policies for rural development under the *Tam Nong* policy and 'New Village' programme. These new rural development policies promote improving economic, social and spiritual life in rural areas through a multifunctional view (see Figure 13.2 adapted from Decision No. 800/QD-TTg 2010). Rural development is expected to involve not only agriculture, but also industry, tourism, culture, social services and many aspects of economic, social and cultural life. These new perspectives suggest that even in rural areas the priority is no longer seen to be farming alone.

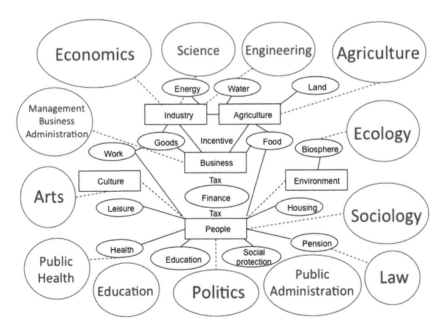

FIGURE 13.2 Multidisciplinary framework of the New Rural Development programme

In order to follow the new strategic orientation, a set of policies has emerged (Resolution 26/NQ-TW; Resolution No. 24/2008/NQ-CP; Decision No. 800/QD-TTg; Decision 491/QD-TTg; Decision 342/QD-TTg). The initial focus was on a standard set of infrastructural targets. This has been replaced by a recognition that the contribution of rural development to food security will mean different things in a peri-urban district with good access to a range of employment opportunities and markets, as compared to highland areas where such opportunities are inevitably constrained. Instead of automatically coming up with a list of infrastructure projects, local government has been urged to bring actors together to identify local needs and investment resources. The need to develop organizations that can drive decentralized rural development is noted, including a legal framework to support the establishment and capacity building for groups of farmers and rural people in vertical market linkages (e.g. commodity associations linking production and processing enterprises), in horizontal linkages (e.g. cooperatives, business associations, farmer associations), in local development linkages (e.g. rural development boards, resource management boards), in socio-political associations and in other rural organizations. The contours of these new organizations are still not clear, but they are likely to play a significant role in the future in linking food security, commercialization and climate change adaptation.

These rural development policies assume, and even promote, a massive demographic shift in the coming years from agriculture to services and industry, and with that probable urbanization. As noted above, such a strategy carries with it a different set of risks. In essence, the focus of rural development efforts implies a not unproblematic search for a new balance among production targets, diversification of agriculture and access to wage labour, each of which involves a different array of risks and opportunities.

Conclusions and implications for the future

The concept of food security in Viet Nam today is fundamentally different from what it was before *Doi Moi*. It has a different meaning for the state, for local government, for market actors, for farmers and for consumers. Among a significant proportion of the rural population, food security is still about rice production, but this is changing. For many today, the focus is more on diversified livelihoods. For some it is about managing climate risk. For others it is getting out of farming, either voluntarily or by being evicted, to allow the construction of factories and urban residential areas. In urban areas concerns about food safety are emerging as greater than concerns about food security *per se*. These changes suggest that the definitions of food security are likely to remain contested, as the meaning of food security is likely to reflect the changing actions and interactions among these different sets of actors over time.

All is not necessarily well with regard to food security (and the entitlements that underpin it) in Viet Nam today, given the recent slowdown of the economy and growing concerns in importing countries about the safety and climate footprint of the products they buy. Pressures on land are creating growing tensions and goal conflicts as it is recognized that expansion of commercial crops, industrialization and urban expansion do not necessarily represent 'win-win' propositions in terms of their impact on rural food security. New policies for multifunctional rural development remain unproven, and scenarios of climate change remain uncertain and contradictory. What is clear, however, is that the Vietnamese state has clearly been able to adapt to changes and challenges over the years, and has found ways to combine pragmatism with vision. Viet Nam has absorbed new ideas, but has not been led by donor fads and demands. Food security policies have been neither mere paper products nor straitjackets. They have been tools for a remarkable transformation in a short period of time.

References

Adger, N. (1999) 'Social vulnerability to climate change and extremes in coastal Vietnam', *World Development,* 27(2): 249–69.

Anh, Tuyet. (2013) *Increasing economic efficiency due to crop/fruit tree diversification in Mekong River Delta,* (in Vietnamese). Online. Available <www.nhandan.com.vn/kinhte/tin-tuc/item/20052402.html> (accessed 25 April 2013).

Belton, B., Haque, M. M., Little, D. C. and Sinh, L. X. (2011) 'Certifying catfish in Vietnam and Bangladesh: who will make the grade and will it matter?' *Food Policy,* 36: 289–99

Boateng, I. (2012) 'GIS assessment of coastal vulnerability to climate change and coastal adaption planning in Viet Nam', *Journal of Coastal Conservation,* 16: 25–36, Online. Available <DOI 10.1007/s11852–011–0165–0> (accessed 25 April 2013).

Cadihon, J-J., Moustier, P. Poole, N. D., Tam, P.T.G. and Fearne, A. P. (2006) 'Traditional vs. modern food systems? Insights from vegetable supply chains to Ho Chi Minh City (Vietnam)', *Development Policy Review,* 24(1): 31–49.

Carew-Reid, J. (2008) *Rapid assessment of the extent and impact of sea level rise in Viet Nam,* Climate Change Discussion Paper 1, Brisbane, Australia: International Centre for Environmental Management (ICEM).

CCCSC (2013) *Adjusting rice cultivation practices to adapt to early floods in lowland areas of Quang Binh province, Vietnam,* Centre for Climate Change Study in Central Vietnam (CCCSC) (Draft).

CNPR (Centre for Nature and People Reconciliation) (2008) *Da Nang: withdrawing 'hanging' lands* (in Vietnamese). Online. Available <www.thiennhien.net/2008/08/02/da-nang-thu-hoi-dat-detreo/> (accessed 11 April 2013).

DARA and the Climate Vulnerable Forum (2012) *Climate vulnerability monitor: a guide to the cold calculus of a hot planet,* second edition, Madrid: Fundación DARA Internacional.

Dasgupta, S., Laplante, B., Meisner, C., Wheeler, D. and Yan, J. (2007) *The impact of sea level rise on developing countries: a comparative analysis,* World Bank Policy Research Working Paper 4136.

Decision No. 100/1981/CT-BBT (1981) Decision on improvement of agricultural product allocation to individuals and labor groups given by the general secretary board, dated 13 Jan 1981.

Decision No. 342/QD-TTg (2013) Decision on modification of some national criteria on New Rural Development given in 491/QD-TTg by the prime minister, dated 20 Feb 2013.

Decision No. 491/QD-TTg (2009) Decision on approval of the national criteria for New Rural Development by the prime minister, dated 16 April 2009.

Decision No. 773-TTg (1994) Decision on exploitation and use of fallow lands, sediment land along river bands and sea coasts by the prime minister, dated 21 Dec 1994.

Decision No. 800/QD-TTg (2010) Decision on approval of the national target programme New Rural Development for the period 2010–2020 by the prime minister, dated 4 June 2010.

Direction No. 100-CT/TW (1981) Decision on improvement of agricultural product allocation to individuals and labor groups given by the general secretary board, dated 13 Jan 1981.

Figuié, M. and Moustier, P. (2009) 'Market appeal in an emerging economy: supermarkets and poor consumers in Vietnam', *Food Policy,* 34: 210–17.

Gainsborough, M. (2010) 'The (neglected) statist bias and the developmental state: the case of Singapore and Vietnam', *Third World Quarterly,* 30(7): 1317–28.

GSO (2010) *Result of the Vietnam households living standard survey 2010,* (in Vietnamese), General Statistics Office, 14.

Hanh P.T.T. and Furukawa M. (2007) 'Impact of sea level rise on coastal zone of Viet Nam', *University of the Ryukyus, Faculty of Science Bulletin,* 84: 45–59.

Hoa, T.T.N., Dung, D. T. and Son, N. H. (2008) 'Vietnam agriculture extension and market and information system', presentation at World Conference of Agricultural Information and IT, Atsugi-Tokyo, Japan.

Hoang Thi Van Anh (2009) 'Rural labour transformation after land withdrawing', (in Vietnamese), paper presented at National Workshop on Land Policy Related to Agriculture, Farmers and Rural Society, Ha Noi.

Le Van Mien, Ton That Phap and Hoang Nghia Duyet (2000) 'Aquaculture – its introduction and development', in V. J. Brezeski and G. F. Newkirk (eds), *Lessons from the lagoon: research towards community based coastal resources management in Tam Giang lagoon Viet Nam,* 115–132, Halifax and Nova Scotia, Canada: Coastal Resources Research Network (CoRR), Dalhousie University.

MARD (2009) *A strategy of agriculture and rural development for period 2011–2020,* Official Correspondence No. 3310/BNN-KH, 12 Oct 2009.

Maruyama, M. and Trung, L. V. (2007) 'Supermarkets in Vietnam: opportunities and obstacles', *Asian Economic Journal,* 21(1): 19–46.

Mergenthaler, M., Weinberger, K. and Qaim, M. (2009) 'The food system transformation in developing countries: a disaggregate demand analysis for fruits and vegetables in Vietnam', *Food Policy,* 34: 426–36.

MONRE (Ministry of Natural Resources and Environment) (2007) 'Wasteful use of land resources'. Online. Available <www.agenda21.monre.gov.vn/default.aspx?tabid=339&idmid=&ItemID=3914> (accessed 31 March 2014).

MONRE (2012) *Scenario on climate change and sea level rise,* Ministry of Natural Resources and Environment.

Moustier, P., Tam, P.T.G., Anh, D. T., Binh, V. T. and Loc, N.T.T. (2010) 'The role of farmer organizations in supplying supermarkets with quality food in Vietnam', *Food Policy,* 35: 69–78.

Nguyen Chu Hoi (2012) 'Economic opportunity in Mekong River Delta', (in Vietnamese). Online. Available <www.probity.com.vn/news/detail/42–0-thuy-san-the-manh-cua-kinh-te-dong-bang-song-cuu-long-pgs-ts-nguyen-chu-hoi-.html> (accessed 25 April 2013).

Nguyen Mau Dung (2011) *Strengthening vocational village, traditional village in order to conserve and develop off-farm and handicraft activities* (in Vietnamese), Department of Agriculture, Forest-Fishery-Salt Processing, MARD.

Nguyen Nguyen Hoai (2011) 'Viet Nam dykes – may become the world heritage'. Online. Available <http://thethaovanhoa.vn/van-hoa-giai-tri/de-viet-nam-xung-dang-la-di-san-nhan-loai-n20110112091830165.htm> (accessed 31 March 2014).

Nguyen Thi Hong Hanh, Nguyen Thanh Tra and Ho Thi Lam Tra (2013) 'Impacts of land withdrawal on local people's livelihood and jobs at Van Lam, Hung Yen', (in Vietnamese), *Journal of Science and Development – Hanoi University of Agriculture,* 11(1): 59–67.

Nguyen Viet Khoa, Vo Dai Hai, Nguyen Duc Thanh (2008) *Cultivation technology in slopping lands,* (in Vietnamese), Ha Noi: Agricultural Publishing House.

Pandey, S., Khiem N. T., Waibel, H. and Thien, T. C. (2006) *Upland rice, household food security and commercialization of upland agriculture in Vietnam,* Manila: International Rice Research Institute.

Rasmussen, O. S. and Tran, C. I. (2005) *Viet Nam fisheries and aquaculture marketing study,* Washington, DC: The World Bank. Online. Available <http://documents.worldbank.org/curated/en/2005/06/7191222/Vietnam-fisheries-aquaculture-marketing-study> (accessed 25 April 2013).

Resolution No. 24/2008/NQ-CP (2008) Resolution of Government on Approval of the Action Plan on Implementing Resolution 26/NQ-TW on Tam Nong.

Resolution No. 26/NQ-TW (2008) Resolution of the 7th Congress issued by the Session X Central Executive Committee on Agriculture, Farmers and Rural Development (Tam Nong).

Resolution No. 63/NQ-CP (2009) The Government Resolution on Ensuring National Food Security.

Truong Huu Quynh, Dinh Xuan Lam and Le Mau Han (2003) *General history of Viet Nam,* (in Vietnamese), Viet Nam Education Publishing House.

Tuan, Le Anh (2010) *Impacts of climate change and sea level rise to the agriculture-aquaculture system in the Mekong River Basin – a case study in the lower Mekong River Delta in Viet Nam,* Viet Nam: Can Tho University.

Vu Duy Giang (2012) 'The use of paddy and broken rice for feeding pig and poultry', (in Vietnamese), presentation at National Workshop on the Use of Low Quality Rice to Replace Corn in Livestock Husbandry, 4 Oct 2012, Ha Noi.

Yu, Bingxin, Zhu, Tingju, Breisinger, Clemens, Hai, Nguyen Manh (2010) *Impacts of climate change on agriculture and policy options for adaptation: the case of Vietnam,* International Food Policy Research Institute Discussion Paper 01015, August 2010.

Zhai, F. and Zhuang J. (2009) *Agricultural impact of climate change: a general equilibrium analysis with special reference to Southeast Asia,* Asian Development Bank Institute Working Paper Series 131, Manila: Asian Development Bank.

14

FOOD SECURITY AND INSECURITY IN AFGHANISTAN

Adam Pain

1. Introduction

On 20 September 2001, nine days after the attack on the United States on 11 September, the World Food Programme issued a press brief[1] stating that Afghanistan was teetering on the brink of widespread famine with nearly a quarter of its people desperately short of food and food aid stocks running out fast. Millions of Afghans were said to have been displaced by 20 years of war, and a severe drought was then in its third year. This was seen to be a catastrophe since, it was argued, some 85 per cent of the population depended on agriculture for survival. Several early warning signs of famine were detected such as reduced food intake, soaring prices, decimated livestock populations and increasing numbers of destitute people. This narrative of famine was carried through into the emergency response after the Taliban had been toppled from power in November 2001, reinforced by reports that dramatized the condition of Afghanistan's rural population (Lautze et al. 2002). Subsequent reports were to emphasize how Western intervention had averted a famine (US Department of State 2002).

Yet this image of collapse and ruin which played to a famine narrative is not well supported by the evidence. While there were claims in a 2001 Draft National Food Security strategy (Sloane 2001) of a complete collapse in rural production followed the destruction of the rural economy by Soviet bombing in the 1980s, other sources pointed to a gradual recovery of the rural economy so that by 1997, wheat production levels were back to 70 per cent of their pre-war level (Fitzherbert 2007). Pinney and Ronchini (2007) showed that the fall in wheat production in Afghanistan as a result of the drought did not lead to a dramatic decline in food supplies. Since the level of commercial wheat imports exceeded those of grain imported for food aid during the drought years (1997–2002) grain prices did not rise. Indeed food aid was estimated to contribute only about 8 per cent of the wheat available in the period 2001–4 (World Bank 2005). The disjuncture between the claims of a rural dependence on domestic wheat production, its collapse due to drought and war and the risks of famine and the evidence of a functioning wheat market clearly supported by demand requires explanation.

There is no doubt that there is evidence of chronic malnutrition in Afghanistan and the food price spikes of 2008 clearly had food security effects. But the diversity of Afghan rural livelihoods and the role of grain markets in the rural economy were poorly understood in 2001. As will be discussed the food securing effects of the rise and fall of the opium economy were largely ignored by the international community. Equally, the thrust of the rural reconstruction agenda has been so guided by a market driven agricultural transformation agenda that, paradoxically, the long decade since 2001 has seen an overall decline in the health of the rural economy rather than its improvement. Thus in 2013 the risks of food insecurity are if anything greater and more entrenched than they were in 2001, driven both by the failure of Afghanistan's state building agenda and the expanded risk environment that many rural households now face due to climate change as well as conflict.

This chapter explores these dynamics of food security in Afghanistan by first setting the context, drawing attention to Afghanistan's history, complexity and geographical variability. It then examines the complex risk environment of the country and the interplay between conflict induced risks, climate based risks and natural disasters. This leads into section 4 and a review of the evidence on food security outcomes which indicates high levels of food and nutrition insecurity. Section 5 explores the national wheat economy and the significant role of international trade in wheat on national wheat availability before discussing in section 6 the evidence on how urban and rural households have sought food security. The final discussion draws attention to the lack of priority given to achieving household food security in the state building exercise.

2. Afghanistan: The context

Afghanistan has been widely described as a failed stated (Ignatieff 2003) and slogans of destruction over 25 years of war, depleted social capital and devastation can be richly harvested from the narratives around Afghanistan.[2] But Afghanistan was never a completely failed state and Cramer and Goodhand point more to a long and episodic but conflict-ridden process of state formation and failure in Afghanistan (Cramer and Goodhand 2002: 886).

Central to this conflict have been the distinctive regional identities (Barfield 2010). There are four major regions, structured around distinct ethnic identities and ancient urban centres: Herat in the west, Kandahar in the south, Balkh (Mazar-e-Sharif) in the north and Kabul in the east, with a fifth region based around Peshawar that was taken by the British and left to Pakistan. These four centres of regional power are central to an understanding of Afghanistan's politics today.

While Abdur Rahman, the emir of Afghanistan from 1880 to 1901 through force may have brought these regions into a nascent Afghan state in the nineteenth century and used enforced settlement of the Pashtuns in the north to assist that process, underlying regional identities and social orders did not disappear. They most strongly reasserted themselves with the emergence of resistance to the Communist government and Soviet occupation after 1978 and consolidated themselves in the period of anarchy that emerged after the fall of President Najibullah in 1992. By the end of the Taliban rule (1996–2001), they had re-established themselves as centres characterized more by their outward economic linkages to neighbouring countries rather than to any centralizing pull of Kabul. That economic orientation has remained since 2001 and has been consolidated by the rise of regional power holders who have remained resistant to the emergence of a strong central government.

Abdur Rahman established centralized economic and political power in Afghanistan, through force. But subsequent history is a testimony to the failure of the state to gain control. There is no stronger evidence for the weakness of the Afghan state up to the 1970s than its inability to make the rural landscape legible (Scott 1998) by establishing cadastral surveys and land taxation systems.

The effects of the conflict from 1978 onwards for Afghanistan's population were various. While their prior experience of the state was its lack of presence apart from a rule of force, starting in 1978 it became the enemy. The war with the Soviet army led to considerable destruction of the rural economy and a major outflow of refugees to Pakistan and Iran. From the period 1978 to 1992 agricultural production declined, but the evidence does not support a picture of complete collapse. As a UNDP report in 1993 noted:

> the agricultural production systems of Afghanistan can only be described as robust and resilient. For fourteen years, from 1978 to 1992, rural production system in Afghanistan continued to support the remaining rural population under conditions of extreme difficulty. Although malnutrition and hunger were reported, this did not degenerate into . . . catastrophic situations.[3]
>
> *(UNDP unpublished report 1993)*

The period after 1992 saw considerable recovery of the agricultural economy stimulated by a large return of refugees and investment in rebuilding the damaged rural infrastructure and a revival in wheat production. The rise to power of the Taliban brought a level of security from 1996 that had been missing, but starting in 1997 a national-level drought gripped the country for nearly five years. This caused considerable hardship, but not a collapse into famine, and wheat markets continued to function. Under the new political conditions following the fall of the Taliban in 2001, households hoped for an economic recovery and that the new political dispensation and the reconstruction effort would provide security and deliver basic services that they had long been deprived of.

But the Afghanistan state building project since 2001 has not delivered (Suhrke 2006; Goodhand and Sedra 2007), highlighting 'the limitations of an orthodox development model' – democracy, good governance and market driven development – as a state building modality (Goodhand and Sedra 2007: 57). The transitions to security, a political settlement and strong socio-economic development have not been achieved and if anything the condition of Afghanistan may be worse than it was in 2001, although different. There is varied opinion as to the reasons for this state of affairs – the limitations of the original Bonn agreement, the confounding of a war on terror with a state building project, the creation of a rentier state through excessive aid (Suhrke's 'more is less' argument [2006)]) – and all are cited as the 'sins of omission and commission' (Goodhand and Sedra 2007: 57). A *de jure* state may have been re-formed with the trappings of democracy, but in effect post-2001 has seen the rise of an informal and shadow state.[4] Informal power relations dominate and are diffused through complex patronage networks and secured through access to an informal economy, of which the opium trade has been a significant dimension. A variable pattern of localized and extended non-state regimes[5] has been created.

There have been improvements in access to basic services funded by aid money and measurable progress has been made in the delivery of access to basic health and education, although delivery has been uneven across the country. While quality of health provision is measurably improving, albeit from a low level, major concerns over education quality persist. Programmes such as the National Solidarity Programme have brought a level of funding and provision of public goods to the village that has never been part of their prior experience, and benefits may have resulted from that both in terms of expectation and appreciation, although the durability of such perceptions may be in doubt. The plethora of other projects and programmes funding public goods provision for roads and other infrastructure has undoubtedly improved access to public goods (Pain 2012).

But the dividends and returns from nearly ten years of aid for Afghanistan have been meagre for many Afghans. Many of the rural and urban poor are certainly no better off than before and for many livelihood security is worse (Kantor and Pain 2011). The lack of employment and work in both the urban and rural labour markets is a testimony to this lack of progress.

3. The risk environment

The political trajectory of Afghanistan and its history of conflict points to an institutional environment of acute risk to rural households. While the state, prior to 1978 was able to maintain a monopoly of force, after 1978 for many it became the enemy and violence by the state against rural populations led to mass destruction, out-migration and a collapse of the rural economy. After the fall of the Najibullah regime in 1992, the conflict between the warring mujahidin factions and the emergence of a context in which commanders and warlords flourished, there was an expansion of this risk environment which lay the conditions for the rise of the Taliban. While their rule within the parts of the country where they gained physical control brought a measure of security to the rural landscape, in dealing with their enemies they were ruthless. Two notable examples stand out: first the laying waste of the orchards and irrigation systems of the Shomali plains (Coburn 2011) outside Kabul and second their siege of Hazarjat and the economic blockage that they imposed which brought about acute food insecurity in the region (Semple 1998). The Taliban frontline against Massoud also brought a siege economy to Badakhshan between 1998 and 2001 which, compounded by the effects of the drought, led to both inflation in food prices and acute hardship (Pain 2008).

During the years of conflict, many households were subject to predatory action by local powerholders who would not hesitate to seize land and livestock. Much of this predatory action was driven by ethnic based divisions and focused around the seizure of natural resources. For example the village of Khilar in Jurm district, Badakhshan, populated by a minority Ishmali population, lost its village pasture at this time to a neighbouring Sunni village (Pain and Kantor 2010: 24). In Faryab during the time of the Taliban, upstream Pashtun villages dammed the Andkhoy River thereby depriving downstream populations of their customary water entitlements. Similar actions and the creation of new irrigation canals by populations in districts upstream of the Balkh irrigation system have reinforced chronic water insecurity downstream (Pain 2007).

A war economy also drove a stripping of natural assets, in particular of trees and forests. Hardship forced individual households to harvest increasing amounts of trees for sale as fuel. In the south the illegal timber trade between Laghman and Pakistan has led to a substantial reduction in forest area over the last two decades (CPHD 2011).

The failure of the post-2001 Afghanistan state to establish a rule of law, let alone deliver a degree of social protection, has meant that the use of power often backed by the gun has persisted and insecurity is pervasive. As a consequence most households draw on social relationships as the one means by which some degree of security can be achieved. But that can carry severe costs since many households exist in a state of dependent security on more powerful patrons (Kantor and Pain 2012), and remain subject to acute uncertainty.

The sources of risk and uncertainty due to the actions of others are layered over a risky physical environment that has also a history of unpredictability. With rising populations and a rural economy in poor health, the effects of natural hazards and risks have become increasingly severe. Rainfall variability is a characteristic of semi-arid environments such as that of Afghanistan and there has been a long history of major or nationwide droughts. Afghanistan has limited storage capacity for water (CPHD 2011) and the winter rainfall recharges the mountain snow cover. This in turn feeds the rivers and groundwater resources that are critical for irrigated agriculture.

Four major country-wide droughts have been experienced in the last century – between 1898–1905, 1944–5, 1970–2 and 1998–2002. The drought of 1970–2, coupled with a hard winter which caused a major depletion of the livestock population, led to acute food insecurity and famine, and it was the lack of government response that has been attributed as a key causal factor in the coup that deposed King Zai Shah in 1972 in favour of his cousin, Daud Khan (Ruttig 2013). More localized droughts are frequent and occur every 3–5 years (CPHD 2011: 5). Annual wheat production is closely correlated with annual rainfall and has a high coefficient of variation given the significance of rainfed wheat to national production. In above average rainfall years such as in 2003, which produced the highest national level of wheat production on record, the higher production was largely because of the productivity of the rainfed crop.

Paradoxically, Afghanistan also experiences relatively high risks of floods, particularly in the north, which occur when high spring rainfalls generate flash floods. In the 19 years from 1992 to 2011 eight serious floods were reported that led to the displacement of more than 240,000 people (CPHD 2011: 76). Although the risk of flooding in Afghanistan is seen to be comparable[6] with that of the United States (CPHD 2011: 78), the rate of flood-related deaths is reported to be among the highest in the world, with an average of over 600 deaths per year due to flooding reported between the years 1980–2000 (CPHD 2011: 76). These high rates are attributed (CPHD, op.cit.) to a combination of poor or absent infrastructure and minimal disaster preparedness.

Climate change effects are likely to compound the risks of drought and floods and their impacts. Already there is evidence of shrinking snow cover in winter (glaciers are confined to the narrow eastern Wakhan panhandle of the country and therefore play a minor role in water supply), but the predictions of drier conditions and an increase in temperatures are likely to make rainfed cultivation of wheat even more variable and uncertain. The effects of climate change on irrigated agriculture will depend to a large extent on the degree to which organizational arrangements do or do not improve the equity and efficiency of water sharing and management. Under current governance circumstances where little significant change has been experienced (Thomas et al. 2012), decreasing water supply will exacerbate existing conflicts over water sharing.

As will be discussed in section 5, wheat prices have been remarkably stable since 2001, reflecting the major role that regional wheat markets have played in ensuring supply to Afghanistan. But with the tightening of supply in the region linked to global wheat markets, coupled with a poor growing season in 2008, the global price spikes in wheat were rapidly transmitted to Afghanistan, leading to a major price rise that did not fall in line with the decline in global wheat prices after the price spike. As will be discussed in the next section, this price spike had major food security effects on poor rural households, most of whom even under normal circumstances are net purchasers of grain. While surplus producers of grain will have benefited from the prices, in good years wheat prices have been depressed, limiting the incentives for surplus production. This has been one of the drivers for cultivating opium poppy. But even with opium poppy, which has had market support (credit, farm gate purchase, etc.) provided by opium traders, there have been major price risks, and agricultural commodity markets in general are seen by many rural households as a major source of risk, thus limiting their engagement (Kantor and Pain 2011).

4. Evidence on food security outcomes

The 2007–8 National Risk and Vulnerability Assessment (NRVA) provides the most recent poverty assessment and it estimates (Ministry of Economy and World Bank,

2010) that in 2007–8 the national poverty rate was 36 per cent, with about 9 million Afghans not able to meet their basic needs. Developed out of a national food security assessment established by the World Food Programme (Pinney and Ronchini 2007: 126–7), the NRVA has provided a national-level poverty assessment based on a representative household data set. The NRVA has been carried out over three rounds since 2003 and has progressively expanded its scope, methods and coverage. The third NRVA (2007–8) collected data for the first time on food consumption and dietary diversity in both the summer and winter, and its results point to key regional and seasonal differences in poverty and food security levels. Table 14.1 summarizes the data on food security outcomes as reported by the third NRVA for per capita daily calorie intake (as a measure of access to food), per capita protein intake and dietary diversity.

The evidence points to significant levels of food and nutrition insecurity. Overall some 29 per cent of the population are calorie deficient, 20 per cent have poor dietary diversity linked in part to the major contribution from wheat in the diet but there are also seasonal dimensions to this with higher levels of calorie deficiency (33 per cent) in the lean spring season. Spatial or locational effects are also significant. There are differences between urban and rural populations, with higher levels of calorie deficiency and higher levels of poor dietary diversity in rural areas. There are also differences between the plain, middle altitude and mountain regions, with high altitude populations suffering from significantly poorer food security outcomes. In Badakhshan and Lagman provinces more than 60 per cent of the population are experiencing calorie deficiency, and 46 and 57 per cent respectively are protein deficient.

But there are also major food security differences between the poor and nonpoor, with the poor defined as a household whose total value of food consumption is less than the poverty line. Over half of the poor are calorie deficient and about a third have diets that are poor in diversity and protein deficient. The significance of the different experience of food insecurity between the poor and non-poor can be found through an analysis of their response to the rise in food prices in 2008. The third NRVA caught the impact of this in the survey and found that a doubling of

TABLE 14.1 Food security outcomes in Afghanistan in 2007–8

	All	*Urban-Rural*		*Terrain Type*			*Non-Poor/Poor*	
		Urban	*Rural*	*Plain*	*Plateau*	*Mountain*	*Non-Poor*	*Poor*
Calorie Deficiency (%)	29	24	30	24	31	33	16	54
Protein Deficiency (%)	17	17	17	15	18	20	8	35
Poor Dietary Diversity (%)	20	14	22	16	26	21	14	32

Source: the Ministry of Economy and World Bank 2012: 4.

wheat flour prices led to a 43 per cent drop in food consumption, although calorie intake and dietary diversity showed a smaller decline (18 per cent and 19 per cent respectively) suggesting quality was traded for quantity. However, there were significant differences between households, with better off households reducing quantity more than quality while poor households (those in the bottom 25th percentile of distribution of real food consumption) with little scope to reduce quantity suffering major changes in the quality of their diet.

Finally, the levels of child malnutrition need to be noted. Some 54 per cent of Afghan children under age five experience chronic malnutrition, a level which is one of the highest in the world. About 72 per cent of children under age five also suffer from the deficiency of key micro-nutrients.

In sum, while no major famine has been experienced within the last three decades, for many households food security is precarious and there is little evidence that it has significantly improved since 2001.

5. The wheat economy of Afghanistan

Wheat is the main staple of Afghanistan, providing nearly 60 per cent of the total daily dietary caloric intake, the highest within the region (Chabot and Tondel 2011). Wheat consumption in Central Asia provides 37–60 per cent of the dietary caloric intake, signifying a considerable dependency on just one commodity for food security, which is not the case for other food insecure regions in West and East Africa for example (Chabot and Tondel 2011). The combination of the significance of wheat in the diet and the role of wheat imports in meeting food demand is a key characteristic of the wheat economy of Afghanistan, which will be briefly reviewed with respect to production and imports before the functioning of markets is explored in more detail.

Table 14.2 summarizes the available data on wheat production and imports and the estimates of total availability. The first point to note is that although the 12-year average of production constitutes about 77 per cent of average availability, in 2004 production only met 57 per cent of availability while in the previous year (2003) it provided 89 per cent of requirements. A core characteristic of wheat production in Afghanistan is the high inter-annual variability of production (a Coefficient of Variation of 0.31, Chabot and Tondel 2011: 9), which is the highest in the region. This is due to variability in rainfall conditions (linked to limited water storage capacity) and evidenced in particular by the fluctuation in the contribution from rainfed wheat to the national production. In good years rainfed cultivation can provide nearly 30 per cent of production, as in 2003. In poor years, such as 2008, it contributed only 8 per cent of production. There is marked variability in the production environment within the regions of the country, which is discussed below.

Wheat imports therefore have made an important contribution to overall wheat availability, averaging about 22 per cent of supply over the 12-year period with maximum contributions of 43 per cent of supply in 2004 and about 7 per cent

TABLE 14.2 Production, imports and total availability of wheat in Afghanistan

	Production (per thousand MT)			Imports (per thousand MT)			Availability
	Total	Irrigated	Rainfed	Imports	Commercial	Aid	
2000	1,469	1,329	140	888	650	239	2,357
2001	1,597	1,514	83	1,220	908	312	2,817
2002	2,686	2,110	576	860	500	360	3,546
2003	4,361	3,017	1,345	550	300	250	4,912
2004	2,293	1,867	426	1,712	1,300	412	4,005
2005	4,265	2,728	1,537	1,000	900	100	5,265
2006	3,711	2,902	809	1,064	980	84	4,775
2007	4,484	2,878	1,606	499	325	174	4,983
2008	2,623	2,406	217	1,614	1,402	212	4,237
2009	5,115	3,433	1,682	361	143	218	5,476
2010	4,532	3,082	1,450	607	525	82	5,139
2011	3,388	3,067	321	1,430	1,430	0	4,818
Average	3,377	2,527	849	984	780	222	4,361
Share %	100	75	25	100	79.3	20	
Share	77.4			22.5	17.9	5	100

Note: years shown in the table correspond to the calendar year of harvest.
Source: Central Statistics Office, Kabul.

of supply in 2009. What is evident is that food aid has played a relatively minor role (an average of 5 per cent over the 12-year period) in total wheat availability, although it contributed just over 18 per cent in 2004.

Given the limited wheat milling capacity within Afghanistan (Chabot and Dorosh 2007) a significant amount of the wheat is imported as flour. However, the sources of wheat grain and flour are linked to the regional production characteristics of Afghanistan and five broad wheat market regions have been identified. Only one of these regions is normally in surplus – the north, based around Mazar-e-Sharif. This wheat surplus from the north and international imports are traded both within the region and in the regional markets in western and central parts of Afghanistan. Although the wheat is imported from Uzbekistan, it is mainly sourced from Kazakhstan. All other regions of Afghanistan are generally wheat deficit and the major source of imports is from Pakistan: from Peshawar in the eastern part and from Quetta in the south (see Table 14.3). There is probably some trade of Iranian wheat to Herat when Iran is in surplus.

With respect to the north, Kazakhstan is Central Asia's largest wheat exporter meeting demand from the Commonwealth of Independent States in general and in particular from Tajikistan, Uzbekistan, Kyrgystan and Russia. The two

TABLE 14.3 Regional wheat markets in Afghanistan

Regional Market	Major Cities	Surplus/Deficit	Domestic Market Links	International Market Links
North	Mazar-e-Sharif	Surplus	To West and Central	From Uzbekistan through Mazar
West	Herat	Deficit	From Northeast and Southeast	From Iran
Central/East	Kabul, Jalalabad	Deficit	Weak links	From Pakistan (Peshawar)
Central–West Highlands	None	Deficit	Weak links	No trade
South	Kandahar	Deficit	To West	From Pakistan (Quetta)

Source: Chabot and Dorosh 2007: 346.

key additional countries that import Kazakh wheat are Afghanistan and Iran. In years of surplus Kazakhstan exports have been able to meet this demand but as a significant part of the Kazakh production is rainfed, its production fluctuates with rainfall availability. In 2008 when poor rains severely reduced Kazakh wheat yields, and also affected wheat production in the region and Afghanistan, supplies tightened and there was, in line with world prices, a significant rise in wheat grain prices. Although the Kazakh government exerts control over wheat production, marketing and exports, it draws on public stocks of wheat to keep flour and bread prices stable.

In contrast, the Pakistan wheat market, which is a major supplier to southern and central/eastern Afghanistan, is subject to heavy regulation by the Pakistani government to support farmers' income and generate stocks for consumers. Internal trade is controlled and prices internally have been relatively stable. Because of the relative importance of irrigated wheat to domestic production, Pakistani wheat production levels have been more stable than in Kazakhstan, but nevertheless in years of poor production and rising prices, as in 2004 and 2008, the movement of wheat internally and exports to Afghanistan have been restricted. In good years the government has subsidized exports. The effect of these interventionist policies has been to create disincentives to wheat production and milling in Afghanistan (Chabot and Tondel 2011: 20) since Pakistani prices of wheat and flour can be easily manipulated to undercut those prevalent in Afghanistan. However, this regulation has also encouraged an informal cross-border trade in flour.

A key question that has exercised analysts of these grain markets is the extent to which these regional grain markets are integrated, both between the grain exporting countries and major importing urban centres in Afghanistan, and internally within Afghanistan. High levels of integration are seen to reflect efficiency in the market and a lack of monopoly behaviour. The tool of analysis to assess whether markets are well integrated is that of price, and the degree to which prices in different locations move together or not is assessed through correlation coefficient measures. Figure 14.1 provides details on wheat price movements in the period 2005–10 in Kokshetau, (northern Kazakhstan), Dushanbe (Tajikistan) and Afghanistan, showing clearly the transmission of the 2008 commodity price spike, and the fact that prices peaked more strongly in Afghanistan and Tajikistan and did not fall in synchrony with those of Kazakhstan. Note should also be made of the rising prices since 2010 on account of a poor harvest in Kazakhstan.

Assessments of the correlation of wheat grain prices between the four trading centres for the period 2005–10 are provided in Table 14.4. Based on the fact that in US, Canadian and Argentine grain markets correlation coefficients are in the range of 0.92 to 0.95, Chabot and Tondel (2011: 25) concluded that the level of integration in the study markets was reasonably strong. Further they argued that if transportation costs and other expenses were taken into account, price differentials could be explained and that non-competitive market behaviour was unlikely. They noted evidence of rising transport prices in the region, which was likely to increase price differentials between locations.

A similar analysis of the degree of market integration between Pakistan and Afghan wheat grain prices pointed to a poorer degree of integration with lower correlation coefficients, which were attributed to the effects of Pakistan's trade policies.

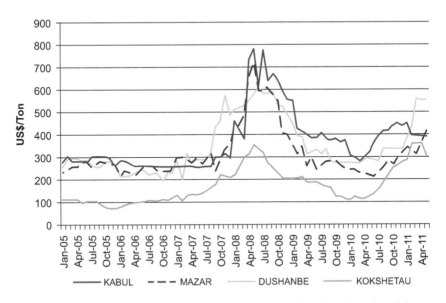

FIGURE 14.1 Wheat grain prices in Kabul, Mazar-e-Sharif, Dushanbe and Kokshetau (current USD/ton, 2005–10)

Source: FEWS Net 2011: 24; prices in Afghanistan and Tajikistan were sourced from WFP; KazAgroMarketing provided prices for Kazakhstan. Prices in local currencies were converted to US dollars using exchange rates from <www.oanda.com>.

TABLE 14.4 Correlation of wheat grain prices in Kabul, Mazar-e-Sharif, Dushanbe and Kokshetau

	Kabul	*Mazar-e Sharif*	*Dushanbe*	*Kokshetau*
Kabul	1	0.86	0.77	0.73
Mazar-e-Sharif	0.86	1	0.83	0.72
Dushanbe	0.77	0.83	1	0.86
Kokshetau	0.73	0.72	0.86	1

Source: FEWS Net 2011: 25.

TABLE 14.5 Correlation of wheat grain prices in four markets in Afghanistan

	Kabul	*Mazar-e Sharif*	*Herat*	*Kandahar*
Kabul	1			
Mazar-e-Sharif	0.89	1		
Herat	0.94	0.95	1	
Kandahar	0.96	0.84	0.92	1

Source: Chabot and Tondel 2011: 29.

In contrast, assessments of subnational market integration using correlation coefficients were used as evidence of a higher degree of internal integration of wheat markets within Afghanistan (see Table 14.5). An earlier study by FEWS Net (Schulte 2007) based on an analysis of the traders in the wheat and flour markets of Mazar-e-Sharif and Herat and their market share and profit margins, also concluded that wheat markets were relatively competitive and was supportive of the conclusions of Chabot and Tondel (2011).

However, a focus on efficiency within markets, as assessed by the use of correlation coefficients in prices, is an extremely poor measure of market performance and ignores the social institutions that underpin marketing systems (Ali Jan and Harriss-White 2012: 42). Even within an approach that prioritizes an efficiency role of markets over other roles, the use of correlation coefficients as a measure of efficiency is seen to be limited. Chabot and Dorosh (2007), who used correlation coefficients to assess wheat market integration in Afghanistan between 2002 and 2005, also argued that markets were relatively well integrated although the values of their correlation coefficients were lower for this earlier period. However, they noted the limitations of using correlation coefficients as a measure of market integration, as they do not take account of transaction costs and price variations in spatially integrated markets. Accordingly they also used co-integration analysis to examine whether spatially separate markets move together. On the basis of their results they concluded that they did but also noted limitations of the method and the lack of analysis of transaction costs.

Other analyses of market in Afghanistan (Lister and Pain 2004; Pain 2007; Pain 2011), which do not privilege efficiency and price correlations but pay more attention to the social institutions that underpin markets – those of geography, social identity and gender – point to the regulation of markets by social factors. Critically, given the almost universal role of informal credit underpinning relations between transacting partners, it is far from the clear that the assessment of price provides a robust assessment of conditions of competition. If the political market place is subject to heavy forms of regulation and rent seeking, so is the economic market place likely to be.

Be that as it may the implications of the critical role of regional trade in ensuring food availability in Afghanistan (as evidenced in 2008 from the effects of tightening the regional supply) are that Afghanistan's food security is vulnerable to the effects of shortfalls in production to the north, which are likely to correlate with poor years of production within the country, and also to the interventionist policies of Pakistan to the south. National-level food availability will not necessarily be secured by regional markets and trade in the future.

6. Household economies and food security

The NRVA (Ministry of Economy and World Bank 2010) points to the persistence of deep levels of poverty and food insecurity but provides little insight into how households have constructed their lives and trajectories of change.

For many, the previous two decades (1978–2001) had been marked by experiences of war, destruction and migration but not a complete collapse of the rural economy. During the first half of the 1990s there was considerable economic recovery until a major drought hit in 1998. This lasted until 2002 and had severe consequences, depleting the reserves of many households and driving further migration and diversification. It did not, however, lead to chronic food insecurity and grain markets did not fail. From 2002 onwards, years of good rains and the expansion of the opium economy brought a rural transformation with an emerging economy driven by reconstruction resource flows that supported the growth of cities. Since 2006, with the decline of opium, recurring droughts, high grain prices (in 2008) and rising insecurity spreading from the south, economic conditions for the poor in both rural and urban areas have declined.

Within this broader landscape of change, how have households fared and how have they addressed food security concerns? The evidence that exists is diverse, ranging from a very limited body of longitudinal studies, more numerous context- (and un-contextualized) and time-specific case studies, to broader cross-sectional data (notably from the NRVA), which usefully complements the more case-specific material. We first examine the evidence drawn from a longitudinal study to identify some key themes and issues, which are then used to drive an exploration of the wider evidence around these.

Urban livelihood trajectories

The cities of Afghanistan have grown significantly since 2001 with a high rate of urbanization driven by returning refugees, conflict and a move out of rural areas because of declining economic opportunities. The urban population is now estimated to constitute 23–30 per cent of the country's population. The rate of urbanization has challenged the limited capacities of municipalities to provide basic services to growing urban populations. In addition, and driven in part by a reconstruction boom and the recycling of opium profits from major drug traders, cities have been subject to significant land grabs by powerful people[7] and escalating land prices (Esser 2009; Pain 2011). Indeed much of the economic growth of Afghanistan's cities since 2001 evidences more the political obstacles to growth (rent-seeking behaviour, appropriation of public resources – see Pain 2011 and the later discussion of these issues for Mazar-e-Sharif) and the limited benefits that have accrued to the poor from this growth.

The effect of this, a 2006 study found (Beale and Schutte 2006), was that a majority of the sample households lived on the margins of cities in their informal and non-legal spaces. An estimated 60 per cent of Kabul residents lived in informal settlements compared to 30 per cent in Mazar-e-Sharif and 25 per cent in Herat (Islamic Republic of Afghanistan 2004), although these are likely to be minimum values. These households had limited access to basic services and social support, with adverse consequences for their ability to earn sufficient income and achieve livelihood security. While access to education was also difficult – in part because of the need for children to work to contribute to the household income – health issues were more severe. Poor availability of and access to health services, combined with environmentally unsafe living conditions – polluted water and the absence of sanitation – imposed a direct cost on households in terms of health problems and time and money devoted to trying to secure clean water.

For most, the major source of income came from working in the informal sector in irregular and insecure conditions. This came with a lack of protected rights and marked seasonal dimensions, indicating the need to create one's own marginal economic opportunities in the informal economy (Beale and Schutte 2006: 34). However, greater economic activity is to be found in Herat, given its location in relation to trade with Iran, which provides more opportunity for work. Only 5–6 per cent of workers had regular employment, between 20 and 60 per cent were self-employed[8] (the greater value was from Jalalabad) and 32–38 per cent worked as casual labourers,[9] with 20–36 per cent in Kabul and Herat respectively working at home (but less than 1 per cent in Jalalabad).[10] About 70 per cent of workers were employed for less than 200 days per year, but those with access to regular employment had a substantially higher number of workdays. Regular employment in Herat generated significantly higher income in contrast to the other two cities, while home-based and self-employed status was associated with correspondingly low income. Casual and self-employed workers in Herat – which included a high percentage of women (43 per cent) – worked more days per year, reflecting the lower average wage rate.

Few individual workers earned a sufficient wage to feed their households (these average seven members) and, as a result, women and children also had to seek income. Access to male labour was the crucial determinant of per capita income levels of households rather than whether the head of household was female or male. Many were self-employed, with home-based work by women particularly significant in Herat and Kabul, but not in Jalalabad. Erratic income hindered savings and the ability to plan and most had to borrow through informal sources in order to smooth consumption, with income often not meeting basic food consumption needs. The majority of households were constantly in debt.

Most households had strong social networks from which they borrowed, effectively providing their only source of social protection. But these horizontal networks based on reciprocity and in a context of shared poverty limit the extent that they can provide an effective safety net. Further, in some contexts where there is a majority ethnic group (e.g. Ismailis in Puli-Khumri or Uzbek in Mazar-e-Sharif) living with small pockets of other ethnic groups (e.g. Hazara), the minority ethnic groups can be excluded from the social networks. Variability in the economic fortunes of different households could be largely accounted for by differences in household structure and composition and in particular the availability of male labour. In sum, what characterized the lives of poor urban households was the multitude of livelihood-related risks to which they were exposed, associated with insecurity of land and housing tenure, lack of service provision, erratic and saturated urban labour markets and the seasonality of employment, making them particularly vulnerable to lack of income during the winter period.

Rural livelihood trajectories

In 2002–4, a short term longitudinal study followed the lives of 390 rural households in 21 villages in seven districts from seven contrasting provinces over an 18-month period. The findings pointed to strong evidence of livelihood diversification, the critical role of non-farm labour for the poor, the significance of labour migration, the presence of widespread debt and significant levels of household expenditure on health (Grace and Pain 2004).

In 2009, a sub-sample of households (64) were revisited to build understanding of the changes in the lives of the case-study households and the factors that had driven these. The study drew attention to both the contrasting provincial and village contexts and the relevance of context to understanding the constraints and opportunities that these households had faced. While Badakhshan has remained relatively secure, both Sar-e-Pul and Faryab had become increasingly insecure during the period of study as the insurgency spread to northern Afghanistan (Guistozzi and Reuter 2011). In Kandahar, insecurity is a daily feature of life.

The evidence showed that many households had gained improved access to basic services since 2002 but this was highly variable between provinces and between villages within provinces (Kantor and Pain 2011: 11). In Kandahar – characterized by both high levels of insecurity, strongly hierarchical and unequal social relations and social norms that restrict women's access to the public sphere and girls' access to education – access to schooling was limited to sons of wealthier households and women did not go to school. In Kandahar City, health facilities were available and households had access to safe drinking water through communal or private borehole wells. By contrast, in Badakhshan – with more equitable social structures and a longer history of education – one of the three villages had a long history of schooling and all boys and girls went to school; in a second village there was poorer access to school for girls because of geographical distance and in a third village, access to education was relatively recent. Health facilities were more available but variable in distance from the villages. The Sar-e-Pul patterns of access to basic services reflected those of Badakhshan, but overall access had improved.

While access to basic services had improved in most villages (but not in the more remote villages in Sar-e-Pul and Badakhshan), the key conclusion was that for a majority of the households economic security had declined significantly, and particularly for households in Sar-e-Pul and Badakhshan. For most, there had been a relative period of prosperity associated with the opium economy, but with the enforced closure of that in Badakhshan and Sar-e-Pul and a series of dry years, the rural economy had declined. Of the 64 sub-sample households, 13 had experienced had some improvements in livelihood security, 10 of these coming from the Kandahar villages. Forty-five of the households (70 per cent) were worse off than in 2002 with just 6 maintaining their economic status.

For those households that had prospered, only one had done so through agriculture (a Badakhshani household), and the basis for this was the rise of the opium economy combined with available household labour. Two other Badakhshani households had also prospered, having managed to leverage education into formal employment with a non-governmental organization. In Kandahar, an area well endowed for agriculture, improvement in household economic circumstances had come through the urban and not the rural economy. For the minority of land-rich households, inherited wealth and strong informal social connections had been the route into a triple diversification – employment with key power holders, access to construction contracts and overseas trade. For those without land, the vibrant but turbulent Kandahar urban economy provided opportunities on the margins for them, more so if the household was labour-rich.

For the static and declining households, household-specific events (e.g. health events), resource constraints of land or labour (or both) in a context of a declining rural economy, and off-farm employment opportunities had led to a number of responses, including able-bodied labour migration to Pakistan and Iran (both seasonally and longer term), enrolment in the Afghan National Police or Army and/or diversification into low-return and unreliable off- or non-farm activities. Where ill health or lack of labour precluded such a response, informal credit, debt and reduced food consumption were the result.

In summary, where agricultural conditions were best, the greatest opportunities for the poor had come through the urban and not the rural economy. In more agriculturally marginal areas that lacked a growing urban economy and had a weak or declining rural economy, being tied to the land has been a cause of impoverishment rather than a possibility for improvement. The one market opportunity there has been in agriculture and to which many responded – opium poppy cultivation – has been closed off, although the lessons that can be drawn from this are returned to later. The rural non-farm economy into which many have been forced to diversify has, for most, provided at best a means of survival but no more. The lessons from this study point not only to the significance of location in relation to access to basic services, but also to the role of urban economies, migration and employment as key dimensions of understanding how households have responded to changing circumstances over this last decade.

7. Discussion

Much of the discussion on Afghanistan's rural economy has been premised on the assumption of the critical role that agriculture has for the rural 80 per cent of the country's population and a hope for the role that a dynamic agriculture would play as an engine of growth for the country. The evidence however points to a flight out of agriculture, a high degree of diversification into the non-farm rural economy, or migration, and the fact that the majority of households are net purchasers of grain. Access to food remains the major challenge and nutrition outcomes are poor.

The one example of dynamic rural growth that can be found is in the spread of opium poppy during the first five years after 2001. It generated positive effects on household poverty and food security, created employment and had multiplier effects in the rural economy given the rise in household incomes and the resulting demand for other goods and services. It is the example that makes the case for the possibility of an agriculturally driven rural transformation. To simply define it in terms of illegality misses the point.

Agricultural transformations in the recent past, as with the Green Revolution in Asia, point to the need for a synergy between rising farm productivity; rising demand with strong state support for rural commodity markets through subsidies on prices; and credit and inputs to help reduce the risks around markets for smallholders and to thicken markets. All this should be based on a long history of prior investment in core public goods, including irrigation infrastructure, roads and electricity (Dorward et al. 2004). To the preconditions of adequate investment in public goods can be added two further core groups of factors closely associated with growth, although these are not necessarily the causal conditions for growth (Williams et al. 2011). The first is confidence in and security of property rights; the second is competitive, suitably regulated but internationally open markets operating under stable macroeconomic conditions.

Self-evidently, many of these basic preconditions do not exist in Afghanistan. There are severe limitations in the basic stock of public goods although these have improved over the last decade; macroeconomic conditions have been far from stable; there is little sentiment within donor policy for subsidies on prices, credit or inputs; the deep level of poverty in rural areas is a major constraint on demand and there is little evidence that there is a growing urban demand. Further, the risk environment, both climate and conflict induced, remains substantial.

But remarkably low priority has been given to ensuring household food security in the state building exercise and there has been resistance to ideas of national self-sufficiency by donors (Pain 2009). Comparative evidence supports a view that ensuring national food security might be a critical first step in state building and economic development (Chang 2009). This has left Afghanistan and its population increasingly vulnerable to food insecurity. With a global rise in staple prices likely; regional and domestic production susceptible to climate change effects of drier periods; poorer supply of irrigation water; and the likely control of exports by neighbouring countries under conditions of shortfall; food insecurity is likely to intensify. As the reconstruction effort falters and international withdrawal gathers pace, the legacy is a state with not much developmental competence. More likely (Goodhand 2012) there will be a consolidated oligarchy of uncertain stability or a descent into a durable disorder. The food security implications of this are grim and a likely outcome is the resurgence of the one source of food security that has been on the agenda over the last decade – opium poppy.

Notes

1 <www.afghanistannewscenter.com/news/2001/september/sep19z2001.html> accessed 21 January 2013.
2 'More than two decades of conflict and three years of drought have led to a widespread human suffering and massive displacement of people in Afghanistan. Many parts of the country are vulnerable to famine, the infrastructure base has been destroyed or degraded and human resources have been depleted. State institutions have become largely non-functional and the economy increasingly fragmented. The social fabric has been weakened considerably and human rights undermined, with women and minorities being the principal sufferers' (Asian Development Bank et al. 2002: 1).
3 United Nations Development Programme, Kabul, Unpublished Report, 1993.
4 Conceptualized here for the context of civil war where the state is wholly informal (Wood and Gough 2006), it has also been applied to more formal states (Harriss-White 2003: 89) where it 'comes into being because of the formal state and it coexists with it'.
5 Regimes, either formal or informal in the sense that Wood and Gough (2006: 1698) apply the term, are 'rules, institutions and structured interests that constrain individuals'.
6 Although there are major discrepancies between the sources that report this (CPHD 2011: 78).
7 The example of Sherpur in central Kabul is widely cited (Schutte and Bauer 2007: 3).
8 Comprising petty trade, manual card pulling, trading recyclable materials, selling phone top-up cards or tonka/rickshaw driving, etc.
9 Unskilled labour, goods loading, brick making, car washing, etc.
10 Washing clothes, weaving carpets, tailoring, embroidery work, spinning wool, domestic cleaning etc.

References

Ali Jan, M. and Harriss-White, B. (2012) 'The roles of agricultural markets. A review of ideas about agricultural commodity market in India', *Economic and Political Weekly,* 47(52): 39–52

Asian Development Bank, United Nations Development Programme and the World Bank (2002) *Afghanistan: Preliminary Needs Assessment for Recovery and Reconstruction,* Draft.

Barfield, T. (2010) *Afghanistan. A Cultural and Political History,* Princeton and Oxford: Princeton University Press.

Beale, J. and Schutte, S. (2006) *Urban Livelihoods in Afghanistan,* Synthesis Paper Series, Kabul: Afghanistan Research and Evaluation Unit (AREU).

CPHD (2011) *Afghanistan Human Development Report 2011. The Forgotten Front: Water Security and the Crisis in Sanitation,* Kabul: Centre for Policy and Human Development (CPHD).

Chabot, P and Dorosh, P. (2007) 'Wheat markets, food aid and food security in Afghanistan', *Food Policy,* 32: 334–53.

Chabot, P. and Tondel, F. (2011) *A Regional View of Wheat Markets and Food Security in Central Asia with a Focus on Afghanistan and Tajikistan,* Famine Early Warning Systems Network, World Food Programme and the United States Agency for International Development.

Chang, H.-J. (2009) 'Rethinking public policy in agriculture – lessons from history, distant and recent', *Journal of Peasant Studies,* 36(3): 477–515.

Coburn, N. (2011) *Bazaar Politics: Power and Pottery in an Afghan Market Town,* Stanford, California: Stanford University Press.

Cramer, C. and Goodhand, J. (2002) 'Try again, fail again, fail better? War, the State, and the 'post-conflict' challenge in Afghanistan', *Development and Change,* 33(5): 885–909.

Dorward, J., Kydd, J., Morrison, J. and Urey, J. (2004) 'A policy agenda for pro-poor agricultural growth', *World Development,* 32(1): 73–89

Esser, D. (2009) *Who Governs Kabul? Explaining Urban Politics in a Post-War Capital City,* Crisis States Working Paper No. 43, London School of Economics.

FEWS NET (2011) *A Regional View of Wheat Markets and Food Security in Central Asia with a Focus on Afghanistan and Tajikistan,* Kabul: USAID, Department for International Development (DfID) and World Food Programme (WFP).

Fitzherbert, A. (2007) 'Rural resilience and diversity across Afghanistan's agricultural landscapes' in Pain, A. and Sutton, J. (eds), *Reconstructing Agriculture in Afghanistan,* 29–48, London and Rome: Practical Action and Food and Agriculture Organization (FAO).

Goodhand, J. (2012) *Contested Transitions: International Withdrawal and the Future State of Afghanistan,* Oslo: Norwegian Peacebuilding Resource Centre.

Goodhand, J. and Sedra, M. (2007) 'Bribes or bargains? Peace conditionalities and "post-conflict" reconstruction in Afghanistan', *International Peacekeeping,* 14(1): 41–61.

Grace, J. and Pain, A. (2004) *Rethinking Rural Livelihoods in Afghanistan,* Synthesis Paper Series, Kabul: Afghanistan Research and Evaluation Unit (AREU).

Guistozzi, A. and Reuter, C. (2011) *The Insurgents of the Afghan North: The Rise of the Taleban, the Self-Abandonment of the Afghan Government and the Effects of ISAF's 'Capture-and-Kill Campaign',* Kabul: Afghanistan Analysts Network.

Harriss-White, B. (2003) *India Working: Essays on Society and Economy,* Cambridge: Cambridge University Press.

Ignatieff, M. (2003) *Empire Lite: Nation-Building in Bosnia, Kosovo and Afghanistan,* London: Vintage.

Islamic Republic of Afghanistan (2004) *Securing Afghanistan's Future: Urban Technical Annex,* Kabul: Islamic Republic of Afghanistan.

Kantor, P. and Pain, A. (2011) *Running out of Options: Tracing Rural Afghan Livelihoods,* Synthesis Paper Series, Kabul: Afghanistan Research and Evaluation Unit (AREU).

Kantor, P. and Pain, A. (2012) 'Social relationships and rural livelihood security in Afghanistan', *Journal of South Asian Development,* 7(2): 161–82.

Lautze, S., Stites, E., Nojumi, N. and Najimi, F. (2002) 'Qaht-e-pool: a cash famine: food insecurity in Afghanistan, 1999–2002', Feinstein International Famine Centre, Medford, MA:

Tufts University. Online. Available <http://famine.tufts.edu/research/natsios.html> (accessed 15 January 2013).

Lister, S. and Pain, A. (2004) *Trading in Power: The Politics of 'Free' Markets in Afghanistan,* Briefing Paper Series, Kabul: Afghanistan Research and Evaluation Unit (AREU).

Ministry of Economy and World Bank (2010) *Poverty Status in Afghanistan: A Profile Based on National Risk Vulnerability Assessment (NRVA) 2007–8,* Kabul: Islamic Republic of Afghanistan and World Bank.

Ministry of Economy and World Bank (2012) *Poverty and Food Security in Afghanistan: Analysis Based on the National Risk Vulnerability Assessment of 2007–8,* Kabul: Islamic Republic of Afghanistan and World Bank.

Pain, A. (2007) *Water Management, Livestock and the Opium Economy: The Spread of Opium Poppy Cultivation in Balkh,* Case Study Series, Kabul: Afghanistan Research and Evaluation Unit (AREU).

Pain, A. (2008) *Opium Poppy and Informal Credit,* Issues Paper, Kabul: Afghanistan Research and Evaluation Unit (AREU).

Pain, A. (2009) *Policymaking in Agricultural and Rural Development,* Briefing Paper, Kabul: Afghanistan Research and Evaluation Unit (AREU)

Pain, A. (2011) *Opium Strikes Back: The 2011 Return of Opium in Balkh and Badakhshan Provinces,* Case Study Series, Kabul: Afghanistan Research and Evaluation Unit (AREU).

Pain, A. (2012) *Livelihoods, Basic Services and Social Protection in Afghanistan,* Secure Livelihoods Research Consortium Working Paper 3, London: Overseas Development Institute (ODI).

Pain, A. and Kantor, P. (2010) *Understanding and Addressing Context in Rural Afghanistan: How Villages Differ and Why,* Issues Paper, Kabul: Afghanistan Research and Evaluation Unit (AREU).

Pinney, A. and Ronchini, S. (2007) 'Food security in Afghanistan after 2001: from assessment to analysis and interpretation to response', in Pain, A. and Sutton, J. (eds), *Reconstructing Agriculture in Afghanistan,* 119–56, London and Rome: Practical Action and Food and Agriculture Organization (FAO).

Ruttig, T. (2013) *How It All Began: A Short Look at the Pre-1979 Origins of Afghanistan's Conflicts,* Kabul: Afghanistan Analysts Network.

Schulte, B. (2007) *Northern Wheat Trader Survey and Afghan Food Security,* Special Report by the Famine Early Warning Systems Network FEWS NET, Kabul: USAID and World Food Programme (WFP).

Schutte, S. and Bauer, B. (2007) *Reduce High Levels of Risks on Poor Families Coping to Make a Living in Afghan Cities,* Policy Note, Kabul: Afghanistan Research and Evaluation Unit (AREU).

Scott, J.C. (1998) *Seeing Like a State: How Certain Schemes to Improve the Human Condition Have Failed,* New Haven and London: Yale University Press.

Semple, M. (1998) *Strategies for Support of Sustainable Rural Livelihoods for the Central Highlands of Afghanistan: A Study to Identify Opportunities for Effective Assistance to Reinforce People's Livelihood Strategies,* Draft, Islamabad: Pattan Development Organization, Mimeograph.

Sloane, P. (2001) 'Food Security for Afghanistan', revised version of paper presented at the International Conference on Analytical Foundations for assistance to Afghanistan, UNDP and the World Bank, Islamabad, Pakistan, 5–6 June, Mimeograph.

Suhrke, A. (2006) *When More Is Less: Aiding State-Building in Afghanistan,* Working Paper 26, Madrid: FRIDE.

Thomas, V. with Mumtaz, W. and Azizi, M.A. (2012) *Mind the Gap? Local Practices and Institutional Reforms for Water Allocation in Afghanistan's Panj-Amu River Basin,* Case Study Series, Kabul: Afghanistan Research and Evaluation Unit (AREU).

US Department of State (2002) Transcript: 'USAID chief says famine averted in Afghanistan'. Online. Available <http://cryptome.org/af-no-famine.htm> (accessed 13 August 2013).

Williams, G., Duncan, A., Landell-Mills, P. and Unsworth, S. (2011) 'Politics and growth', *Development Policy Review,* 29(S1): S29–S55.

Wood, G. and Gough, I. (2006) 'A comparative welfare regime approach to global social policy', *World Development,* 34(10): 1696–712.

World Bank (2005) *Enhancing Food Security in Afghanistan: Private Markets and Public Policy Options,* Washington, DC: World Bank.

15

FOOD SECURITY AND INSECURITY IN INDIA

Amit Mitra

Introduction

> Between one half to two thirds of India's people do not get food of the right type, and between one quarter to one third do not even get enough quantity to eat in order to sustain a healthy active life. Add to this the staggering magnitude of malnutrition and actual hunger, the accelerating growth of population which even on a conservative assumption is expected to reach 625 million by 1980 and the one billion mark by 2000, and one has the colossal dimension of India's food problem.
>
> *(Sukhatme 1965: 90)*

This somewhat Malthusian view of India's food security challenge, from the then director of FAO's Statistics Division, did not foresee the Green Revolution, which through an inter-linked series of agro-technological innovations in the 1960s and 1970s led to the widespread introduction of High-Yielding Varieties (HYVs) of cereal crops. The Green Revolution transformed the supply of food grains in India, placing the nation in a much better position to feed itself. More than four decades later India has achieved food self-sufficiency and was in 2012 the world's largest rice exporter (Pritchard et al. 2013: 41; Chandrasekhar 2012). At the same time, food stocks held by the Food Corporation of India stood at 76 million tonnes, 138 per cent more than the stipulated 39 million tonnes buffer stock.

But while national food security has been achieved, many households remain food insecure. Food insecurity and malnutrition in India co-exist with bulging domestic food grain facilities and strong export markets. There are more malnourished people in India than in any other country in the world. In 2012, 217 million or 17.5 per cent of India's population was malnourished, leading Pritchard et al. (2013: 19) to comment: '[Based] on these statistics, if the undernourished people in

India constituted a single country, it would be the fifth most populous country in the world'.

India's economic growth over the last two decades has been phenomenal, leading to images of 'India shining'. But the high economic growth rates have not been accompanied by growth in employment and labour absorption into industry, pointing to a failed agrarian transformation. Poverty and hunger have not been reduced to the extent that might be expected. As Pritchard et al. (2013: 2) note, '[S]een in international terms, India's GDP growth is contributing too anaemically to the reduction of poverty and food security. This failure has been termed as the Indian "enigma"' (see also Ramalingaswami et al. 1996; Headey et al. 2011; Gillespie et al. 2012; Walton 2009). As Pritchard et al. point out (2013: 2), '[F]or most developing countries the prevalence of children underweight for age (a good measure of under-nutrition) falls by roughly half the rate of GDP growth. Had international trends applied to India, the nation's average annual growth of 4.2 per cent between 1990 and 2005 should have reduced the prevalence of underweight children by 2.1 per cent annually or over 27 per cent in total over this 15-year period. In fact, this measure of food insecurity fell by just a meagre 10 per cent between 1990 and 2005.'

According to the 2011 *Global Hunger Index Report* (IFPRI 2011), India continues to be in a category of those nations where hunger is 'alarming'. The hunger index between 1996 and 2011 rose from 22.9 to 23.7, while 78 of the 81 developing countries studied – including Pakistan, Bangladesh, Viet Nam, Kenya, Nigeria, Myanmar, Uganda, Zimbabwe and Malawi – have all succeeded in reducing hunger. The 2011 hunger and malnutrition report (HUNGaMA 2011) shows that the proportion of children that are malnourished in the 100 poorest districts of India continues to be over 40 per cent (cf. Saxena 2012).[1]

On 12 September 2013, after prolonged debate between political parties, civil society organizations and activists, both houses of the Indian Parliament passed the National Food Security Act (NFSA), which is seen to enshrine a right to food (hence it is also known as the Right to Food Act).[2] The NFSA aims to provide subsidized food grains to some 67 per cent of India's 1.2 billion people (TOI 2013a), and eligible households will be able to purchase five kilograms of food grains per person every month at highly subsidized rates.[3] Pregnant women and lactating mothers will be entitled to a daily free meal during the pregnancy and six months after childbirth, and maternity benefits of at least Rs 6000. Provision has been made for free meals for children (6 months to 14 years) as well (NFSA 2013).

India thus represents a series of paradoxes with respect to food security. On the one hand it has achieved national grain self-sufficiency, but the levels of household food insecurity remain high. High rates of economic growth have been achieved but poverty and food insecurity have not been reduced. As will be discussed, the contribution of agriculture to GDP has declined significantly but an agrarian transformation has not been achieved. Although India has now enacted a bill enshrining the right to food, many are doubtful as to whether it will achieve what it claims. This chapter investigates these paradoxes leading to an assessment of what the

NFSA might achieve. But it starts with a review of the state of food and nutritional insecurity in India, pointing to significant regional and social variability in these dimensions.

Patterns of food and nutritional insecurity in India

Despite the impressive growth in output of food grain, the per capita availability of cereals and pulses has declined over the last 15 years (Mukherjee 2007). Average daily food consumption should provide about 2,400 kcals per person (FAO 2008), but for the poorest 30 per cent of the Indian population food intake provides only about 1,600 kcals. Even though more than a third of Indian men and more than half of Indian women are underweight, over 15 per cent of the population is estimated to be overweight and obese (IIPS 2009). India thus faces a double burden of under- and over-nutrition (Gillespie et al. 2012).

Women and children are particularly nutritionally insecure. Nearly 52 per cent of married women in India between the ages of 15 and 45 are anaemic, as compared to only 20 per cent in sub-Saharan Africa (Mukherjee 2007). Over 74 per cent of all children less than three years of age suffer from anaemia, with the percentage rising as high as 84 per cent in Haryana. Nearly 53 per cent of India's children under the age of five are malnourished, as compared to 32 per cent in sub-Saharan Africa. The results of the third (2005–6) National Family Health Survey (NFHS-3) (IIPS 2009) showed that almost half (46 per cent) of all children aged less than three are underweight and almost 80 per cent of children in the age group 6–35 months are anaemic. Undernourishment amongst children has been a direct contributor to 54 per cent of all childhood deaths (Arnold et al. 2009; Pritchard et al. 2013). Forty-four per cent of these deaths were connected to 'mild to moderate under-nutrition, making the children more vulnerable to mortality from a range of health problems including measles, malaria, diarrhoea and the risk of infection. Acute under-nutrition, in which the absence of food is linked in a direct way to the cause of death, was estimated as the cause of 11 per cent of child mortality in India' (Pritchard et al. 2013: 23).

Regional patterns

Food insecurity in India has strong regional as well as seasonal dimensions. Using data from the NFHS-3 (IIPS 2009) and the 61st round of the National Sample Survey (NSS) in 2004–5, Menon et al. (2009) constructed a State Hunger Index (SHI). The index was constructed around three inter-linked dimensions of hunger – inadequate food consumption, the proportion of underweight children and child mortality rates. The SHI, calibrated to international indices, showed 'that though India's overall national hunger performance ranks it akin to Burkina Faso and Zimbabwe (66th in a list of developing countries), the best performing state (Punjab) is equivalent to the 33rd-ranked developing country (Nicaragua) and the worst performing state (Madhya Pradesh) is equivalent to the 81st developing country (Chad)'.

The Chennai-based MS Swaminathan Research Foundation (MSSRF) has also computed a Composite Index of Food Insecurity (CIFI) using seven indicators: percentage of women with anaemia, percentage of women with chronic energy deficiency, percentage of stunted children, percentage of population daily consuming less than 7,940 kjoules, percentage of households without access to safe drinking water and percentage of households without toilets on the premises (Athreya et al. 2008: 47). The data for the first five of these indicators were drawn from the NFHS-3 and the last two from the 2001 national census.

Minor differences between the SHI and the CIFI notwithstanding, the overall pattern both indices highlight is a band of states – from Gujarat in the west, Madhya Pradesh and Chattisgarh in the centre and Jharkhand, Orissa and Bihar in the east – that have the worst food security levels in the country. Madhya Pradesh, Chattisgarh, Orissa and Jharkhand all have substantial proportions of scheduled tribe (ST)[4] populations, large forested areas and historically low levels of infrastructure. Bihar traditionally has high levels of unequal resource access, especially of land, and is prone to disasters such as floods (Pritchard et al. 2013: 28).

Within each state, there are some districts that are more vulnerable than others to food insecurity. Drèze and Khera (2012) constructed a Human Development Index (HDI) and 'Achievements of Babies and Children' (ABC) index at the district level. Comparison of the HDI and the ABC showed how the cumulative impact of the living conditions (HDI) affects child-related deprivation (the ABC index). The results suggest that pre-existing patterns of spatial inequality drive unequal life opportunities for the next generation (Drèze and Khera 2012; Pritchard et al. 2013: 30) and enable identification of clusters of inequality in northern and eastern India.

Particularly deprived are eight districts of southern Orissa, known as the KBK region after the three districts (Kalahandi, Bolangir and Koraput) they originally belonged to and which have had a long history of drought. Kalahandi is at the centre of this belt, 'both geographically and in terms of Indian political discourse about hunger and starvation' (Pritchard et al. 2013: 30). Banik analysing starvation in India noted: 'Kalahandi has become synonymous with drought and starvation' (Banik 2007: 5) and that as recently as 1996–7, a drought saw 94 people officially recorded as having died from starvation (unofficially, the estimate was more than 300 [Banik 2007: 6]). Subsequent years saw more reported deaths from starvation (in addition to high levels of undernourishment and distress migration), leading to 'the projection of Kalahandi as India's starvation capital', with 'almost all Prime Ministers of India' over recent years visiting there (Banik 2007: 6, cited in Pritchard et al. 2013: 31). Overlying drought proneness,'Banik argues that a particular combination of features embedded at the district level – a history of maladministration, land ownership inequalities [and] high proportions of ST and SC[5] populations' (Pritchard et al. 2013: 30) contributes to the hunger of the area. Essentially, these drivers hold true for almost all the 'hunger districts'.

The lean periods in the agricultural year reduce rural labour's access to income and reduce its nutritional intake. Food insecurity is not uniform throughout the year (Mukherjee 2007). Although the intensity and duration of the food insecurity periods may differ across states, even under the best of conditions rural households are food insecure for at least three months in a year, and the months when people need the highest energy intake is at the peak of labour demand for crop cultivation. Consumption of food is, therefore, lowest during the period when the poor have to work the *hardest*. Even when food intake in the non-food insecure months is sufficient in quantity, the quality of food and nutritional composition is an area of major concern because of the dominance of starch in the diet. People may not be starving in these months but their nutritional intake is seriously jeopardized by the poor nutritional quality of their diet and limited intake of protein and fibre.

Urban poverty

About 377 million (31 per cent) of India's 1.2 billion population (Office of the Registrar General and Census Commissioner 2011) live in urban areas and this is increasing at a faster rate than its total population. By 2030 it is estimated that 41 per cent (575 million) of the population will be living in cities and towns. Economic development and urbanization are closely linked. In India, cities contribute over 55 per cent to GDP, and urbanization has been recognized as an important component of economic growth (cf. Mukhopadhyay and Revi 2009). Over 80 million poor people were living in the cities and towns of India and an estimated 61.80 million people lived in slums in 2001 (UNDP 2009: 2) with limited access to basic water and sewerage facilities. Around 25 per cent of the urban population nationally is below the poverty line and there is evidence that urban poverty is also deepening, linked to limited opportunities for unskilled workers.

However, as a number of sources have noted (Athreya et al. 2010; Chatterjee et al. 2012), urban areas have been neglected in food security studies. Yet there is evidence that there is rising urban inequality which, combined with a neglect of investment in urban public health and increasingly insecure employment, is giving rise to a serious deterioration in food security for many of the poorest households. Food insecurity is likely to be greater in smaller towns than larger cities. About half of the women in urban areas are estimated to be anaemic and suffering under-nutrition (Athreya et al. 2010: xiv) and the extent of child under-nutrition remains high. Nationally over 37 per cent of urban children aged 6–36 months are stunted, with urban areas in Maharashtra recording levels of 40 per cent compared to 27 per cent in Kerala (Athreya et al. 2010: 72). Chatterjee et al. (2012) found that nearly 60 per cent of urban slum households in Mumbai were severely food insecure, and households where the woman was the main earner were particularly likely to experience severe food insecurity.

As Athreya et al. (2010) note, the key issues that characterize urban food insecurity are those of access and absorption. There is as yet no urban equivalent of the National Rural Employment Guarantee scheme and the dominance of casual, poorly paid work in the informal sector is a critical constraint to food access for the urban poor. Equally, the lack of access to safe drinking water and the lack of availability of health services contribute to poor absorption of food.

Social patterns

Underlying the regional patterning of food insecurity are social patterns of inequality. The scheduled tribes (ST) and scheduled castes (SC) of India comprise nearly 99 per cent of the nation's 37.2 per cent of households below the poverty line (GOI 2009) and are characterized by intergenerational poverty and food insecurity. It is rare to see an upper caste household in India, even when landless, suffering from food insecurity.

The levels of exclusion and caste discrimination that these social groups experience are well shown by the findings of the Indian Institute of Dalit[6] Studies survey conducted in 531 villages of five states: Rajasthan, Uttar Pradesh, Bihar, Andhra Pradesh and Tamil Nadu in 2003. The government of India, as will be discussed below, has long been running a mid-day meal scheme (MMS) and public distribution system (PDS) for food. But an assessment of SC access to these government programmes revealed significant patterns of inter-state variation, where greater levels of dalit access to MMS and PDS corresponded with a lower incidence of exclusion and caste discrimination. Higher proportions of dalits in the village populations correlated with their greater access to these government programmes (Thorat and Lee 2005: 4201).

The physical quality of life indicators of the scheduled tribes are also poor. This includes high maternal, neo-natal and infant mortality rates. It is rare to find men above 60 and women above 70 in India's tribal districts (except for the north-eastern states). Rai (2005) found that of 1,000 ST households surveyed in 40 villages in Rajasthan and Jharkhand, 99 per cent faced chronic hunger. Another 25.2 per cent of households had faced severe food rationing during the week prior to the survey. A further 24.1 per cent of households had rationed consumption throughout the month prior to the survey. Over 99 per cent of the ST households had lived with one or another level of chronic hunger and food insecurity during the whole previous year. Of the 500 sample ST households surveyed in Rajasthan, not a single one had secured two meals in one day for the whole previous year (Rai 2005).

Scheduled tribes have also been fast losing their lands through forced displacement and their forest rights as a result of developmental activities (big dams, special economic zones) and the enforcement of conservation areas. Annually, over 500,000 people are displaced by infrastructure construction, including hydroelectric and irrigation projects, mines, industrial complexes and super-thermal[7] and nuclear power plants (Kothari 1997). Estimates for the numbers displaced by large projects since independence in India range from 21 million to 50 million (Hemadri et al. 2000), but it is estimated that STs comprise some 40 per cent of the displaced, though they account for only 8 per cent of India's population (Fernandes and Thukral 1989; Mitra and Rao 2009; Mehta 2009). Even in states like Kerala that have had well-known land reforms, the struggle for restoration of alienated lands remains a losing battle for its 0.321 million ST population, 80 per cent of whom are landless (Sreerekha 2010).

Gendered differences

Hunger is related to impoverishment, and factors such as access to education and health also have a significant impact on the status of nutrition. A major link in the intergenerational cycle of impoverishment is the low status of women. As the National Family Health Survey 2009 shows, half of women aged 15–49 years suffer from low levels of nutrition and education, which can lead, as will be discussed below, to malnutrition among children. In turn, as Rampal and Mander (2013: 51) argue, '[U]ndernourishment seriously affects school participation and is linked to lower cognitive growth and poor school performance, leading to school drop-outs. More significantly, while early under-nutrition and cognitive-social impairment is preventable, it is irreversible after the age of two years.' The outcomes are, as noted earlier, that young children in India suffer from the highest levels of stunting, underweight and wasting observed in any country of the world (Arnold et al. 2009). Gendered inequalities thus can have direct intergenerational effects.

In patriarchal India gender discrimination is prevalent in rural areas. Women are often excluded from major institutions, public and private. Within the household, despite their reproductive, care giving chores and productive roles, they may often not be involved in major decision-making processes. In terms of development indicators such as education and literacy, women lag far behind men. Female labour force participation rates remain well below that of men. Even where women work there is discrimination in wages, with one study showing women receiving only about 60 per cent of the wages for equivalent work done by men (Rengalakshmi et al. 2002). Women's work is often not recognized in the household or even by the women themselves, let alone in the official data (Mitra 2008a; 2008b).

India passed a National Rural Employment Guarantee Act (NREGA) in 2005 drawing on the experience of the Maharashtra Rural Employment Guarantee scheme (Sjoblom and Farrington 2008). The NREGA, in providing 100 days of employment a year at the rate of Rs 100 a day, is a significant step. It is likely to have contributed to food security for the poor, and there may have been gendered benefits. On the face of it, the scheme seems to have ensured better wages for rural women. According to a NSS survey report (TOI 2013b), the disparity in wages between male and female workers had declined between 2004–5, when there was no national rural employment guarantee programme and 2007–8, when the scheme was introduced. Compared to an average daily wage of Rs 48.50 for a woman in urban areas, women in the NREGA scheme got Rs 79 daily in rural areas. With the introduction of this scheme, the wage difference between men and women has been further reduced, while in urban areas, where there is no such scheme, the differential has increased. While the scheme ensures higher wages to workers, the NSS survey report highlights that wages being given are still less than the stipulated Rs 100 per day, indicating major problems of implementation and access to the full entitlement offered by law.

However, despite the increasing understanding of the importance of gender in relation to food security and nutrition in particular, it is not well researched

(Pritchard et al. 2013: 161). Gillespie et al. (2012: 13), in a critical review of the literature on India, found just 11 studies that explored linkages between agriculture, female employment, female socio-economic power, household expenditure (linked to food and health), intra-household allocation of food and nutritional outcomes, but only one study actually assessed the nutritional outcomes. Their review confirms that social and cultural norms define gender roles and dictate that rural women and girls assume home production responsibilities in rural households (Gurung 1999). The intra-household decisions on allocation of labour often are biased and women are relegated to domestic tasks. Increasingly, faced with economic pressures, gender roles have become more flexible to enable women to engage in work traditionally regarded as belonging in the male domain. Conversely, rigid gender role definitions dictate that men should not perform household tasks. However, women are taking on new roles in farm production, off-farm production and community production to ensure the family's access to food and household resources. Where women can command greater income this can have effects on expenditure on foods but this is not guaranteed.

The literature on the links between agriculture, female employment, caring capacities and practices and nutrition outcomes is unclear (Gillespie et al. 2012: 13–14), pointing to ambiguous effects depending on the nature of female employment. No clear evidence either way of employment of women having nutritional consequences for children is available. Equally, the limited literature on the linkages between female employment in agriculture, energy expenditure and the nutritional and health outcomes for the female worker provide little clarity, and Gillespie et al. (2012: 15) concluded that 'general rural livelihood characteristics matter for optimal maternal and nutrition outcomes more than occupations per se'.

However, what is clear is that there has been an increasing feminization of agriculture (Binswanger-Mkhize 2013). To understand why this has come about it is necessary to explore what Binswanger has called the stunted structural transformation of the Indian economy where 'workers have moved primarily from the agriculture sector to the rural non-farm sector rather than to more secure jobs with pension and health benefits in the urban economy' (Binswanger-Mkhize 2013: 5).

India's failed agricultural transformation

Three key phases can be identified in relation to agricultural trends in India (Gillespie et al. 2012). The first was pre–Green Revolution (before 1965) characterized, as the opening quote of this chapter makes clear, by an imbalance between production and domestic needs. The second phase was the period of the Green Revolution (mid 1960s to late 1980s) during which India's agriculture sector had an impressive long-term record of taking the country out of serious food shortages despite rapid population increase. This was achieved through a favourable interplay of infrastructure, technology, extension, and policy support backed by strong political commitment that increased productivity. This resulted in a quadrupling of food grain yields, and food grain production increased from 51 million tonnes in 1950–1 to 217 million tonnes in 2006–7 (GOI 2008: 3). Production of oilseeds, sugarcane and cotton have also increased more than four-fold over the period, reaching 24 million tonnes, 355 million tonnes and 23 million bales respectively, in 2006–7 (GOI 2008: 3).

The Green Revolution transformed the supply/demand balance of food grains in India, placing the nation in a much better position to feed itself. However, at the same time, it also altered the social relations between the production and consumption of food. By encouraging a concentration of production in more favourable environments, the Green Revolution created new economic and environmental dependencies (Pritchard et al. 2013: 42). Strong contrasts have appeared between the agriculturally dynamic regions of the Punjab, Haryana, Kerala, West Bengal and Tamil Nadu amongst others, while states such as Chattisgarh, Jharkhand and the interior of Orissa have languished.

The Green Revolution produced a diverse range of interrelated direct effects (improvements in the welfare of poor farmers who adopted these innovations) and indirect effects such as reduced food prices, changes in employment arrangements in agriculture, and employment and income effects in non-agricultural sectors (Pritchard et al. 2013: 43). However, the Green Revolution also created new forms of social exclusion and 'small farmers were sometimes displaced by large farmers, tenants by workers, workers by labour saving innovations and producers in marginal areas by those in better endowed environments' (De Janvry and Sadoulet 2002: 2). Thus although the new varieties were seen to represent an evolutionary technique adopted first by less risk averse farmers with better access to information and inputs, 'when introduced into an entrenched power structure . . . [they were] used so as to benefit the powerful' (Lipton and Longhurst 1989: 401, cited in Pritchard et al., 2013: 45). Thus the Green Revolution had clear distributional effects, increasing inequality between regions and farms (Freebairn 1995: 265, cited in Pritchard et al. 2013).

In the early 1990s a third phase set in with a shift in policy which liberalized trade in agriculture, leading both to a decline in the relative agricultural growth rate and a rise in the relative prices of cereals despite increasing buffer stocks. A key area of withdrawal of central government has been budgetary allocations for the agricultural sector. Agriculture had a 14 per cent share in the budget during the First Plan (1951–6); in the Sixth Plan (1980–5) this fell to 5.8 per cent and to 4.9 per cent during the Ninth Plan (1997–2000).

The allocation to agriculture and allied sectors in the central plan was substantially increased from Rs 21,068 crore (Rs 210,680 million) in the Tenth Plan to Rs 50,924 crore (Rs 509,240 million) in the Eleventh Plan. However, as percentage of the total central plan the share of agriculture and allied sectors was around 2.4 per cent in both the plans, and this increased to around 3 per cent in 2007–8 (GOI 2010a: 64, Table 4.4).

Average state level expenditure on agricultural infrastructure has also declined since 1997 and was only 5 per cent of the average state allocation in 2001–2 (Vasavi 2012: 77). Institutionally, this has meant the virtual withdrawal of direct involvement by the central government from agriculture (Vasavi 2012), although certain subsidies have remained in place. In economically backward districts agricultural policy supports from the state are more defined by their absence than presence. Interestingly, since 2003–4, public investment in the agriculture sector has accelerated, leading to a higher share of public sector gross capital formation. Public investment in the agriculture sector has increased from 17 to 28 per cent over this period. Gross capital formation in agriculture also increased as a proportion of agricultural GDP after 2003–4, from 11.1 per cent to 15.1 per cent in 2008–9 (GOI 2010a: 64).

The effect of this is that the importance of India's agriculture sector to the economy has declined, with its share of GDP falling from 30 per cent in 1990–1 to 14.5 per cent in 2010–11. This has not been accompanied by a matching reduction in the share of agriculture in employment. About 52 per cent of the total workforce is still employed by the farm sector (NSS 66th Round, July 2009–June 2010), although the share of income in the rural economy from non-farm activities has increased (Jatav and Sen 2013).

In the classic agrarian transition, agriculture provides raw materials and labour for industrial development, with labour being fed by surplus cheap food from commercialized agriculture and globalized commodity markets (Byres 1991). Labour moves out of agriculture but rising rural incomes provide an important home market for industrial development. In turn, rising profits from both the industrial sector and agriculture lead to investment in agriculture, giving rise to further increases in agricultural productivity.

But, although GDP from agriculture has more than quadrupled, from Rs 10837.40 billion in 1950–1 to Rs 4859.37 billion in 2006–7 (both at 1999–2000 prices), the increase per worker has been rather modest. GDP per agricultural worker was around Rs 2,000 per month in 2006–7, only about 75 per cent higher in real terms than in 1950, compared to a four-fold increase in overall real per capita GDP. The main failure has been the inability to reduce the dependence of the workforce on agriculture significantly through the creation of sufficient non-farm opportunities to absorb the surplus labour in rural areas; this includes a failure to equip those in agriculture to access such opportunities, or to equip the rural population with the education and skills to migrate to better opportunities. Half of those engaged in agriculture are still illiterate and just 5 per cent have completed higher secondary education (12 years of schooling). Incomes and education are the lowest among agricultural labourers. Farming members of small family-operated farms are twice as likely to be illiterate as non-farming members. Ensuring food security and farmer welfare thus require support systems to extend technology and scale benefits to an existing workforce in agriculture that lacks non-farm skills and that is ageing and increasingly feminized.[8]

Overall, the average size of operational holdings in India has diminished progressively from 2.28 ha in 1970–1 to 1.55 ha in 1990–1 to 1.23 ha in 2005–6. As the 2005–6 Agriculture Census shows (Agricultural Census Division 2007), the proportion of marginal holdings (area less than 1 ha) has increased from 61.6 per cent of total holdings in 1995–6 to 64.8 per cent in 2005–6. About 18 per cent are classified as small holdings (1–2 ha), about 16 per cent as medium holdings (more than 2 but less than 10 ha) and less than 1 per cent as large holdings (10 ha and above).

What appears to have happened is that agriculture has not delivered the capital surplus to drive the transformation in non-agricultural sectors. Instead, as Lerche et al. (2013: 344) argue, 'liberalization and globalization has enabled wider non-agrarian sourcing of capital and new urban and international markets have developed for industrial outputs'. Thus the high levels of economic growth seen in India have taken place despite weaker agricultural growth and have not absorbed labour from agriculture. This stunted structural transformation of the Indian economy (Binswanger-Mkhize 2013) will continue to drive an overall decline in farm size, a retention of labour in rural areas and a search for rural non-farm income. This in turn will contribute to the feminization of agriculture and the rise of part-time farming which, without rising non-farm income sources in rural areas, is likely to contribute to increasing immerization.

Since 2003–4, and following the downturn in growth rates in GDP in agriculture, which between 1997–8 and 2002–3 averaged 0.5 per cent per annum, growth rates have revived, and between 2003–4 and 2011–12 they have averaged 2.9 per cent (Lerche 2013: 392). However there are strong regional dimensions to these trajectories, and within regions, differences between irrigated and non-irrigated areas. Some states (e.g. Gujarat, Rajasthan and West Bengal) show growth rates of above 2 per cent while others have had negative growth rates (Orissa, Kerala and Tamil Nadu). Certain states – notably Chattisgarh, Jharkhand and Orissa – remain characterized by low levels of irrigation and adoption of new crop varieties or HYVs.

Increasing environmental risks

Two key threats to agricultural production are becoming increasingly important, both linked to the environment and natural resources. The first is the dependence of agricultural production on groundwater irrigation given the fact that more than 60 per cent of India's agriculture is rainfed. It is estimated by some sources (World Bank 2013) that 15 per cent of India's groundwater resources are overexploited and there is widespread evidence of rapidly declining water tables and increasing levels of salinity (Reddy and Mishra 2009). In part this can be linked to continuing subsidies for irrigation water and electricity for pump sets (Vasavi 2012; Pritchard et al. 2013), which have encouraged cultivation of water-intensive crops, over-use of water, ground water depletion/salinity and waterlogging in many areas. A subsidy for nitrogen fertilizer has resulted in nitrogen/phosphorous/potassium imbalance and acted as a disincentive for the use of organic manures. As a result, the linkages between crop and livestock production have been reduced. Despite the liberalization, some of the earlier subsidy arrangements remain in place to placate rural voters (Pritchard et al. 2013: 74–5; Vasavi 2012: 75–6), but with clear environmental consequences. Despite the rhetoric on removal of subsidies in agriculture, 'input subsidies for fertilizers, power and irrigation have largely remained intact and ben-efitted either the fertilizer and pesticide industries or large farmers especially those in the northern and central wheat belts and in the sugarcane and commercial crops of Maharashtra' (Vasavi 2012: 75–6).

Superimposed on declining water resources are the effects of climate change (GOI 2010b; World Bank 2013). Key risks that are seen to affect Indian agriculture are increases in temperate and extremes of heat, changes in the monsoon pattern on which Indian agriculture so strongly depends and an increasing intensity of extreme weather events. While much will depend on the extent to which global tempera-tures rise there is already evidence of rising temperatures in India, and unusual and unprecedented spells of hot weather are expected to occur far more frequently and cover much larger areas.

There is evidence of a decline in the monsoon rainfall since the 1950s and an increasing frequency of heavy rainfall events. Dry years are expected to be drier and wet years wetter, triggering more frequent droughts as well as greater flooding in large parts of India. Droughts have already had major effects in 1987 and 2002–3 on more than half of India's crop area, resulting in a large decline in crop produc-tion. The frequency of droughts is expected to rise significantly in north-western India particularly in the states of Jharkhand, Orissa and Chattisgarh (World Bank 2013: 111).

Already there is direct evidence of effects on agriculture. Although rice yields have increased, there is clear evidence (World Bank 2013: 126) that rising tem-peratures and lower rainfall at the end of the growing season have already caused losses in Indian rice production. In addition, research on wheat yields shows that they peaked in India around 2001 and that high temperature extremes (greater than 34° C in northern India) have reduced wheat yields (World Bank 2013).

With upward pressure on food prices both globally and within India, and with a more risky environment in relation to production, the prognosis for security of food supply from Indian agriculture and the ability of poor households to access food is not encouraging. It is in light of these risks and the existing challenges of access to food that the Right to Food Act must be considered.

The National Food Security Act 2013

In September 2000 India committed itself to the Millennium Development Goals (MDGs), including those on hunger.[9] There was an optimism based on the high economic growth rates that these goals, especially those relating to eradication of hunger, could be easily achieved. But the 2005 mission to India by the UN Special Rapporteur on the Right to Food was sceptical and concerned:

> Starvation deaths have not been fully eradicated, nor has discrimination against women and against lower castes. Corruption and a wide range of violations including forced labour, debt bondage and forced displacement (destroying people's access to productive resources) remain serious obstacles to the realization of the right to food. In the current transition to a more

liberalized, market oriented economy, the poorest are more disproportionately bearing the costs. . . . Despite the progress made in the progressive realization of the right to food in India since independence, the Special Rapporteur is concerned that there are signs of regression particularly amongst the poorest. In monitoring progress towards the MDGs, the Planning Commission has noted that India was not currently on track to achieve the goals set in relation to malnutrition and undernourishment.

(Special Rapporteur on the Right to Food 2006, cited in Pritchard et al. 2013: 21)

It is against this background that the National Food Security Act was passed. The Indian Ministry of Agriculture's Commission on Agricultural Costs and Prices (CACP) referred to it as the 'biggest ever experiment in the world for distributing highly subsidized food by any government through a "rights based" approach' (Gulati and Jain 2013: 8). The bill extends coverage of the pre-existing Targeted Public Distribution System, India's principal domestic food aid program, to two thirds of the population, or approximately 820 million people. Initially, the Lok Sabha Standing Committee (2013) estimated that a 'total requirement of food grains, as per the Bill would be 61.55 million [metric] tons in 2012–13'.[10] The CACP calculated in May 2013 that ' . . . the requirement for average monthly Public Distribution System (PDS) offtake is calculated as 2.3 MT for wheat (27.6 MT annually) and 2.8 MT for rice (33.6 MT annually)' (Gulati and Jain 2013: 10–11). When volumes needed for the Public Distribution System and 'Other Welfare Schemes' were aggregated, the CACP estimated rice and wheat requirements to total an 'annual requirement of 61.2' million metric tonnes (Gulati and Jain 2013: 10–11). However, the final version of the bill signed into law includes an annex, 'Schedule IV', which estimates the total food grain allocation as 54.926 million metric tonnes (NFSA 2013: 18).

There has been substantial debate since the bill was passed as to the costs. The Lok Sabha Standing Committee estimated that the value of additional food subsidies (i.e. on top of the existing Public Distribution System) 'during 2012–13 works out to be . . . Rs 2409 crores',[11] that is, 24.09 billion rupees, or about $446 million USD at the then-current exchange rate, for a total expenditure of 1.122 trillion rupees (or between $20 and $21 billion USD).[12] However, the CACP calculated, '[c]urrently, the economic cost of FCI [Food Corporation of India] for acquiring, storing and distributing foodgrains is about 40 per cent more than the procurement price' (Gulati et al. 2012). The commission added,

The stated expenditure of Rs 120,000 crore annually in NFSB is merely the tip of the iceberg. To support the system and the welfare schemes, additional expenditure is needed for the envisaged administrative set up, scaling up of operations, enhancement of production, investments for storage, movement, processing and market infrastructure etc. The existing Food Security Complex of Procurement, Stocking and Distribution – which NFSB perpetuates – would increase the operational expenditure of the Scheme given its creaking infrastructure, leakages & inefficient governance.

(Gulati et al. 2012)

The commission concluded that the total cost for implementation of the bill '. . . may touch an expenditure of anywhere between Rs 125,000 to 150,000 crores', i.e. 1.25 to 1.5 trillion rupees (Gulati et al. 2012). There has been a more public debate where opponents of the bill have in the view of Sinha (2013) attempted to inflate the costs in order to argue that the bill is unaffordable. Some have pointed, for example, to distributional leakages around the existing PDS, which in some states can rise to 50 per cent or more through corruption. Himanshu and Sen (2013:70) cite estimates of 40 per cent of the food distributed through the PDS not reaching users in 2009–10, although these are national figures and hide not only significant recent improvements in the PDS distribution system but also inter-state variability. There are also the costs of wasted food stocks but as Sinha notes in citing the Annual Report of the Department of Food and Public Distribution (2013: 33), there is evidence of significant decline in wastage from 0.1 per cent of offtake in 2007–8 to 0.004 per cent in 2012–13.

But the more significant criticism of the bill has come from those who consider that the measures of the bill were significantly watered down and that by incorporating a targeting rather than a universal approach (Himanshu and Sen 2011), the bill will not effectively improve food security for the poor. Civil society groups like the Right to Food Campaign have argued that the NFSA does not 'provide a commitment to provide everyone in India with a legal guarantee to a full, nutritionally appropriate diet' (Pritchard et al. 2013: 156). Indeed only 67 per cent of the population is provided for and the monthly food grain ration potentially will satisfy less than half of their needs. The cereal requirements of an individual have been assessed to be 10 to 14 kg a month (Parsai 2013), and targeted households will have to go to the market for the remainder of their needs. M. S. Swaminathan, the eminent agricultural scientist, has pointed out that the NFSA addresses only the issue of inadequate consumption arising from lack of purchasing power. It ignores other correlates of food security such as the availability of food in the market (related to production) and lacks any commitment to adequate stocks for everyone at all places and at all times. Issues of improving utilization and ensuring for example the absorption of food in the body – which is determined by the availability of clean drinking water, sanitation, toilets, primary healthcare and nutritional literacy (Swaminathan 2013) – are goals simply to be achieved. The bill also does not, in Swaminathan's view, address the issue of food and nutrition security holistically. Under-nutrition in a cereal-based diet is a major cause of malnutrition and while the bill considers under-nutrition resulting from calorie deprivation, the protein-hunger arising from inadequate consumption of protein-rich foods (like pulses, milk, eggs, etc.) and hidden-hunger caused by the deficiency of micronutrients in the diet (like iron, iodine, zinc, vitamin A and vitamin B12) are not sufficiently addressed.

However, despite the criticism, there is compelling evidence that the PDS system, on which the NFSA is superimposed, has not only improved its performance in many if not all states in recent years, but it is also increasingly important as a means of income support and social protection in rural India. The analysis of Himanshu and Sen (2013) points to the fact that the PDS and the mid-day meal scheme (MMS) have significantly increased the contribution of in-kind transfers to both poverty reduction and nutrition. They argue that their analysis shows that during 2009–10, which was also a year of severe drought and high food prices, the effect of these schemes was to lift 55 million above the poverty line. Drèze and Khera (2013) agree on the significance of the PDS, but also point to the significant variability between states with respect to its effects. Their analysis shows that the effects of the PDS on rural poverty[13] reduction ranges from the relatively small (7 to 16 per cent) as in Bihar, West Bengal and Uttar Pradesh to the significantly high (56.8 to 83.4 per cent) in the states of Chattisgarh and Tamil Nadu. This is primarily because of the functioning of the PDS in these different states. The significance is that, as they put it, 'the PDS is now an important source of economic security for poor people in many states' (Drèze and Khera 2013: 59). However there are also a significant number of states in India where the poor governance of the PDS system deprives the poor of access to the PDS.

Conclusions

In one respect it could argued that the good news to be drawn from the functioning of the PDS is that the National Food Security Act, even though it is far from perfect, will contribute significantly to the right of the poor to access food. The more sobering conclusion that could be drawn concerns the scale of the challenge that India faces and the levels of food insecurity still prevalent in the country that have to be addressed. There are few grounds for optimism that the structural conditions that have given rise to the Indian 'enigma' will be resolved. The progressive informalization of labour and the slow growth of urban employment will keep the rural labour force in rural areas under conditions when farm size continues to shrink, particularly in more marginal agricultural areas. There is an optimistic view (Binswanger-Mkhize 2013) that would see the possibility of an economy-wide growth with an agricultural transformation that is technology-driven, more resource efficient, diversified and productive. This would support the rise of the rural non-farm economy and stronger rural–urban linkages, with income benefits leading to better food security and nutrition outcomes. But there are many who would point to the deep structural constraints in India's economy that make such a transformation wildly improbable and the need for public action to address food and nutritional insecurity as an enduring requirement for the central government.

Notes

1 The districts for the HUNGaMA survey were selected using the Child Development Index developed in 2009 by Indicus Analytics for UNICEF India. The HUNGaMA survey covers the 100 rural districts that ranked at the bottom of the index – and 12 districts ranked near the top. These 12 top districts were selected to represent a spread of examples across India; six of them are the top ranking rural districts in the six states (one district per state) of the 100 focus districts; the remaining six are the top ranking rural districts in Himachal Pradesh, Kerala and Tamil Nadu (two districts per state), the three states whose

rural districts led the all-India index ranking. http://naandi.org/HungamaBKDec11LR.pdf

2 http://pib.nic.in/newsite/erelease.aspx?relid=99309; see http://en.wikipedia.org/wiki/National_Food_Security_Bill,_2013 and the references cited therein for summaries of the debates.

3 This includes rice at Rs 3 per kg; wheat at Rs 2 per kg and millets at Rs 1 per kg.

4 ST = scheduled tribes.

5 SC = scheduled caste.

6 Dalit refers to the excluded castes and includes all the scheduled ones. They still face untold forms of discrimination in India.

7 The Indian government has established a series of what it calls SuperThermal Power Stations to address the country's power deficit.

8 Eleventh Five Year Plan cf: http://planningcommission.nic.in/plans/planrel/fiveyr/welcome.html

9 Goal 1 of the MDGs is Eradication of Extreme Poverty and Hunger and Target 2 (within Goal 1) was to halve between 1990 and 2015 the proportion of people who suffer from hunger.

10 Standing Committee on Food, Consumer Affairs and Public Distribution (2012–13), Fifteenth Lok Sabha, Ministry of Consumer Affairs, Food and Public Distribution (Department of Food and Public Distribution) (January 2013), *The National Food Security Bill, 2011, Twenty Seventh Report,* available http://164.100.47.134/lsscommittee/Food,%20Consumer%20Affairs%20&%20Public%20Distribution/Final%20Report%20on%20NFSB.pdf

11 A crore is equivalent to ten million (10,000,000).

12 Standing Committee on Food, Consumer Affairs and Public Distribution (2012–13), Fifteenth Lok Sabha, Ministry of Consumer Affairs, Food and Public Distribution (Department of Food and Public Distribution) (January 2013), *The National Food Security Bill, 2011, Twenty Seventh Report,* available http://164.100.47.134/lsscommittee/Food,%20Consumer%20Affairs%20&%20Public%20Distribution/Final%20Report%20on%20NFSB.pdf.

13 There are at least three different poverty lines used in India and considerable debate as to the strengths and weaknesses of each.

References

Agricultural Census Division (2007) *Agricultural Census of India, 2005–6,* Ministry of Agriculture, Government of India. Online. Available <http://agcensus.nic.in/> (accessed 01 December 2013).

Arnold, F., Parasuraman, S., Arokiasamy, P. and Kothari, M. (2009) *Nutrition in India – National Family Health Survey (NFHS-3), India, 2005–06,* Mumbai: International Institute of Population Sciences, and Calverton, Maryland, USA: ICF Macro.

Athreya, B. V., Bhavani, R. V., Anuradha, G., Gopinath, R. and Velan, A. S. (2008) *Report on the State of Food Insecurity in Rural India,* Chennai: MS Swaminathan Research Foundation and World Food Programme.

Athreya, B. V., Rukmani, R., Bhavani, R. V., Anuradha, G., Gopinath, R. and Velan, A. S. (2010) *Report on the State of Food Insecurity in Rural India,* Chennai: MS Swaminathan Research Foundation and World Food Programme.

Banik, D. (2007) *Starvation and India's Democracy,* London: Routledge.

Binswanger-Mkhize, H. P. (2013) 'The stunted structural transformation of the Indian economy: agriculture, manufacturing and the rural non-farm sector', *Economic and Political Weekly,* 48(26–27): 5–13.

Byres, T. J. (1991) 'The agrarian question and differing form of capitalist agrarian transition: an essay with reference to Asia', in J. Breman and S. Mundle (eds), *Rural Transformation in Asia,* pp. 3–76, New Delhi: Oxford University Press.

Chandrasekhar, C. P. (2012) 'India's triumph in rice', *The Hindu,* 23 December.

Chatterjee, N., Fernandes, G. and Hernandez, M. (2012) 'Food insecurity in urban poor households in Mumbai, India', *Food Security,* 4(4): 619–32.

De Janvry, A. and Sadoulet, E. (2002) 'World poverty and the role of agricultural technology: direct and indirect effects', *Journal of Development Studies,* 38(4): 1–26.

Drèze, J. and Khera, R. (2012) 'Regional patterns of human and child deprivation in India', *Economic and Political Weekly,* 48(39): 42–49.

Drèze, J. and Khera, R. (2013) 'Rural poverty and the Public Distribution System', *Economic and Political Weekly,* 48(45–46): 55–60.

FAO (2008) *FAO Methodology for the Measurement of Food Deprivation: Updating the Minimum Dietary Energy Requirements,* Rome: Food and Agriculture Organization, Statistics Division.

Fernandes, W. and Thukral, E. G. (eds) (1989) *Development, Displacement and Rehabilitation,* Delhi: Indian Social Institute.

Freebairn, D. K. (1995) 'Did the Green Revolution concentrate income? A quantitative study on research reports', *World Development,* 23(2): 265–79.

Gillespie, S., Harris, J. and Kadiyala, S. (2012) *The Agriculture-Nutrition Disconnect in India: What Do We Know?* IFPRI Discussion Paper 01187, Washington, DC: International Food Policy Research Institute.

GOI (2008) *Eleventh Five Year Plan (2007–2012), Vol 3,* Planning Commission, Government of India, New Delhi: Oxford University Press. Online. Available <http://planningcommission.nic.in/plans/planrel/fiveyr/11th/11_v3/11th_vol3.pdf> (accessed 12 December 2013).

GOI (2009) *Report of the Expert Group to Review the Methodology for Estimation of Poverty,* Planning Commission, Government of India, November. Online. Available <http://planningcommission.nic.in/reports/genrep/rep_pov.pdf> (accessed 12 December 2013).

GOI (2010a) *Eleventh Five Year Plan as Approved by the National Development Council,* Chapter 4: Agriculture. Online. Available <http://planningcommission.nic.in/plans/mta/11th_mta/chapterwise/chap4_agri.pdf> (accessed 12 December 2013).

GOI (2010b) *Climate Change and India: A 4 x 4 Assessment. A Sectoral and Regional Analysis for 2030s,* Executive Summary, Delhi: Ministry of Environment and Forests, Government of India.

Gulati, A., Gujral, J. and Nandakumar, T. (2012) *National Food Security Bill, Challenges and Options,* CACP Discussion Paper 2, December, New Delhi: Ministry of Agriculture, Government of India. Online. Available <www.prsindia.org/uploads/media/Food Security/CACP Report on Food Security Bill.pdf> (accessed 31 January 2014).

Gulati, A. and Jain, S. (2013) *Buffer Stocking Policy in the Wake of NFSB: Concepts, Empirics and Policy Implications,* CACP Discussion Paper 6, May, New Delhi: Ministry of Agriculture, Government of India.

Gurung, J.D. (1999) 'Searching for women's voices in the Hindu-Kush Himalayas', in Gurung, J.D. (ed.), *Searching for Women's Voices in the Hindu-Kush Himalayas,* pp. 1–36, Kathmandu: International Centre for Integrated Mountain Development.

Headey, D., Chui, A. and Kadiyala, S. (2011) *Agriculture's Role in the Indian Enigma – Help or Hindrance to the Under-Nutrition Crisis,* IFPRI Discussion Paper, 00911, Washington, DC: International Food Policy Research Institute.

Hemadri, R., Mander, H. and Nagraj, V. (2000) 'Dams, displacement, policy and law in India', prepared for *Thematic Review: Displacement, Resettlement, Rehabilitation, Reparation and Development,* World Commission on Dams (WCD). Online. Available <www.dams.org/docs/kbase/contrib/soc213.pdf> (accessed 31 May 2010).

Himanshu and Sen, A. (2011) 'Why not a universal food security legislation?' *Economic and Political Weekly,* 46(12): 38–47.

Himanshu and Sen, A. (2013) 'In-kind food transfers – impact on nutrition and implications for food security and its costs', *Economic and Political Weekly*, 48(47): 60–73.

HUNGaMA (2011) *The HUNGaMA Fighting Hunger and Malnutrition*, HUNGaMA Survey Report, Hyderabad, India: Naandi Foundation.

IFPRI (2011) *Global Hunger Index Report: The Challenge of Hunger*, Bonn: International Food Policy Research Institute.

IIPS (2009) *National Family Health Survey (NFHS-3) 2005–6, Nutrition in India*, Mumbai: International Institute for Population Studies.

Jatav, M. and Sen, S. (2013) 'Drivers of non-farm employment in rural India – evidence from the 2009–10 NSS round', *Economic and Political Weekly*, 48(26–27).

Kothari, S. (1997) 'Whose independence? The social impact of economic reforms in India', *Journal of International Affairs*, 51(1): 85–116.

Lerche, J. (2013) 'The agrarian question in neoliberal India: agrarian transition bypassed?' *Journal of Agrarian Change*, 13(3): 382–404.

Lipton, M. and Longhurst, R. (1989) *New Seeds for Poor People*, Baltimore: John Hopkins University Press.

Mehta, L. (ed.) (2009) *Displaced by Development – Confronting Marginalization and Gender Injustice*, New Delhi: Sage Publications.

Menon, P., Deolalika, A. and Bhaskar, A. (2009) *India State Hunger Index: Comparison of Hunger Across States*, Washington, DC: International Food Policy Research Institute.

Mitra, A. 2008a 'Gender Caste and Growth Assessment – West Bengal Sub-National Study', (Mimeo).

Mitra, A. 2008b 'Social and gender aspects in the target area', Working Paper No. 1, India, Convergence of Agricultural Interventions in Maharashtra's Distressed Districts, International Fund for Agricultural Development, (Mimeo).

Mitra, A. and Rao, N. (2009) 'Displacing gender from development: a view from SantalParganas, Jharkhand', in Mehta, L. (ed.), *Displaced by Development – Confronting Marginalization and Gender Injustice*, New Delhi: Sage Publications.

Mukherjee, A. (2007) *Micro-Level Food Insecurity in Contemporary India: Perspectives of the Food Insecure*, Bangkok: Poverty and Development Division, United Nations Economic & Social Commission for Asia and the Pacific. Online. Available <http://enrap.org.in/PDFFILES/Food%20Security%20in%20Contemporary%20India.pdf> (accessed 10 December 2013).

Mukhopadhyay, P. and Revi, A. (2009) 'Keeping India's economic engine going: climate change and the urbanization question', *Economic and Political Weekly*, 44(31): 59–70.

NFSA (2013) *The National Food Security Bill, 2013*. Online. Available <www.thehindu.com/multimedia/archive/01404/National_Food_Secu_1404268a.pdf> (accessed 9 December 2013).

Office of the Registrar General and Census Commissioner (2011) *2011 Census Data*, Delhi: Ministry of Home Affairs, Government of India. Online. Available <http://indiafacts.in/india-census-2011/urban-rural-population-o-india/> (accessed 6 January 2014).

Parsai, G. (2013) 'Food Bill in a political quagmire', *The Hindu*, New Delhi edition, 6 June.

Pritchard, B., Rammohan, A., Sekher, M., Parasuraman, S. and Choithani, C. (2013) *Feeding India – Livelihoods, Entitlements and Capabilities*, London and New York: Routledge, Earthscan.

Rai, P. (2005) *Political Economy of Hunger in Adivasi Areas*, New Delhi: Centre for Environment and Food Security, (Mimeo).

Ramalingaswami, V., Jonson, U. and Rohdel, J. (1996) 'Commentary – the Asian enigma', in *The Progress of Nations*, New York: UNICEF.

Rampal, A. and Mander, H. (2013) 'Lessons on food and hunger: pedagogy of empathy for democracy', *Economic and Political Weekly,* 48(28): 50–7.

Reddy, D. N. and Mishra, S. (2009) 'Agriculture in the reforms regime', in D.N. Reddy and S. Mishra (eds), *Agrarian Crisis in India,* New Delhi: Oxford University Press.

Rengalakshmi, R. and Research team (2002) *Rural and Tribal Women in Agrobiodiversity Conservation: An Indian Case Study,* RAP Publication 2002/08, Chennai: MS Swaminathan Research Foundation and Bangkok: FAO Regional Office for Asia and the Pacific.

Saxena, N. C. (2012) 'Hunger and malnutrition in India,' *IDS Bulletin* 43(S1), July 2012.

Sinha, D. (2013) 'Costs of implementing the National Food Security Act', *Economic and Political Weekly,* 48(39): 31–34.

Sjoblom, D. and Farrington, J. (2008) *The Indian National Rural Employment Guarantee Act: Will it reduce poverty and boost the economy?* Project Briefing No. 7, London: Overseas Development Institute.

Sreerekha, M. S. (2010) 'Challenges before Kerala's landless: the story of Aralam Farm', *Economic and Political Weekly,* 45(21): 55–62.

Sukhatme, P. V. (1965) *Feeding India's Growing Millions,* Bombay: Asia Publishing House.

Swaminathan, M. S. (2013) 'What the food bill does not consider', *Asian Age,* 12 September. Online. Available <http://archive.asianage.com/columnists/what-food-bill-does-not-consider-231> (accessed 9 December 2013).

Thorat, S. and Lee, J. (2005) 'Caste discrimination and food security programmes', *Economic and Political Weekly,* 4198–4201.

TOI (2013a) 'Government defers promulgation of ordinance on Food Security Bill', *Times of India,* New Delhi, 13 June 2013. Online. Available <http://articles.timesofindia.indiatimes.com/2013–06–13/india/39950536_1_food-security-bill-ordinance-route-opposition-parties> (accessed 9 December 2013).

TOI (2013b) 'Disparity in wages between male, female workers in rural India reduced', *Times of India,* Delhi, 28 May 2010.

UN Commission on Human Rights, *Report of the Special Rapporteur on the Right to Food: Mission to India,* 20 March 2006, E/CN.4/2006/44/Add.2, available at: <www.refworld.org/docid/45377b210.html> (accessed 4 August 2014).

UNDP (2009) *India – Urban Poverty Report,* Summary of Chapters. Online. Available <www.undp.org/content/dam/india/docs/india_urban_poverty_report_2009_related.pdf> (accessed 9 December 2013).

Vasavi, A. R. (2012) *Shadow Spaces – Suicides and the Predicament of Rural India,* Gurgaon, India: Three Essays Collective.

Walton, M. (2009) 'The political economy of India's malnutrition puzzle', *IDS Bulletin* 40(4): 16–24.

World Bank (2013) *Turn Down the Heat: Climate Extremes, Regional Impacts, and the Case for Resilience,* a Report for the World Bank by the Potsdam Institute for Climate Impact Research and Climate Analytics, Washington, DC: World Bank.

16

SOUTH KOREA

Food security, development and the developmental state

Anders Riel Müller

Abstract

South Korea is one of the major economic successes of post-war reconstruction and development. Much has been written about the successful role of the state in designing industrial policies and keeping tight control of business and finance. But the state not only tightly managed strategic industrial sectors, it was and still is deeply involved in food supply management. This is an often overlooked and perhaps less successful aspect of the South Korean developmental state. This chapter outlines how government policies for food and agriculture have changed over time, highlighting the strong role of the state, the challenges of securing food supplies, and the framing of food supply management in national (food) security terms. The findings are that food security is a concept that in effect has been used as a matter of national security concern by the state to mobilize the people, but that the particular challenges that food security policy tried to address have changed significantly.

Introduction

South Korea has been one of the most economically successful post-colonial states. Its economic success has often been attributed to it being an effective developmental state, in addition to favourable world market conditions and strong financial support from the US and Japan. The South Korean government's management of industrial inputs such as energy and raw materials, and of food, has been an area that has received less attention. Food supply management has been a significant area of concern for the South Korean state since its early efforts to industrialize following the end of the Korean War. Since the 1950s South Korea has depended on food imports to a larger or lesser degree and management of this has as such been a quite significant aspect of state-led development. Thus food security and economic development polices have been interconnected (Timmer 2005). The ways in which the South Korean state has sought to manage these interconnected processes are interesting because the country has relied heavily on food imports throughout most of its modern history. However, this has received little attention in the literature on South Korea's development since the early 1980s. This is not unique to studies of South Korean development but rather a symptom of a greater marginalization of studies on food and agriculture in contemporary capitalist development. As Timmer argues (2005), the mainstream assumption is that food security ceases to be a problem once a country becomes wealthy enough. Nevertheless, food remains central to economic development as one of the raw materials on which industrialization depends (McMichael 2009b; McMichael 2013; Friedmann and McMichael 1987). That South Korea, one of the richest nations in the world, should be food insecure may seem absurd to many and data would seem to support that. According to the Global Food Security Index from the Economist Intelligence Unit, South Korea ranks as the 18th most food secure country in the world.

The question then arises, that if the assumption is that food security decreases when economic wealth increases, why has food security policy remained an important and sensitive issue in South Korean politics even today? This most likely has to do with differing definitions of what food security means. That food insecurity is eliminated with wealth creation can only be true if the definition of food security remains the same. But even global definitions of food security have changed over time (Maxwell 1996). Food security as a concept, whether political or academic, is difficult to pin down, and subject to multiple interpretations. For the purpose of discussion here, it is more useful to begin from Carol Bacchi's observation that concepts are rarely descriptive of anything, but rather are prescriptions for where we ought to go from here (Bacchi 2000: 45). Policy concepts are rarely static but are designed for particular purposes and redefined to fulfill other purposes. As Bacchi has also noted, government policies take on particular shape and understanding in relation to issues of policy concern (Bacchi 2009). This is also the position here. Food security is a concept that changes in definition in relation to what it seeks to do for policy purposes (Hettne 2010). The objective is therefore to identify how and why the South Korean government has used food security policies to address issues of food supply management by framing policies as a matter of food security within a broader agenda of political and economic objectives.

The analytical framework relies on theories of the developmental state (Woo 1991; Burmeister 2000; Burmeister 1990; Kohli 2004; Woo-Cumings 1999; Amsden 1989) and international food regimes (McMichael 2009a; Araghi 2010; Friedmann and McMichael 1987; McMichael 2009b; McMichael 2013). Theories of the developmental state are useful for understanding the direct role of state-led development in South Korean history. As Jung-En Woo convincingly argues, the developmental state in South Korea strictly managed the strategically important finance and economic sectors according to national development objectives. Companies and industrial sectors were launched through strategic allocation of state-controlled financial capital (Woo 1991). This tight control of finance enabled the state to promote or suppress particular sectors and companies according to the state's will. These policies could be economically as well as politically motivated, but one overarching driver for development and the developmental state was the issue of national security because of the continued military threat from North Korea. Meredith Woo-Cumings argues, along with a number of other scholars, that the economic success of many developmental states can be attributed to the ability to mobilize industry and labour in a common project against an external threat (Woo-Cumings 1998; Doner et al. 2005; Halabi 2011; Zhu 2001). The security imperative provides a very plausible explanation for why the South Korean state considered rapid industrial transformation a priority. Also, by framing political and economic projects in national (in)security terms, the state was able to both mobilize industry, labour and consumers for the national project while keeping dissent at a minimum despite the high social cost of developmentalism. These costs include among others, suppression of political freedom, low wages and low welfare spending. Food supply management, framed as food security policy, is one area in which the government repeatedly has used security language in justifying changes in food policy in the name of the national development project. The agricultural and food sectors are rarely considered in developmental state literature, but as Larry Burmeister points out, the food and agricultural sectors have been among the most tightly regulated and managed sectors by the state in South Korea (Burmeister 1990: 199). However, there are also differences between the food sectors and other industrial sectors. South Korean agriculture, food processing and retail still have a large share of small farmers and small businesses − very different from the few large conglomerates that dominate other economic sectors. Therefore, the state has developed a series of large state agencies, state cooperatives and public corporations through which it has managed food and agriculture especially during the 1970s, the apogee of the developmental state. This we will return to later. The point to be made here is that the state has played and still plays a very significant role in managing food supply, and it has done so, and still does, by framing food supply management in food security terms. That is, food supply management is a matter of national survival. One weakness in developmental state theory is its primary focus on the relationship between the state and business, that is, on domestic state–business relationships. The relationships in the South Korean food and agriculture sector are quite different from the finance and industrial sectors, the usual focus of developmental state theories, because food production and retail until recently were dominated almost entirely by small producers and retailers. This different composition of the food and agriculture sector compared to many of the manufacturing sectors needs to be taken into account analytically.

South Korea's food supply management has oscillated between self-sufficiency strategies and import strategies. Emphasis on either self-sufficiency or food imports has in turn been based on domestic economic and political concerns, on one hand, and trends in the international food system on the other. Food regime theory provides a complimentary analytical lens that enables understanding of the trajectories of food and agriculture within a broader global capitalist system. Food regime analysis problematizes linear representations of agricultural modernizations and emphasizes the central role of food and agriculture in the trajectories of the capitalist world economy and the state system (McMichael 2009b; Friedmann and McMichael 1987). Food regime theory historicizes the global and national food system as a series of crises, transformations and transitions in the world economy and the hegemonic role of a few nations or corporations in shaping the world food system. There has been a serious critique of food regime theory for relying too heavily on theories developed for manufacturing industries, thus throwing into question the periodization of specific food regimes (Goodman and Watts 1994; Goodman 1997). This criticism called into doubt the relevance of food regime theory in the late 1990s. Food regime theory's primary focus on identifying particular regimes of hegemony in the global food system also reduced food system transformations to the logic of capitalist accumulation, thus understating the agency of the state. Nevertheless, food regime theory continues to provide a useful analytical perspective for understanding the transformations in the South Korean food system in connection to broader transformations in the global food system. Specifically, food regime theory enables us to contextualize food supply policy decisions by the Korean state in a global political economy of food. By historicizing the concepts of food security, development and the state, this chapter intends to show shifting policy approaches to food supply management, framed as food security, over time in South Korea and identify the drivers that created shifts in policy priorities. By combining developmental state theory and food regime theory, this chapter identifies shifting regimes of food supply policy and the political and economic drivers, domestic and international, influencing state food supply management policy. These shifts occurred in the space for negotiation between South Korea's position in the global economic order and domestic political and economic circumstances. It is these shifts and their internal and external drivers that are the focus of this chapter.

The chapter is organized as follows: first, a historical overview of the food security–developmental state nexus in South Korea will contextualize how the two concepts have been articulated historically up to the beginning of the 2007–2008 world food crisis. Three shifts in food security policy are highlighted: the rebuilding of the South Korean economy and early industrialization where food security was a matter of upholding a supply of US food aid. This period was followed by a shift to self-sufficiency strategies as US food aid was withdrawn and new needs of the domestic economy and political establishment emerged by the early 1970s. The third period of slow trade liberalization began in the early 1980s due to pressure from trading partners and significant shifts in the diet towards a more animal-protein-rich diet. The second part of this chapter analyses the government's formulation of food security strategy since the 2007–2008 food crisis, with particular emphasis on the recent decision to obtain food security through large-scale overseas land acquisition and overseas agricultural activities in production, marketing and logistics. The chapter ends with an analysis of the changing configurations of food security, economic development and the developmental state, investigating what transitions in the domestic political and economic environment and the global food system have been decisive in shaping the food policies devised by the state in particular periods.

Post-war South Korean state formation and US food aid

For centuries, the Joseon[1] dynasty had isolated itself from the outside world, but changes in the global political economy in the nineteenth century altered the structure of Korean agriculture, turning the kingdom into a major exporter of agricultural commodities to Japan. The first export boom came after the 1876 Gangwha treaty when Japanese gunboat diplomacy forced the Hermit Kingdom[2] to open up to foreign trade. Rice exports generated profits, which in turn were used to import foreign manufactured goods and build infrastructure such as railroads and electricity generation and distribution (Chung 2006). The second boom came under Japanese occupation and colonization, especially following the 1918 rice riots in Japan when high rice prices led the Japanese administration to develop a thirty-year plan for expansion of Korean rice production and export 90 per cent of production to Japan (Chung 2006: 233). The export-oriented agro-industrial infrastructure developed by the Japanese during the interwar period was almost completely dismantled following the Korean War. The physical infrastructure was destroyed by the war itself, and the Japanese imperial-capitalist/Korean landlord alliance of agricultural institutions was dismantled by the completion of land reforms in South Korea, effectively eliminating the capitalist landlord class (Kay 2002; Hsiao 1981; Shin 1998; Kohli 1999; Cumings 2005). South Korean agriculture became characterized by a so-called minifarm structure (Burmeister 1992) where the majority of farm households owned less than one hectare (Hsiao 1981: 81).

In the years following the Korean War, after an armistice agreement was signed in 1953, South Korean policy focused primarily on political stability, military security and industrial import substitution. Political stability was the main priority and purging of civil unrest from what the government perceived to be pro-north and pro-unification agitators. The government perceived these agitators to be associated with communist and peasant movements, and neutralization happened through violence and coercion but also by accommodating some demands such as land reform. Nevertheless national food deficiency was a serious issue exacerbated by the influx of approximately 2.5 million refugees from North Korea (Hsiao 1981). Hunger and famine were widespread in the years after the war. During this period of post-war recovery, political stabilization and industrialization, South Korean food security and development were underwritten by food aid from the United States under the Food for Peace program (PL480). The PL480 provided both food commodities and agricultural inputs such as fertilizer to increase domestic agricultural productivity. The PL480 program incorporated South Korea into the US food surplus regime, in which US agricultural surpluses were distributed to countries important to US foreign interests (Friedmann and McMichael 1987). Food aid provided food to the urban population and kept domestic agricultural prices low. Food security through US food aid also had significant impacts on South Korean food culture, introducing new food items into the diet such as processed meat products, dairy and bread, especially in urban areas. One author describes the lines of hungry people waiting to receive their daily ration of porridge made from US flour and milk powder, products entirely unfamiliar to most Koreans at the time (Lee 2013). Most Koreans were lactose intolerant and many suffered from diarrhoea caused by the unfamiliar food, but, with nothing else to eat and with the aid of US supported dietary education, people were taught that flour and dairy were superior products, an assumption that still prevails to this day.

During these early years of the new republic, South Korea remained an unstable state with high levels of corruption in the administration of President Syngman Rhee. The country lagged behind its North Korean foes in almost every aspect of economic development despite massive economic and military aid from the US. The situation changed in 1960–61. President Syngman Rhee was forced to resign in April 1960 because of popular protests caused by what many believed were rigged elections. Rampant corruption in his administration, political repression, a stagnant economy and poverty all contributed to his downfall. A democratically elected government was installed for the first time in South Korean history and the Second Republic was established with Chang Myon as president. The new government put strong emphasis on economic development through industrialization, but this government only lasted eight months. General Park Chung Hee took power in a military coup on 16 May 1961. The Park Chung Hee era is usually associated with the birth of the developmental state, effectively shifting focus from import substitution to export-oriented industrialization, guided capitalism and rapid industrialization. At this time, North Korea was ahead in both industrial and military power and Park Chung Hee's drive for industrialization can be understood in this context of continued danger of a potential war with North Korea. Thus, industrialization and economic growth in order to build military capacity were vital to upholding national security. The Park regime was heavily dependent on loans, grants and investment from Japan and the US during the first decade. Economic growth between 1962 and 1971 averaged 9.7 per cent and the gross national product grew from USD 2.5 billion in 1961 to USD 12.4 billion in 1971 (Boyer and Ahn 1991: 32).

However, industrialization led to a shifting of balance in urban–rural relations. Despite the fact that food security through food-self-sufficiency was stated as a target in the first two five-year plans (1962–66 and 1967–71) under Park Chung Hee, economic development policies favoured industrialization at the expense of agriculture[3] (Hsiao 1981; Boyer 2010). Food security remained underwritten by PL480 through the first phases of rapid industrialization. In fact US food aid increased steadily from 669,000 metric tons in 1965 to 3.6 million tons in 1972 or one-fourth of South Korean grain consumption (Hsiao 1981: 242). In the first period of Park Chung Hee's rule, the massive mobilization for industrialization remained underwritten by cheap food imports from the US surplus food regime to feed the urban population and growing ranks of industrial workers coming from the countryside to the urban centres looking for economic opportunities.[4]

Food security and the 'Big Push' for rice self-sufficiency

By the early 1970s, shifts in US aid priorities, the broader global food system and domestic political concerns led to a major shift in how to secure food supplies. The last year that the US provided PL480 sales in local currency was 1970, thus creating a greater burden on the government's balance of payments. The PL480 Title II program, which provided direct donations of food aid, also ended in 1970 (Hsiao 1981). South Korea could no longer rely on PL480 for its food supply needs. President Park also had domestic concerns. The squeezing of the agricultural sector to finance[5] industrialization had led to an increase in urban to rural income disparities. By 1966 urban labourers' incomes eclipsed that of rural farmers for the first time since the founding of South Korea (Hsiao 1981). The lack of economic opportunities in rural areas caused millions of rural residents to migrate to the cities to look for work, leading to shortages of rural labour (Boyer and Ahn 1991). The stagnation of rural economic growth was a problem for Park Chung Hee, as the rural population was an important political pillar for the regime whose legitimacy was in large part based upon a notion of growth with equity. The mobilization of the countryside for food self-sufficiency thus made sense both politically and economically. The population in 1975 was still 50 per cent rural and self-sufficiency would improve the balance of trade so vital to the government's industrialization strategy. In 1975, for example, grain imports amounted to USD 689 million or 14 per cent of the country's foreign exchange earnings (Young and Dong 1981: 51).

In 1970, President Park Chung Hee announced a set of new and ambitious agricultural and rural modernization programs to increase domestic agricultural production and improve rural livelihoods. The most famous of these programs was the Saemaul Undong[6] program (New Community Movement). Saemaul Undong was to revitalize rural areas through a spiritual awakening of the rural population and modernize agriculture through massive investments in rural and agricultural infrastructure (Boyer and Ahn 1991; Brandt 1979; Burmeister 2006). The program, in its early years, was successful in raising material living standards in the countryside through the construction of modern houses made from cement, new roads, storage facilities, mechanization and irrigation infrastructure. The program, however, also allowed the authoritarian regime to extend its political control all the way down to village level where Saemaul Undong leaders functioned as a parallel authority to the public administrative system (Moore 1984). The South Korean minifarm structure, the result of land reforms two decades earlier, was mobilized to achieve food security through food self-sufficiency. The government quadrupled expenditure on large-scale infrastructure projects such as dams, reservoirs and irrigation (Boyer and Ahn 1991).

Green Revolution policies were also implemented by introducing new cultivation techniques, seed varieties and machinery, and significant protectionist measures were established. Consumers on the other hand were encouraged to buy Korean produce through campaigns that made rice consumption a matter of national duty. High tariffs and import bans on agricultural commodities were implemented alongside the rise of the statist agricultural system (Kim 2006). Direct subsidy programs such as a rice-purchasing program that guaranteed government-fixed prices helped to boost income. Meanwhile consumer prices were kept low by selling rice to consumers below the purchasing prices. As such, the rice-purchasing program subsidized both the rural and urban family economy in order to allow industrialization plans to continue. By the late 1970s South Korea became self-sufficient in rice for the first time since World War II. Saemaul Undong and other development programs also deepened the political penetration and control by the Park regime. State-controlled organizations such as the National Agricultural Cooperative Federation and Saemaul Undong were not only institutions for development. They were also tools of political control that went all the way down to village level throughout the country (Burmeister 2006; Moore 1984; Boyer and Ahn 1991).

By the end of the 1970s, average rural income was again higher than average urban income and the country had become self-sufficient in its most important food staple, rice. The agricultural modernization also had wider economic benefits to the country. Surplus capacity in construction and manufacturing industries found new domestic markets through the rural modernization program, supplying construction materials and consumer goods to rural consumers and providing direct benefits to industry as well. As such, the Saemaul Undong period of the 1970s was a time in which food security, rural development and national development had clear synergetic linkages. The Saemaul Undong period however proved not to be sustainable either economically or politically. At the national level, the government accumulated massive deficits in its nationalized grain and fertilizer operations by the late 1970s at a time when the market liberalization agenda was emerging. With the assassination of Park Chung Hee in 1979, the Saemaul Undong program lost its economic momentum. The organization itself continued mostly as a political tool for President Chun Do Hwan, the military general who seized power following Park Chung Hee's death, while the economic importance of the agricultural sector declined and the government reoriented its focus on export industries.

The decade was thus characterized by an export-led industrial strategy and an import substitution food supply strategy. This was the consequence of both greater changes in external economic relations with the US and the need to secure domestic political legitimacy. The food security strategy in the 1970s was distinct from the previous period because food security became closely linked to economic development policies through a food self-sufficiency strategy in which growth in the industrial economy financed economic growth and development in the rural sector.

Food security, (free) trade and the rise of feed imports

Just as South Korea managed to raise agricultural productivity to unprecedented levels, changes in the larger global economy were already in motion to dismantle statist and protectionist food security strategies. The assassination of Park Chung Hee led to political instability, and the wider international economic recession of 1979–80 signalled a new economic and political order. Moreover the US, South

Korea's largest trading partner, was no longer ready to accept the heavy protection of South Korean industries from foreign imports. The economic crises of the 1970s had put the Keynesian mode of economic governance into question and the US had already abandoned the Bretton Woods agreement in 1970.[7] Free trade and *laissez faire* economics were back in vogue, especially in the US, and South Korea had no choice but to begin dismantling the developmental state because of anti-dumping measures from major trading partners. The domestic agricultural sector, which for a decade had become the main mechanism for achieving self-sufficiency, could no longer be protected to the same extent as before. South Korea was forced to open up markets for US rice, wheat, tobacco and cereals (Woo 1991). Otherwise, South Korean industrial exports would have been subjected to penalties because farm protectionism was seen to limit export markets for surplus agricultural producers such as the US, and threaten the international free trade regime (McMichael and Kim 1994).

South Korea gradually liberalized its agricultural sector (Francks et al. 1999). Import restrictions on some agricultural commodities were lifted and tariffs were reduced. The pressure to further reduce protectionism was strengthened in the 1990s. During the Uruguay Round of the General Agreement on Tariffs and Trade (GATT) and the Agreement on Agriculture of the World Trade Organization (WTO), South Korea sought to reduce trade restrictions for important export-oriented industries such as heavy manufacturing, electronics and textiles. However, from the GATT/WTO perspective South Korea was required to eliminate agricultural protectionism. As a compromise, South Korea pursued a 'bifurcation' strategy of a heavily protected national market for rice and livestock while opening up to international commodity relations for other agro-food circuits such as animal feed and processed flour goods (McMichael and Kim 1994). South Korea thus faced a contradictory situation in which large corporate export-oriented industries co-existed with a small-scale highly protected farm sector and a relatively under-capitalized food processing industry (OECD 1999).

The 1980s and 1990s also drastically changed South Korean diets. Rice consumption per capita has been declining since the early 1980s. Total rice consumption per capita was 132.2 kg in 1980, declining to 69.8 kg in 2011 (Statistics Korea 2012). Animal-protein consumption, on the other hand, increased significantly as consumers became wealthier. The average meat consumption per person was only 5.2kg in 1970 but reached 41.1kg in 2011 (Statistics Korea 2011). To meet demand the government encouraged commercial livestock farming, replacing existing small-scale farmers' subsistence livestock, and many farmers were eager to enter this new and more profitable sector. Since the early 1980s the production of domestic beef has increased steadily with a short-term decrease following the 1998 Asian financial crisis (Korea Rural Economic Research Institute 2010). However, the number of livestock producers gradually declined. During the past twenty years the number of livestock farms decreased from 620,000 in 1990 to 170,000 in 2009, with average herd sizes increasing from 2.6 heads per livestock farm to 15.1 heads in 2009 (Korea Rural Economic Research Institute 2010). Accordingly, the rise in meat consumption sharply increased South Korea's dependence on imported grains such as corn and soybeans because of the ecological barrier to feedstuff production within the country caused by the high level of utilization of agricultural lands for rice farming. As a result, South Korea is now the third largest importer of corn and the fifth largest wheat importer in the world (United States Department of Agriculture, Foreign Agricultural Service 2013). The shift in diet also posed new challenges to government food security policies. Food security policy in recent years shifted its focus from rice production to securing animal feed and hence became more about feed security (see a parallel example of this in Chapter 13, Viet Nam); feed security is the central element of post-2007–2008 food security policy.

The 1980s saw a gradual decline both in terms of national food self-sufficiency and economic security of rural livelihoods, leading to the emergence of pro-democracy rural movements that played a crucial role in the democracy struggles of the 1980s (Abelmann 1996). The introduction of democracy did little to change the tide of food import dependence. With the completion of the Uruguay Round of the GATT, South Korea agreed to gradually remove all import restrictions on agricultural commodities except rice. Food self-sufficiency between 1980 and 2000 declined from 69.6 per cent to 55.6 per cent of national requirements and from 56 per cent to 29.7 per cent if one includes feed grain. This decline in food self-sufficiency clearly reflects the rapid change in diet towards animal protein, shifting the food security challenge from one focused on rice to one increasingly related to meat consumption. The economic importance of the agricultural sector also continued to decline *vis-à-vis* strong growth in South Korea's export sector. In 1980, agriculture contributed 13.8 per cent to GDP but only 4.2 per cent in 2000. For two and a half decades South Korean food security and economic development were underwritten by a global food regime with relatively low and stable prices benefiting the overall economy, but to the detriment of domestic agriculture.

The array of developmentalist agricultural institutions were forced to partly redefine themselves as the statist period of agricultural policy was gradually dismantled (Kim 2006; Burmeister 2000). Throughout this process of liberalization, state institutions, public corporations and state-controlled cooperatives that previously controlled almost every aspect of agriculture had to redefine their roles in some areas, but they also retained control of certain sectors such as rice production and financial credit. While rice and meat sectors remained protected, overall agricultural policy shifted to protecting the agricultural household as international competition began to severely depress the rural economy. The declining importance of the domestic agricultural sector in food security policy and competition from overseas producers led to reduced income levels and higher indebtedness among Korean farm households (Korea Rural Economic Research Institute 2010: 40). Since 1995 farm household debt has increased by 300 per cent while farm household income has only increased by 50 per cent. In 1995 urban household income was only slightly higher than rural income, but by 2009 farm household income was only 60 per cent of urban household income (Korea Rural Economic Research Institute 2010: 103). Policies to address this included debt relief, campaigns to encourage consumers to buy Korean, expanding agricultural and non-agricultural income opportunities through rural industrialization projects, and improving living conditions (Korea Rural Economic Research Institute 2010). With the conclusion of the Uruguay Round of the GATT in 1995, South Korea lifted all import restrictions on agricultural products except rice. The further liberalization of the South Korean food sector deepened the 'defensive' role of agricultural policy and state-controlled institutions in seeking to protect and shelter farmers from global competition rather than increasing productivity.

For two decades, national food security moved away from self-sufficiency strategies to an increasing dependency on agricultural imports. External pressures, the declining economic and political importance of national agriculture, and changing diets have been influencing factors. As Burmeister stated in 1992, the South Korean minifarm structure of small-scale rice and livestock producers moved from a relatively articulated relationship with national economic development in the 1970s to one of relative disarticulation in the following decades (Burmeister 1992). The linkages between food security, domestic agriculture and national economic development were weakened as export-oriented industrialization led to increasing integration into the world market. Furthermore, the more meat-centric diet put pressure on the production capacity of South Korea's limited agricultural land that was already reserved primarily for rice production. The rapidly growing livestock industry had to rely on imported feedstuffs. Food security policy instead shifted to a dependence on the global grain markets in what McMichael has termed the corporate food regime controlled by northern trading companies (McMichael 2009a). Domestic agricultural policy in turn became more about managing the decline of South Korea's farm sector. But from the government's development perspective, the reliance on foreign grain imports was in many ways beneficial because of two decades of declining food commodity prices (FAO 2012b). This was all about to change in the new millennium.

Current status of South Korea's food import dependence

Today South Korea is the world's fifth largest grain importer, importing more than 13 million tons annually from the global market (Lee 2011); however, it does not have any significant control over global grain supply chains. South Korea is primarily dependent on imports from the US, Australia and Brazil for soybeans, wheat and corn and the trade is controlled by four major trading companies – Cargill, Archer Daniels Midland, Bunge and Louis Dreyfus Commodities Group (LDC) – which make up 56.9 per cent of the total South Korean grain trading volume (H. Park 2011a). If one includes a handful of Japanese traders such as Mitsui, Mitsubishi and Marubeni, then 79 per cent of total imports of the three major grains[8] are controlled by foreign trading companies (H. Park et al. 2012). Of the three major grains, large parts of soy and corn imports are for animal feed. Approximately 75 per cent of imported corn is used for feed and in 2009 almost 17 million tons of feed were consumed in total by Korean livestock and broiler operations (Ting et al. 2012). There is clear evidence that food supply concerns are now about securing animal feed. South Korea's food supply is thus sourced from a few countries, and a limited number of companies is considered to leave South Korean food industries and livestock producers particularly vulnerable to price fluctuations by influential policymakers from both sides of the political spectrum, government officials in the Ministry of Food and Agriculture and influential Korean agricultural economists (Chung 2011). With no large domestic agricultural trading firms, the South Korean government cannot stabilize prices through futures markets and other mechanisms. The potential impacts of rising and volatile world grain prices on the overall economic performance of the economy and political stability are major worries for South Korean policymakers. According to Organisation for Economic Co-operation and Development (OECD) inflation statistics, consumer price inflation in South Korea has been above the average of OECD countries from 2008 to 2011, especially, food price inflation. It has been two to five times higher than the average OECD level except in 2008. Food price inflation in Korea was 8.1 per cent in 2011, which is the second highest rate among all OECD countries (OECD Stats 2013). In most periods manufacturing products and services make up the lion's share of consumer price inflation. But in May 2009 and around September and October 2010 food price inflation's contribution to total inflation was 35% and 50% respectively. The contribution of food price inflation to total consumer price index is of concern to lawmakers, who fear voter backlash and negative impacts on the national economy.

The Korean Food Import Price Index (KFIPI 2005 = 100) is heavily influenced by global raw material prices and the way it behaved in the years immediately after the 2007–2008 food crisis caused concern among Korean lawmakers. The KFIPI rose from 140 in January 2008 to 179 in October 2008. In 2009 the KFIPI went down again to approximately 155, only to rebound sharply in the second half of 2010, reaching almost 240 in January 2011 (Bank of Korea 2012; FAO 2012a). Korea came out of the 2007–2008 food crisis without significant economic consequences, but it left many experts and policymakers concerned about the high import dependency and its long-term consequences for the economy. Food price increases can affect inflation and hence impact on national economic strategy. Therefore, the Korean government deemed it necessary to come up with solutions to long-term instability in the global food markets.[9] State intervention into South Korean food and agriculture, as a result of global grain price fluctuations reflect not only a continued developmentalist stance of the government, but also an expectation from agricultural producers, food industries and the general public that the state should intervene as it has done in the past to protect Korean economic interests. It was no longer sufficient to rely on global trade and 'free markets' governed by the WTO and controlled by a few large trading companies. The question was how to address this challenge. Here again the linkages between food security and economic development become apparent, but in new configurations. First, as stated earlier, diets had become more rich in animal protein, accounting for most of the surge in grain import dependence. Food security was thus not about absolute measures of access to sufficient calorie intake, but rather about how to maintain certain diets in ways that did not negatively affect the overall competitiveness of the economy. Thus, the main question of food security became how to secure a buffer supply of feed grains for price stabilization. The government had to consider various ways to stabilize food supply and prices, taking into account the limitations imposed by the GATT and WTO both in terms of regulation and market power imbalances in the global food market. The government set out to identify new strategies for food supply management for the twenty-first century.

Post-2007–2008 food crisis and neo-mercantilist food/feed security strategies

In an interview with the newspaper *Korea Herald,* senior researcher Kim Yong-Taek from the government research centre Korea Rural Economics Research Institute (KREI) explained the perceived weakness of Korea's food system: 'We have long lacked a control system for agricultural commodities. Despite tough conditions surrounding the issue, it's a timely decision given the necessity and a global trend' (Shin 2011). In a concerted effort to address the perceived weakness in the Korean food system, the Ministry of Food, Agriculture, Forestry and Fisheries[10] (MIFAFF) announced a new strategy for food security in 2008 under the name '10-year Comprehensive Plan for Overseas Agricultural Development'. The ten-year strategy set targets for overseas production and provided a framework for supporting overseas agricultural development in order to gain greater control over overseas agricultural production and trade. The strategy had two major components: i) establish overseas trading companies which could secure commodities and stabilize prices through the futures market and ii) support overseas agricultural production, processing and logistics.

The first part of the strategy sought to establish agricultural trading firms in key markets. The first firm was set up in 2009 in Chicago, the site of the Chicago Board of Trade (CBOT), under the name At Grain Company (At Grain). The Chicago commodities exchange is the single most important exchange for trading. At Grain is a joint venture between the public corporation Korea Agro-Fisheries Trade Corporation (55 per cent), Samsung C&T (15 per cent), STX Corporation (15 per cent) and Hanjin Transportation Co., Ltd. (15 per cent) (S. Park 2011a). All three private companies are among the Korean heavyweights in logistics and trade. At Grain planned to invest an initial USD 45 million and up to USD 240 million over a ten-year period. By establishing a firm in Chicago the South Korean government hoped to improve utilization of futures markets to secure stabilization of grain prices and reduce risk exposure. At Grain is also actively engaged in acquiring the necessary infrastructure, such as grain elevators and port facilities, to control the grain commodity chain. The target is to supply up to 30 per cent of Korea's grain needs through this partnership (H. Park 2011b). The government is also considering setting up similar trading firms in key markets such as Brazil, Russia and Ukraine. Despite the involvement of four Korean conglomerates, the government retains a majority share in the company. At Grain is a clear example of what Evans

(1995) calls a demiurge[11] role for the state and it emphasizes the strong role that the government still plays not only in setting policies and regulation, but also as a direct player in food security policy.

The second component of the strategy sought to establish overseas food production, logistics, processing and marketing activities. The target is to secure 385,000 hectares of overseas farmland by 2018 producing 1.8 million tons of wheat, corn and soybeans (S. Park 2011b). This amounts to approximately 10 per cent of total grain imports and the intention is to build a buffer stock of grains. The government so far has provided a budget of USD 197 million between 2007 and 2011 for loans to Korean companies willing to enter into overseas agricultural activities (Ministry for Food Agriculture Forestry and Fisheries 2012). The loans have a low interest rate of 2–3 per cent and are repayable over ten years with a five-year grace period.[12] As a condition for granting the loan, companies are required to bring government-mandated amounts of their products back to Korea in case of food emergencies. It is, however, uncertain under what circumstances this clause can and will be activated. The strategy was revised in 2011 and made into law as the 'Overseas Agricultural Development Cooperation Act' in January 2012. Apart from becoming law, the strategy also includes additional measures to ensure more effective overseas food production. The existing '10-year Comprehensive Plan for Overseas Agricultural Development' from 2008 will continue, but measures will be taken for more systemic support of company activities overseas. Second, the 'Council of Overseas Agricultural Development' chaired by the vice minister of MIFAFF will be established to monitor and review policies and projects. Third, the 'Association of Overseas Agricultural Development', a private industry association, was founded in the spring of 2012 with the support of the state company Korea Rural Community Corporation (KRC).[13] Lastly, new specialized funds to support overseas agricultural investment were made possible by the new act.

The main agency for implementation of overseas support services is the Overseas Agricultural Development Service (OADS), a division under KRC operating under a mandate from MIFAFF. OADS provides a comprehensive support system for private enterprises wanting to invest in offshore food production. OADS is responsible for evaluating loan applications and monitoring the performance of companies given loans. The money provided as loans has to be spent within a year and the company can submit subsequent loan applications in following years. OADS also provides general investment environment surveys for different countries as well as customized surveys for companies granted loans at no additional cost. Survey services include analysis on labour supply, agricultural and rural infrastructure surveys, etc. OADS also visits the overseas projects receiving government loans to assist companies in enhancing agricultural productivity. Thus, while the agricultural operations themselves are entirely in private hands, government involvement is apparent in pre-screening and site valuation, in training and technical assistance of company staff and in evaluation. These support policies are implemented because most Korean private companies have little experience and expertise in overseas agriculture. Given that overseas agricultural development is a high-risk and capital-intensive investment, the government is providing a broad range of services to encourage more Korean companies into this new area of business.

At the end of 2012, more than 100[14] companies were active in 20 different overseas countries. Twenty-nine companies had been granted loans from the government. Loans were granted to both private companies as well as semi-public companies often owned by local governments as shareholders; 42,300 hectares of farmland had been leased or acquired and 171,000 tons of wheat, corn and beans had been produced by the end of 2011 (Ministry for Food Agriculture Forestry and Fisheries 2012). According to OADS data from early 2012, 74 of the total of 85 declared overseas agricultural investments by Korean companies are concentrated in Asia – particularly in Cambodia, the Philippines, Laos and Viet Nam – and Far East Russia (Korea Overseas Agricultural Development Service 2012). Korean companies seem to prefer regional investments closer to home for logistical reasons. According to OADS, more than 65 per cent of all private investors[15] produce grains such as maize, wheat and soybean, Korea's main import grains. So while the Korean government has been engaging in offshore agricultural investments directly through state-owned companies in trade, it is trying to follow a more indirect role in promoting South Korean–controlled production overseas. The companies receiving loans from the funds provided through the Overseas Agricultural Development Cooperation Act are mainly small and medium sized companies, but larger Korean conglomerates are also involved either directly or as shareholders in some of the smaller companies and it is expected they will expand activities higher in the value chain in the future. The government hopes that as expertise is accumulated, Korean companies will begin to invest in more profitable areas of the global agro-industrial value chain, but it seems so far that the larger companies are taking a more cautious approach.[16]

International criticism regarding land grabbing is a cause for concern. By only being indirectly involved through loans and support services, the government seeks to shelter itself from potentially bad international publicity. In the 2008 case when the negotiations of the South Korean company Daewoo Logistics with the Madagascan government for the lease of 1.3 million hectares became international headline news (Blas 2008; Financial Times 2008; Walt 2008), this was seen to reflect negatively on South Korea despite the fact that the government was not involved in the deal.[17] In an interview, the director of OADS stressed that they require companies receiving loans from the government to respect international codes of conduct for foreign land investment as well as national laws in host countries. There is however limited or non-existent monitoring of whether the companies comply.[18] The government is also seeking to reduce criticism and opposition in host countries by integrating investments with other aspects of Korean overseas relations such as Official Development Aid (ODA). MIFAFF seeks to strengthen linkages between ODA projects and private corporations' investments to reduce potential local resistance and establish friendly relations with host countries, as well as to raise the rate of success for South Korean companies. Thus, to reduce investment risk, ODA activities such as technology transfer and knowledge dissemination are being strengthened.

The Rural Development Administration (RDA), another agency under MIFAFF, is actively conducting agricultural ODA projects. For example, in 2010 the RDA established an agricultural research centre in Cambodia. So far research activities have centred on developing corn varieties suitable for Cambodian conditions. Interestingly enough, this is also the crop that South Korean companies operating in Cambodia are the most interested in growing. Through overseas agricultural technical centres, called KOPIA (Korea Project on International Agriculture),[19] RDA provides local technical support on cultivation techniques, seed breeding and machinery. It also implements projects including technology transfer, agricultural resource development, training programs and infrastructure development.

The new food supply strategy is a significant departure from earlier periods. First of all, while it still relies on overseas grain resources, the Korean government is taking both a direct and indirect role in establishing some degree of control over the feed grain supply chain.[20] Second, new sectors of the Korean economy are being mobilized for this new food security strategy including large shipping companies, trading companies, feed companies and small to medium sized companies and entrepreneurs with no previous experience in agricultural production, logistics or marketing. There are some clear linkages to previous developmental state efforts in the industrial sectors, but also significant differences. The government does not have the tools to force companies into new areas of business through coercion as in earlier eras. Instead, the government provides incentives such as cheap loans and stresses the economic potential in overseas agriculture as a new profitable sector that Korean companies can enter. At a conference in the summer organized by the Overseas Agricultural Development Association, a private industry organization,[21] government officials from South Korea, Brazil, Indonesia and Russia presented promising investment prospects for Korean companies. Government food security policy is thus being integrated with the interests of domestic companies searching for new lucrative business opportunities overseas, which, it is hoped, will benefit economic development domestically.

The 2007–2008 food crisis called for new configurations. The liberalization of agricultural markets through GATT and the WTO and the corporate grain oligopolies were now deemed detrimental to food security and continued economic development. Liberalized markets could no longer be trusted to provide a stable supply of cheap grains. The overseas agricultural development plan is building on old business relations, but it is entering a new sector. Despite consolidation in the domestic agricultural sector, the minifarm structure persists to this day, and government agencies and large government-controlled public corporations still play a central role in managing food supply and agriculture even though some of their leverage has been curbed by liberalization. The agricultural institutions that played a central role during the 'Big Push' in the 1970s are once again mobilized, but this time for an entirely different project. This time food security is to be secured through acquisition of overseas agricultural resources and infrastructure, either directly via government state-owned enterprises or indirectly via private companies tempted into the business by favourable loans and possibilities in new and potentially profitable business areas. Rather than mobilizing the domestic agricultural sector in another 'Big Push', the government has turned to the corporate sector and entrepreneurs who are always looking for new economic opportunities. The role of domestic agriculture remains unclear in this context of new food security priorities.

Conclusion

Neither food security, economic development nor the developmental state are static concepts. They are given particular meanings and purposes depending on changing national, international and global contexts. They appear in various configurations over time and each configuration is historically contingent on previous periods as well as perceived new challenges. Food aid backed the fledgling South Korean government's search for political stability following the devastating war with the North. Later, as industrialization became the main focus of the ruling regime, food aid ensured cheap food to the growing ranks of industrial workers while squeezing the countryside through price depression on agricultural commodities. The dynamics of food security, economic development and the developmental state in South Korea have undergone several reconfigurations over time. Food security and economic development are often regarded as mutually reinforcing processes where food security improves as nations get wealthier (Timmer 2005). But, as this chapter suggests, food security is not a static objective. It is a dynamic concept that changes over time according to changing environments and political priorities.

Similarly, the developmental state is not a static entity. Many observers of South Korea's economic development have considered the past three decades as a gradual dismantling of the developmental state, but the developmental state is far from dead. As Joseph Wong observed in 2004:

> The reconfiguration of the developmental state today and into the future does not mean its obsolescence. The developmentally oriented state continues to play important roles in East Asia's economic, social, and political development. One should not equate liberalization, globalization, transnational harmonization, or economic policy convergence with the retreat of the state from the tasks of promoting national development.
>
> *(Wong 2004: 357)*

The developmental state has reconfigured itself for new challenges and is constantly looking for new ways to manage the economy despite heavier constraints. The election of Lee Myung Bak in 2007 saw a re-emergence of developmental state features in the food sector. Price volatility of key imported commodities such as food, energy and industrial raw materials propelled his administration to a renewed emphasis on the need to directly and strategically manage the supply of raw materials, including food and feed.

In the case of South Korea, food supply management has re-emerged as a major priority for the government in recent years. The framing of these policies uses the language of earlier eras but the qualitative content of policy has changed. While food security policy in the years following the Korean War was about adequate food supplies, food supply policy has gone through variations of food security as securing *desired food using desired means*. In the 1970s government food supply policy primarily meant ensuring an adequate supply of rice, using domestic agriculture to boost overall economic growth. In this process, developmental state institutions were the primary agents for facilitating mobilization of the rural population behind the government's development agenda. As diets changed and the global conditions for statist protectionist agricultural measures deteriorated, food security policy hinged on cheap feed grains from global markets. The importance of domestic agriculture for food security diminished and was more or less sacrificed in free trade negotiations as South Korea's economic development depended on the export of manufactured goods. In the twenty-first century the instability and uncertainty of global markets has led to new configurations between food security, economic development and the developmental state institutions. The domestic agricultural sector remains on the margins while the corporate and entrepreneurial sectors have become the agents of change that the government seeks to mobilize for food security, not by producing food within domestic territories, but rather through direct control over agricultural resources overseas. In this respect, South Korea's policy of overseas agricultural investment follows in the footsteps of similar Japanese initiatives that began in the 1970s with most notable success in Brazil and Australia. While many have declared the end of the developmental state in the age of economic liberalization, this chapter has highlighted that the developmental state institutions have endured and re-emerged albeit in a less authoritarian form, but with a decidedly developmental impetus. National development is still high on the political agenda for South Korea. North Korea may still be the primary military threat, but concern about the competitiveness of the national economy in an increasingly tense regional economic and political atmosphere in Northeast Asia is perhaps a stronger motivation today. Resource security, in a resource scarce country, has come to play a prominent role in contemporary South Korean politics. As in earlier periods, the government is seeking to mobilize business and popular support on a national(ist) platform that emphasizes the threat that resource insecurity may pose for national security and maintaining the developmental achievements of the past. State intervention in the food and agricultural sectors remains strong despite 25 years of slow liberalization, but a shift has occurred since 2007. The state gradually reduced its historically strong role in managing food supply and demand from the early 1980s up until the 2007–2008 food crisis. For 25 years, state intervention focused primarily on protecting and preparing domestic agricultural producers and food industries for market liberalization. This changed in 2008 when the state, once again, took on a more aggressive role both directly through its At Grain joint venture in the US and indirectly by providing finance and technical services to Korean companies overseas. This shift in policy and role of the state in the food and agricultural sectors indicates that the government at the time perceived food and agriculture as a potential new cause for macroeconomic instability and hence for new national development goals. The business–state alliance that is well known from developmental state theory on South Korean industrialization has historically not been a significant player in food security policy. This appears to have changed in the twenty-first century, as food becomes an attractive area of investment. As big business enters the food supply chain in South Korea, the structure of the food system domestically and overseas may change significantly.

The study of food and agriculture in South Korea highlights some of the dilemmas of the East Asian development model that are rarely discussed. Both South Korea and Japan are economies in which export-oriented industries and foreign trade are dominant, but where food and agriculture policy is a constant matter of concern for the state for both economic and political reasons. In South Korea, many policymakers and experts often cite the weak food and agricultural sectors as reasons why the country is still not on par with the richest countries in the world. That food supply can become the potential Achilles' heel of one of the most significant economic development successes of the twentieth century requires us to rethink the weaknesses of conventional ideas of development and modernization.

Notes

1 The Joseon state was founded in 1392 and lasted until 1897. The state covered the entire Korean peninsula up to the Yalu River. Today the river is the border between North Korea and China.

2 Since the Japanese invasions of Joseon of 1592–98, the kingdom maintained an isolationist policy recognizing Chinese suzerainty, but otherwise severely restricting interaction with the outside world.

3 The early period of Park's rule did see the establishment of state institutions that were to become central in the 1970s food self-sufficiency drive including the National Agriculture Cooperative Federation, the Rural Development Administration, and the Union of Land Improvement Association, which later would be renamed the Korea Rural Community Corporation.

4 South Korea experienced a population boom especially between 1955 and 1966. Because of the lack of economic opportunities and land in rural areas, many people had few options but to migrate to urban centres looking for employment in the nascent industrial sectors and education. Between 1960 and 1975, rural to urban migration averaged 445,000 people a year. The urban population grew from 28 per cent to 50.9 per cent during this period (Jang 2005: 19).

5 The extent to which the 'squeezing' of the agricultural sector through taxation and depression of food prices was a significant source of finance for industrialization is debatable. See for example Kay 2002 and Hsiao 1981.

6 Saemaul Undong is often hailed as a success story in rural development, but also attracts widespread criticism. The failure of Saemaul Undong to deliver sustainable improvements in rural livelihoods was for example an important factor in the rise of peasant movements during the 1980s. Saemaul Undong was also used for political control and corruption, and remains to this day a controversial rural development program (Moore 1984; Boyer and Ahn 1991; Kang 2002; Abelmann 1996).

7 South Korea entered the GATT (General Agreement on Tariffs and Trade) in 1967 but restricted imports of many agricultural commodities important to domestic production such as rice, barley, maize, etc.

8 The soybean is technically not a grain but for the sake of simplicity, it is defined as a grain in this chapter.

9 Japan, which is similar to Korea in terms of population density, geographic environment for cultivation, agricultural structure, and dietary life, embarked on intensive investment in the global grain trading business in the mid 80s. It focused on securing grain collectors, storage facilities or elevators near ports in the US. Through this process, Japanese trading transnational corporations (TNCs) such as Marubei, Itochu, Mitsui and Mitsubishi success-fully entered the global grain trade business. As a result, even though Japan is highly reliant on food imports, it is estimated that the Japanese feed themselves reliably by their own channels away from the influence of major grain traders (Kim 2010). This Japanese model was recommended to Korea by many opinion leaders after the 2007–2008 food crisis.

10 The ministry changed its name in early 2013 to Ministry of Agriculture, Food and Rural Affairs.

11 By 'demiurge' Evans refers to the state as a direct producer and provider of services.

12 In the initial ten-year overseas agricultural development plan, the grace period was three years and the repayment period was seven years. The loan terms were changed in 2012 to allow companies more time to establish their businesses.

13 KRC is a state-owned company under the Ministry for Food, Agriculture, Forestry and Fisheries. The company's main area of activity is rural infrastructure development. The company has operated overseas agricultural cooperation projects for decades and is the main implementing agency for the overseas agricultural development program.

14 Every Korean private and public company investing in overseas agriculture has to report it to the government. Thus, the data reported to OADS includes all Korean companies that have completed negotiations and made actual investments in host countries.

15 According to Korean law, any Korean company engaged in agricultural activities overseas with the intent of exporting back to South Korea has to declare their activities to the government.

16 Interview with the vice chairman of the Korea Overseas Agricultural Development Asso-ciation.

17 The Madagascar case, however, never really made headline news in South Korea and the majority of the general public are unaware of the failed attempt by Daewoo Logistics.

18 In the director's opinion, ethical oversight of investments was the responsibility of the host government and not of the Korean government.

19 KOPIA centres are located in Asia (Viet Nam, Myanmar, Cambodia, the Philippines, Thailand, Sri Lanka), Community of Independent States (CIS) and South America (Uzbekistan, Paraguay, Brazil, Bolivia, Ecuador), and Africa (Kenya, Democratic Republic of Congo, Algeria, Ethiopia).

20 Establishing overseas agricultural production is not such a new idea. South Korea also initiated efforts in the 1970s and the 1980s but both attempts failed. There is however very little information about these earlier efforts because they were quite small and unco-ordinated.

21 The Overseas Agricultural Development Association is on paper a private industry orga-nization, but it was initiated by MIFAFF, the offices are hosted by the public corporation KRC, and the chairman and vice chairman are both senior KRC managers.

References

Abelmann, N. (1996) *Echoes of the Past, Epics of Dissent: A South Korean Social Movement,* Berkeley: University of California Press.

Amsden, A.H. (1989) *Asia's Next Giant: South Korea and Late Industrialization,* New York: Oxford University Press.

Araghi, F. (2010) 'Food regimes and the production of value: some methodological issues', *Journal of Peasant Studies,* 30(2): 41–70.

Bacchi, C. (2000) 'Policy as discourse: What does it mean? Where does it get us?', *Discourse: Studies in the Cultural Politics of Education,* 21(1): 45–57.

Bacchi, C. (2009) *Analysing Policy: What's the Problem Represented to Be?* Frenchs Forest, Australia: Pearson.

Bank of Korea (2012) *Economic Statistics System, Bank of Korea.* Online. Available <http://ecos.bok.or.kr/> (accessed 17 September 2012).

Blas, J. (2008) 'S Korean group to lease farmland in Madagascar', *Financial Times,* 19 November 2008. Online. Available <www.ft.com/cms/s/0/22ccaa98-b5d9–11dd-ab71–0000779fd18c.html#axzz1ZtUwI4le>.

Boyer, J. (2010) 'Food security, food sovereignty, and local challenges for transnational agrarian movements: the Honduras case', *Journal of Peasant Studies,* 37(2): 319–51.

Boyer, W. W. and Ahn, B.-M. (1991) *Rural Development in South Korea: A Sociopolitical Analysis,* Cranbury: University of Delaware Press.

Brandt, V.S.R. (1979) 'Rural development in South Korea', *Asian Affairs,* 6(3): 148–63.

Burmeister, L. L. (1990) 'State, industrialization and agricultural policy in Korea', *Development and Change,* 21(2): 197–223.

Burmeister, L. L. (1992) 'Korean minifarm agriculture: from articulation to disarticulation', *The Journal of Developing Areas,* 26(2): 145–68.

Burmeister, L. L. (2000) 'Dismantling statist East Asian agricultures? Global pressures and national Responses', *World Development,* 28(3): 443–55.

Burmeister, L. L. (2006) 'Agricultural cooperative development and change: a window on South Korea's agrarian transformation', in Y.-S. Chang and S.H.L. Lee, (eds), *Transformations in Twentieth Century Korea,* pp. 64–86, Oxon and New York: Routledge.

Chung, J.-Y. (2011) 'High food and commodity prices and policy implications', *Samsung Economic Research Institute Weekly Insight,* 16(11): 9–13.

Chung, O.-I. (2006) *Korea Under Siege, 1876–1945: Capital Formation and Economic Transformation,* Oxford: Oxford University Press.

Cumings, B. (2005) *Korea's Place in the Sun: A Modern History,* New York: W.W. Norton.

Doner, R.F., Ritchie, B.K. and Slater, D. (2005) 'Systemic vulnerability and the origins of developmental states: Northeast and Southeast Asia in comparative perspective', *International Organization,* 59(02): 327–61. Online. Available <www.journals.cambridge.org/abstract_S0020818305050113> (accessed 5 March 2013).

Evans, P. (1995) *Embedded Autonomy: States and Industrial Transformation,* Princeton N.J.: Princeton University Press.

Food and Agriculture Organization of the United Nations (2012a) *FAO Food Price Index.* Online. Available <www.fao.org/worldfoodsituation/wfs-home/foodpricesindex/en/> (accessed 9 December 2012).

Food and Agriculture Organization of the United Nations (2012b) *The State of Food and Agriculture 2012,* Rome: Food and Agriculture Organization of the United Nations.

Financial Times (2008) 'Food security deal should not stand', Editorial, *Financial Times,* 19 November 2008. Online. Available <www.ft.com/cms/s/0/20cf7936-b670–11dd-89dd-0000779fd18c.html#axzz1KcarljSc> (accessed 26 April 2011).

Francks, P., Boestel, J. and Kim, C.H. (1999) *Agriculture and Economic Development in East Asia,* Abingdon, UK: Taylor & Francis.

Friedmann, H. and McMichael, P. (1987) 'Agriculture and the state system: the rise and fall of national agricultures, 1870 to the present', *Sociologia Ruralis,* 29(2): 93–117.

Goodman, D. (1997) 'World-scale processes and agro-food systems: critique and research needs', *Review of International Political Economy,* 4(4): 663–87.

Goodman, D. and Watts, M. (1994) 'Reconfiguring the rural or fording the divide? Capitalist restructuring and the global agro-food system', *Journal of Peasant Studies,* 22 (1): 1–49.

Halabi, Y. (2011) 'Protracted conflict, existential threat and economic development', *International Studies,* 46(3): 319–48.

Hettne, B. (2010) 'Development and security: origins and future', *Security Dialogue,* 41(1): 31–52.

Hsiao, H.-H.M. (1981) *Government Agricultural Strategies in Taiwan and South Korea: A Macrosociological Assessment,* Taipei: Institute of Ethnology Academia Sinica.

Jang, S. (2005) *Land Reform and Capitalist Development in Korea,* Gyeongsan: Gyeongsan National University.

Kang, D.C. (2002) *Crony Capitalism,* Cambridge: Cambridge University Press.

Kay, C. (2002) 'Why East Asia overtook Latin America: agrarian reform, industrialisation and development', *Third World Quarterly,* 23(6): 1073–1102.

Kim, C. (2006) 'The rise and decline of statist agriculture and the farmers movement in South Korea', *Korea Observer,* 37: 129–47.

Kim, Y.-T. (2010) 'Realities and challenges of offshore agricultural development', in *Outlook for Agriculture 2010: Green Growth and New Way Out of Agriculture and Rural Communities,* pp. 373–403, Seoul: Korea Rural Economic Research Institute.

Kohli, A. (1999) 'Where do high-growth political economies come from? The Japanese lineage of Korea's "Developmental State"', in M. Woo-Cumings (ed.), *The Developmental State,* pp. 93–136, New York: Cornell University Press.

Kohli, A. (2004) *State-Directed Development: Political Power and Industrialization in the Global Periphery,* Cambridge: Cambridge University Press.

Korea Overseas Agricultural Development Service (2012) *Status of Overseas Agricultural Expansion,* Korea Overseas Development Service. Online. Available <www.oads.or.kr> (accessed 23 August 2012).

Korea Rural Economic Research Institute (2010) *Agriculture in Korea 2010,* Seoul: Korea Rural Economic Institute.

Lee, C.-H. (2013) *Food War 2030,* Seoul: Korea Food Security Foundation.

Lee, G. J. (2011) 'War with global four majors: food security a matter of life and death', *Segyeilbo,* 9 September 2011 (in Korean).

Maxwell, S. (1996) 'Food security: a post-modern perspective', *Food Policy,* 21(2): 155–70.

McMichael, P. (2009a) 'A food regime analysis of the "world food crisis"', *Agriculture and Human Values,* 26(4): 281–95.

McMichael, P. (2009b) 'A food regime genealogy', *Journal of Peasant Studies,* 36(1): 139–69.

McMichael, P. (2013) 'Land grabbing as security mercantilism in international relations', *Globalizations,* 10(1): 47–64.

McMichael, P. and Kim, C. (1994) 'Japanese and South Korean agricultural restructuring in comparative and global perspective', in P. McMichael (ed.), *The Global Restructuring of Agro-Food systems,* pp. 21–52, Ithaca NY: Cornell University Press.

Ministry for Food Agriculture Forestry and Fisheries (2012) *Four Years of the Lee Myung-Bak Government Agricultural Achievements and Key Promotional Tasks of 2012,* Seoul: Ministry of Food, Agriculture, Forestry and Fisheries.

Moore, M. (1984) 'Mobilization and disillusion in rural Korea: the Saemaul movement in retrospect', *Pacific Affairs,* 57(4): 577–98.

OECD (1999) *Review of Agricultural Policies in Korea,* Paris: OECD.

OECD Stats (2013) *Consumer Prices Food 2000–2012.* Online. Available <http://stats.oecd.org> (accessed 19 March 2013).

Park, H. (2011a) 'New food security strategies in the age of global food crises', *Samsung Economic Research Institute Monthly Focus,* Seoul: Samsung Economic Research Institute.

Park, H. (2011b) 'Rising agro prices and responses', *Samsung Economic Research Institute Weekly Insight,* 16(30): 9–13.

Park, H., Jung, H., Kim, H. and Chae, S. (2012) 'Korea needs own global resource majors', *Samsung Economic Research Institute Weekly Insight,* 9(12): 9–13.

Park, S. (2011a) 'South Korea starts grain venture in Chicago to secure supply', *Bloomberg,* 29 April 2011. Online. Available <www.bloomberg.com/news/2011–04–29/south-korea-starts-grain-venture-in-chicago-to-secure-supply.html> (accessed 23 July 2012).

Park, S. (2011b) 'South Korea to increase overseas farming on record food costs', *Bloomberg,* 10 July 2011. Online. Available <www.bloomberg.com/news/2011–07–10/south-korea-to-expand-overseas-farming-on-rising-food-costs.html> (accessed 20 March 2011).

Shin, G.-W. (1998) 'Agrarian conflict and the origins of Korean capitalism', *American Journal of Sociology,* 103(5): 1309–51.

Shin, H. (2011) 'Korea strives for agricultural security', *Korea Herald,* 24 March 2011. Online. Available <www.koreaherald.com/business/Detail.jsp?newsMLId=20110324000671> (accessed 24 March 2011).

Statistics Korea (2011) *Social Indicators in 2011,* Seoul.

Statistics Korea (2012) *Per Capita Food Grain Consumption in Korea,* Seoul.

Timmer, C.P. (2005) 'Food security and economic growth: an Asian perspective', *Asian-Pacific Economic Literature,* 19(1): 1–17.

Ting, M. K., Choi, S. and Smith, G. H. (2012) *Grain and Feed Annual 2012,* Washington, DC: USDA Foreign Agricultural Service.

United States Department of Agriculture, Foreign Agricultural Service (2013) *World Wheat, Flour, and Products Trade,* Washington, DC. Online. Available <www.fas.usda.gov/psdonline/> (accessed 19 March 2013).

Walt, V. (2008) 'The breadbasket of South Korea: Madagascar', *Time Magazine.* Online. Available <www.time.com/time/world/article/0,8599,1861145,00.html> (accessed 5 May 2011).

Wong, J. (2004) 'The adaptive developmental state in East Asia', *Journal of East Asian Studies,* 4: 345–62.

Woo, J. (1991) *Race to the Swift: State and Finance in Korean Industrialization,* New York: Columbia University Press.

Woo-Cumings, M. (1999) *The Developmental State,* Ithaca: Cornell University Press.

Woo-Cumings, M. J.-E. (1998) 'National security and the rise of the developmental state in South Korea and Taiwan', in Rowen, H.S. (ed.), *Behind East Asian Growth,* pp. 319–37, London and New York: Routledge.

Young, H.K. and Dong, S.B. (1981) 'Food policies in a rapidly developing country: the case of South Korea, 1960–1978', *The Journal of Developing Areas,* 16(1): 47–70.

Zhu, T. (2001) *Threat Perception and Developmental States in Northeast Asia,* Canberra: Australian National University Department of International Relations.

17

FOOD SECURITY, NICARAGUA

Food purchase or food production, which is best for the rural poor?

Francisco J. Perez, Arthur H. Grigsby and Sandrine Fréguin-Gresh

Introduction

Nicaragua's rural areas are dominated by smallholders who rely on their agricultural production to meet a significant proportion of their household consumption needs.[1] However, most producers do not cover their annual consumption of staples such as maize and beans. This chapter looks critically at whether efforts to support increased production are the best way to increase the food security of these smallholders and for Nicaragua as a whole. To understand the implications of food security policy choices, this chapter analyses trade-offs, opportunities and risks facing smallholders as they look for the best way to enhance their food security by either trying to maintain or increase their subsistence production, pursue off-farm income sources or combine both activities.

Farming is a risky business in Nicaragua. Over the past 23 years agriculture in Nicaragua has suffered various shocks caused by changes in international coffee prices (2001–2); weather phenomena (El Niño); major hurricanes (Mitch 1998 and Felix 2007); drastic reductions in public expenditure (1990–6/2001–2); increasing input, energy and fuel costs (2006–7); food price spikes (2007–9); the global financial crisis (2009–11); and a domestic cattle price crisis (2010–13). All of these factors have influenced the food security level through reducing incentives to basic food production, and by affecting employment, income and prices for urban consumers.

In the last two decades, the main attempts to address food insecurity by public institutions, international agencies and NGOs have changed little in Nicaragua. In the peak of the crises, food has been distributed for food insecure populations, and in periods of recovery, food production for subsistence has been supported. Significant resources have been invested in technical assistance, seed distribution and input distribution, but average yields have scarcely improved. According to FUNICA (2012), average maize yields in Nicaragua are only 0.95 TM/ha for maize,[2] while those for beans are 0.53 TM/ha. By comparison, maize yields in El Salvador average 3.23 TM/ha and 0.59 TM/ha for beans.

Thus, the idea of investing funds to promote food production by households primarily oriented towards subsistence raises questions of the efficient use of scarce resources allocated to this sector. Focusing food security strategies for rural households on agricultural production might not be the most effective way to enhance their food security. Finding alternatives for employment and income to provide food access could have greater impact. Using a case study, this chapter draws on data from a national survey to investigate whether allocating resources for increasing smallholder subsistence production is the best option for enhancing the food security of rural households or not.

This is not just a technical issue. The policy implications of such a shift are profound, and indeed could be seen to imply a redefinition of the role of the state in Nicaragua. Historically Nicaraguan development has been defined by its agricultural sector: cattle and indigo producers before 1890, coffee producers since 1890, agro-industrial producer and exporter (bananas, cotton, coffee and sugar) after the 1950s. Contrasting economic models have been applied, including modernization paradigms (1950–79), state-centred models (1980–9) and a free-market-oriented economy (1990–2006). Currently, the focus is on promotion of traditional exports, led by strong cartel groups with a strong social agenda for reducing rural poverty. According to the Central Bank Annual Report 2012 (BCN 2013) and the World Bank (WB 2013), the agriculture sector remained the most important economic sector given its contribution to GDP (18.6 per cent), employment (31 per cent) and share of total exports (89 per cent of merchandise). The share of agriculture in total exports in the period 2009–12 (88 per cent) is evidence of the significance of the role of agriculture in Nicaragua's economy.

However, the conditions for a continued focus on agriculture are changing. Nicaragua is in a demographic transition. The rural population is shrinking, falling to 44 per cent of the total population, and 40.4 per cent of the total economically active population (see Figure 17.1). Since 2002 there has been a clear trend of decline in proportion of the population working in agriculture.

The Agrarian Census 2001 (INEC 2002) found that there were 200,000 farms, of which 59 per cent were subsistence families (those who depend on rural labour markets), 31.7 per cent peasants (those who hire temporary workers), 6.6 per cent farmers (those who hire permanent labour) and 2.7 per cent agrarian capitalists (those who do not live on their farm). The 2011 census reported 262,546 farming households but 15 per cent of them controlled 74.5 per cent of the total land (INIDE 2013). The proportion of small farmers is growing but at a higher rate than their proportion of total land holdings. Thus, small farmers with less than 7 ha increased from 47.4 to 60 per cent of total farming households, but their control of land increased at a lower rate from 4.3 to 5.6 per cent of total land.

Expenditure on the agricultural sector represents 6–7 per cent of total public expenditure. Nicaragua promoted a national strategy for rural development based on developing and enhancing specific value chains using Porter's competitive

Nicaragua: Agriculture employment generation (% of Total employment)
60.0
50.0
40.0
30.0
20.0
10.0
0.0
1990
1991
1992
1993
1994
1995
1996
1997
1998
1999
2000
2001
2003
2004
2005
2006
2007
2008
2010
y = -0.0892x² + 1.0971x + 37.415
R² = 0.6419
Agriculture (% of Total employment)
Manufacturing (% of Total employment)
Services (% of Total employment)
Polinómica (Agriculture (% of Total employment))

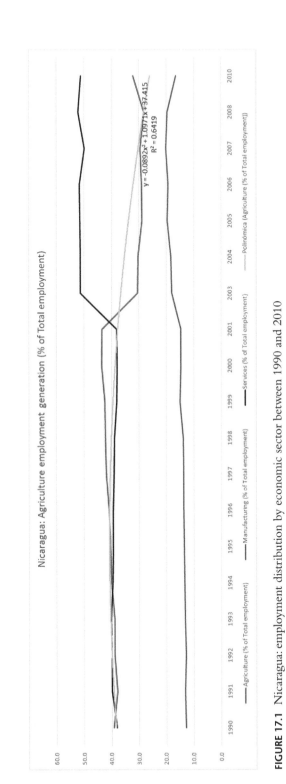

FIGURE 17.1 Nicaragua: employment distribution by economic sector between 1990 and 2010

Source: based on World Development Indicators 2013.

clusters approach. The main goal was to enhance Nicaraguan comparative advantages on food production by reducing transaction costs through vertical integration of all the processes in a value chain. For vertical integration, the key export products (the most competitive ones) were selected: coffee, peanut and soybean, dairy, meat, vegetables, grains and shrimp. Using this approach, the Nicaraguan state aims to increase agriculture's contribution to GDP, employment and exports as the main way to reduce rural poverty.

However, the food security policy was not clear. Up to 2006, this was a tacit policy, and it was expected that the drivers of economic development (agricultural exports and export processing zones) would generate enough employment for the Nicaraguan population; however, by 2005 the Living Standards Measure Survey (LSMS) report shows an increase in the number of families under the extreme poverty line (from 15.2 to 17.2 per cent of total population). This situation opened a new discussion on how to improve Nicaragua's food security status, and with a new government and a different approach regarding the role of the state, a new food security policy was designed and released in 2009.

This chapter presents part of the national debate about what are the main tools for reducing food insecurity in rural areas. Using value chains and agricultural production system approaches, the authors analyse rural household responses to the changes in the governance of value chains stimulated by the economic liberalization process, with emphasis on the food crisis. A key theoretical assumption is that families use an economic rationale for responding to structural changes and policy incentives. Thus, there is not a global answer, but rather specific ones for every segment of an agrarian structured society.

Food security response to volatile food prices

Nicaragua faced a major food price spike between September 2007 and March 2008. Food prices had already been rising since 2003, with short-term increases in 2003, 2005, 2006, leading over time to a virtual doubling of prices, mainly for rice, beans, maize and vegetable oil. The 2007–8 crisis was related to several factors such as oil price increases, climate shocks, international prices, population growth, price expectations and shifts to bio-fuels. Despite signals since 2006 that the world was facing this crisis, the Nicaraguan state only reacted in September 2007, after Hurricane Felix, when poor domestic production and expectations of high prices could no longer be ignored.

The state response to the crisis was basically of two types: trade policy measures and food distribution to food insecure populations. The trade policy measures focused on suspending tariffs for bean imports and subsidizing consumer prices, and as a result rice imports doubled and bean imports increased as well. The government suspended the import tariff (30 per cent) for beans in October 2007 (MIFIC 2007a). This policy was a supply-side tool, aiming to increase the total volume in the market and as a result of the increased supply, achieve a new lower price equilibrium. Between October and December 2007, Ministerio de Industria Fomento y

Comercio [MIFIC][3] (2007b) reported imports of beans as follows: 90.71 MT[4] from Mexico, 78 MT from USA and 503.2 MT from China, totaling around 672.2 MT. Prices did not fall as quickly as expected and in the case of rice, they did not fall to the levels before 2007. The oligopolistic behaviour limited the effects of trade policies by restricting price transmission and retaining the gains of market (De Franco and Arias 2010). A key factor influencing public response to the crisis was the absence of a national food stock linked to a lack of infrastructure for food storage.

The second policy response was a consumer price subsidy ranging from 12 to 30 per cent. This program was focused on urban areas and benefited around 14.5 per cent of all Nicaraguan families, but it did not meet demand throughout the year. The national government bought products from farmers at market prices and sold them to consumers at lower prices than the market. As a result, and depending on the market share of the government sales, this alternative forced intermediaries to reduce their profits and ultimately reduced prices. Through the National Enterprise for Access to Basic Food (ENABAS), and using several distribution points, the government sold packages of basic products, initially just 12 kg of beans per week, but in a second phase, the packages included up to 12 kg of rice, 10 kg of sugar and 4 litres of vegetable oil. These were sold to every family able to buy them with no restriction; families' access to basic products and grains depended on their access to cash and the availability of foodstuffs at their local distribution point. The national government set fixed prices in order to subsidize consumers by around 30 per cent of the highest peak market price. The policy included four basic items: beans, maize, rice and vegetable oil (ENABAS 2008). Because of price variation, by February 2008 the fixed price represented different levels of subsidy depending on the product: rice 20 per cent, beans 15 per cent and vegetable oil 12 per cent (ENABAS 2009).

In November 2007 the food distribution program (ENABAS) started to sell grains and vegetable oil in Managua and Pacific urban areas. In the case of beans, consumer prices were between 60 and 67 per cent of the market price. Managua was the key geographical target area in this program because it has 24.6 per cent of the country's total population, and 39.7 per cent of the total urban population. In 2007, 84 per cent of distribution points and 92 per cent of beneficiaries were in Managua. Based on the assumption that every beneficiary came from a different household, this program may have benefited up to 9.3 per cent of the total population. Limits to coverage were related to the limited infrastructure for collecting and distributing food and the necessary process of creating a neighborhood level of distribution points.

Distributions to food insecure populations have also been developed, starting in 2007, with mechanisms linked to incentives for employment and for maintaining children in school. There is also a program for food production: in this case families receive seeds and fertilizers for producing grains. The crisis was a window of opportunity for developing the food security policy and programs in 2009. Programs for fighting hunger in rural areas, which have been implemented since 2007, were integrated into this policy. These include Zero Hunger (provides animals to families who own at least 0.7 ha), Patio Coupon (provides horticulture and medicinal plant seeds for landless families), seeds and coupons for fertilizers (for those families who produce at least 0.7 ha of grains). In the 2009 focus group interviews undertaken by Perez et al. (2010), the members of rural communities recognized the importance and effects of food security and price subsidy programs; however, they pointed out that only a few people benefited from the coupons; the amount of food distributed with consumer price subsidies was small; and there were long periods without procurement and subsidized food being available.

Market integration

Nicaragua is in the early stages of market integration, since most of the producers are not linked to national and international value chains. The potential of a given producer to benefit from these value chains relates to several factors. New and challenging demands of these markets with respect to quality, volume and delivery date are clearly set for some items, while most still maintain traditional standards. The state has yet to assume a clear role in facilitating/arbitrating disputes between producers and industrial groups. Public policies and commitments to facilitate technical change in order to meet product quality requirements (e.g. vaccines and production-related infrastructure) are weak. Integration into dynamic, value-added product markets and value chains has been limited at the national level, since measures to mediate between smallholders and buyers such as contract farming and vertical integration (production, gathering, processing and trading) are weak. Integration is primarily happening at other levels of the marketing channels (wholesale markets, collection centres, other intermediaries).

In the case of grains, the producers have a weak integration to high value chains. In the case of maize, there was an early integration to supermarkets with some small groups of producers beginning to establish maize areas for the purpose of supplying supermarkets. The maize value chain associated with industrial processing is mainly linked to the coffee industry, where maize is a complementary input for obtaining low prices and low quality coffee products. Maize and sorghum are used in coffee mixes in different relationships: 80 per cent of coffee and 20 per cent maize, 70:30 and 60:40. The market for red beans is quite competitive, and cooperatives and producer organizations consider this crop to be a very high-risk activity requiring capital and liquidity, since domestic prices are volatile, and exports are restricted by the government. Supermarkets tend to be big buyers that use traditional and national brokers to buy grains in different regions of the country, according to the production season. Because of low prices for producers, there are few incentives for investing in technology that improves quality of food production. Quality requirements mostly address moisture and foreign material in the first level of transformation associated with cleaning and packing. As a consequence small and medium producers are linked only to low value-added chains.

Because of these various factors, caution is warranted in assessing the potential of rural families to take advantage of commercial markets. Some producers have been integrated into the new commercial chains, but many subsistence producers have been excluded from new commercial markets.

Liberalization changing terms of trade for agricultural production

Trade liberation has opened regional markets, and Central America is the main partner for agricultural exports for Nicaragua. The USA is the second partner and a free trade agreement has been signed. Nicaraguan tariff protection is the second lowest in the region after El Salvador. By 2021 it is expected that the average tariff will be 1.6 per cent. Nicaragua has signed free trade agreements with Mexico and Chile, and it is negotiating with Canada, Taiwan and the European Union.

This is in sharp contrast to the 1980s, when there was a strong public intervention in order to control trade and distribution of products and inputs. The Sandinista government developed several institutions for trading coffee, rice, cotton, meat and sugar. After 1990, price controls were eliminated and market institutions were privatized and as a result commercial chains were reconfigured. Currently, agricultural products are traded by market mechanisms of supply and demand. There are no controlled prices, nor is there direct public intervention for fixing prices.

Prices for commercial agricultural products have been unstable during the last 16 years. Gourmet and organic coffee have offered alternatives with higher and more stable prices. Meat, rice, peanut and sesame prices have been relatively stable, but the general trend for prices of grains for producers[5] has been downward since 1999. The costs of inputs such as nitrogen-based fertilizers correlate with petroleum prices, which are currently rising. Therefore, the terms of trade for agricultural production have been deteriorating over time. Table 17.1 highlights that by 2008 a producer needed 659 kg of maize in order to cover the cost of 181.8 kg of fertilizers. This amount of maize is equivalent to 45 per cent of the national average yield (1,477.3 kg). A similar situation happened in 2012, when producers needed 454.54 kg of maize to cover the cost of fertilizers.

Export product value chains, such as coffee, meat, sugar, sesame and peanuts, tend to be subject to monopoly control either in processing and/or the exporting process. Infrastructure facilities for coffee processing (cleaning, peeling, parchment stage, cutting and packing) tend to be concentrated in a few companies such as CISA Exportadora, the meat industrial group and the dairy industrial group. There is a regional integration process through foreign investment in products such as peanut, sesame and cheese by Salvadorian and Guatemalan companies. There are initial levels of vertical integration to global networks such as Starbucks (coffee), and Cargill (poultry and pork) and Wal-Mart through Hortifruti (fruits and vegetables), since there are several cooperatives selling directly to those firms.

According to Grigsby and Perez (2007; 2009) and Michelson et al. (2012), traditional informal markets tend to dominate domestic trade; however, supermarkets are increasing their share of vegetable markets. Supermarkets are part of a global

TABLE 17.1 Terms of trade between grain and fertilizer prices in Nicaragua

Year	Price of beans for producers (C$/qq)	Price of maize for producers (C$/qq)	Fertilizer prices (N, P, Q) (C$/qq)	Fertilizer prices Urea (N) (C$/qq)	Cost of fertilizers for beans C$/mz)	Cost of fertilizers for maize (C$/mz)	Term of trade for beans	Term of trade for maize
2001	321.2	93.4	132.5	150.0	264.9	564.9	0.8	6.1
2002	311.0	101.3	142.1	146.5	284.1	577.2	0.9	5.7
2003	265.5	94.3	151.7	171.2	303.3	645.7	1.1	6.8
2004	412.7	126.9	202.0	254.4	404.0	912.7	1.0	7.2
2005	491.4	152.0	231.5	296.8	462.9	1,056.5	0.9	7.0
2006	401.8	142.7	267.7	311.5	535.4	1,158.5	1.3	8.1
2007	547.3	216.4	366.8	426.2	733.6	1,586.0	1.3	7.3
2008	901.0	216.3	832.4	732.6	1,664.7	3,129.8	1.8	**14.5**
2009	611.9	341.6	761.8	559.1	1,523.6	2,641.7	2.5	7.7
2010	889.3	279.7	597.5	552.9	1,195.0	2,300.7	1.3	8.2
2011	876.0	302.8	664.1	605.6	1,328.1	2,539.4	1.5	8.4
2012	684.1	274.6	712.9	672.1	1,425.7	2,769.8	2.1	**10.1**
2013	494.1	351.4	710.4	687.6	1,420.8	2,795.9	2.9	8.0

Key

C$ = Cordobas (local currency), qq =4 5.45 kg, mz = 0.7ha

NPQ = Nitrogen, Phosphorus and Potassium; a complete fertilizer

Terms of Trade refers to the amount of maize/beans necessary to cover the cost of fertilizers

The bold numbers indicate that, based on 2008 price of maize, producers would need 14.5 QQ (659 kg) to cover fertilizer costs. This would represent 69 % of average maize yield.

Source: Perez (2011), based on MAGFOR price statistics.

integration with the USA network, mainly through Wal-Mart. Hortifruti, a regional company, tends to monopolize supermarket supply of fruit and vegetables.

Factors impacting on agriculture as a smallholder livelihood

The potential to increase production and engage in markets is related to the ability of smallholders to afford and access inputs. According to Grigsby and Perez (2007), after 1990, the agricultural input market was controlled and there was oligopolistic control of imports and distribution of 70 per cent of seeds, fertilizers and pesticides, since only three companies were involved. Nitrogenous fertilizers and pesticides are the main intermediate imports for agriculture. Financial markets are developing with commercial banks and micro-finance institutions providing approximately USD 220 million in agriculture loans, covering between 15 and 20 per cent of producers. However, credits are mainly for short-term loans and with interest rates higher than 25 per cent (Perez 2011). The government is promoting rural financial markets by creating a produzcamos (bank for agricultural production) however, the procedures are even more bureaucratic than the private sector, and as a result the farmers' organization claims that few small farmers have access to credits.

Nicaragua has a dynamic land market, although there are serious property conflicts. Land prices vary from USD 4–7,000/ha in the well-connected Pacific areas to USD 60–80/ha in the new agricultural areas, and where there is limited infrastructure. In the next decade, because of population pressures and aggressive land concentration, access to land by smallholders is likely to come under pressure, leading rural households to migrate to urban areas.

Agricultural extension and agricultural insurance markets are not developed. Extension programs are subsidized by public funds and international cooperation and reach a minority of smallholders. There are no public policies to foster insurance markets, and this is a serious restriction for a farm-contracting system, since agriculture is considered very risky.

After 1990, international trade and domestic trade were reconfigured, as public institutions were dismantled. International price crises, free market policies, property conflicts, high levels of unemployment and long periods of drought generated a new socioeconomic context for rural families. Nicaragua is now in a transition phase: rural population growth has decreased, but not enough to reduce pressure on natural resources, thus, internal migration will continue. In the short run fertile land and water resources will be scarce. Approximately 30 to 40,000 persons from the rural economically active population struggle unsuccessfully to find jobs in agriculture every year.

Depending on the social sector (rural workers, subsistence families, peasants, farmers or agrarian capitalists), families have adapted their economic units to the new context. For some crops, such as coffee, subsistence and semi-subsistence families sometimes benefit from support by NGOs, international cooperation and public institutions in order to be integrated into alternative markets. For other products such as cheese, smallholders are generally excluded from the quality value chains, since they have no capital to make the investments needed to meet increasingly strict standards. They can access credits, but for short-terms and at high interest rates, a situation that is a disincentive to investment in cattle, for example. These factors suggest that poverty is a serious constraint for rural families in terms of technical change, investment and market integration.

In the past, the poor often migrated by opening up new land at the 'agricultural frontier', i.e. the formerly vast rainforests in the east of the country. Today the only unexploited land still available includes small areas bordering natural reserves, indigenous communities and swamps of the Caribbean areas. Nonetheless, the extensive path of agrarian development is incorporating 106,000 hectares a year into agriculture. In 2012, the deforestation rate was estimated at 70,000 ha/year. The unsustainability of these practices is starting to restrict internal migration. Together with population growth and declining soil fertility in some areas these changes seem likely to trigger an aggressive land concentration process. Subsistence families mainly use extensive agricultural production systems, since they cannot afford technical investments to increase yields. Subsistence families tend to rely on natural soil fertility, and after two to three years of farming newly opened land, yields decrease significantly. There is a call for public policies to restrain this extensive production model with a high level of land concentration; a controversial alternative for changing this situation is the implementation of a progressive tax on land tenure. This option might force the intensification of (especially) cattle production since owning a large area of land will become expensive.

Thus, the potential for overcoming constraints and obstacles lies in intensification, which in turn requires market integration to generate the necessary investment capital. Intensification for smallholders sometimes involves integration into fair trade, quality and organic markets. Technical change can also increase yields, producing higher volumes in the same area.

This is not the only option. Short-term migration has been an important exit for rural families. This allows families to migrate to Costa Rica and/or El Salvador during harvesting season in order to obtain funds for sowing in May. International migration outside the region is also an important economic strategy; however, new regulations in the USA are blocking migrant flows.

Case studies: Whose food security benefits from which pathway?

Nicaragua is facing a divergence of trends in agriculture. On one hand, there is rapid growth of agro-exports; and on the other, rising levels of food insecurity. Pressures to produce for markets, shifts to reliance on wage labour, and declining prospects for subsistence production are changing the relationships between livelihoods and food security. This chapter describes the reliance of rural households on purchasing foodstuffs, with some households buying more than 90 per cent of the total amount of food they consume. These findings have policy implications since most food security programmes directed towards subsistence and extremely poor households are based on the assumption that increased production is an effective way to address their food insecurity. These findings suggest that food production primarily contributes to food security among medium income households.

The evidence from Nicaragua that is presented in this section is based on the results of a national survey of smallholder strategies to respond to globalization's effects (Grigsby and Perez 2009). Territories, municipalities, counties and agro-food value chains were selected to be part of the survey in order to examine the distinct processes of differentiation of rural families and the different levels of integration related to infrastructure and markets. Regions were selected based on the principle of a representative analysis of three different dynamics reflecting different degrees of market integration and productive infrastructure, and were classified as three types of territories:

- 'Winners' from globalization: the municipality of El Cua is located in the coffee zone with specialization and integration processes in organic coffee chains, mainly among smallholder households. Although this municipality has low level of access to infrastructure, it has developed a high level of integration with global coffee value chains through organic, fair trade and conventional markets. The municipality of El Viejo, in the Pacific, represents the area of greatest market and infrastructure integration. It is a territory where pockets of rural land reform smallholders are embedded among large landowners who produce peanut and sesame. It is an area of high investment in roads, mass media and access to seaports.

- '*Marginalized*' territories, far from markets and infrastructure: the municipality of La Libertad has a differentiation process in its dairy value chain. The area is inaccessible, with poor infrastructure subject to seasonal disruption. Its population is dispersed in the rural districts, and many households rely on mules for access to markets.
- '*Intermediate*': in the municipality of Muy Muy, the dairy processing industry exerts an influence on prices and product quality, and in the municipality of Terrabona, in the dry zone, families combine non-agricultural strategies, temporary migration, and production of traditional market grains and vegetables for supermarkets.

The evidence from these contrasting areas is used to explore the differentiation processes as determined by agricultural structures and the dynamics of marginalization and exclusion in different contexts. The distinct characteristics allow an exploration of patterns of household adaptation to the restructuring of rural markets and the role of alternative non-agricultural activities in the reconfiguring of the household economy. A quantitative analysis was performed using a household survey, which allowed us to analyse household strategies in response to globalization's structural effects on agriculture and rural development. Based on the census conducted in the qualitative study, a random sample of households was chosen combining those in which agriculture played different roles. The total number of surveyed households was 1,575 using a safety margin of 10 per cent replacement, eventually including 1,458 in the survey analysis from La Libertad (290), Muy Muy (299), Terrabona (281), El Cua (300) and El Viejo (288). A production system economic analysis was performed. This allowed calculation of the agricultural value-added, using a factor production approach. The analysis also includes non-agrarian income and an analysis of the production efficacy and efficiency, based on the production factor approach: labour, land and investment.[6]

The overall objective of the survey was to analyse the basic assumptions of differentiation, marginalization, exclusion and reconfiguration of rural household economies as consequences of trade liberalization. In order to explore the determinants of the inclusion/exclusion of rural families in relation to dynamic value chains, this analysis included two levels: family and territorial. Specifically it aimed to analyse the relationship between the reorganization of value chains and the segmentation of households in rural areas through the cases of beans, dairy products and vegetables. It also investigated how changes in value chains stimulate economic dynamics in territories with quite different characteristics in terms of infrastructure, production, types and distance from markets. Finally, the segmentation process within rural territories and its relationship to social, economic and ecological vulnerability was explored.

The survey results identify a clear relationship between the contribution of agriculture to the household (in both absolute and percentage terms) and the increase of annual total income. There was also a close relationship between the contribution of income from agriculture to overall household income and food purchasing. A smaller agricultural income led to higher expenditure on food items. There was also a clear relationship between income levels and the proportion of non-agrarian income, with households in the top income quintile having a higher proportion of non-agricultural income; thus subsistence families are not the main users of alternatives such as urban labour markets, small businesses in the service sector or even migration (remittances). This result is important since there are several projects devoted to fostering poor families' incomes through non-agrarian economic activities, promoting small businesses related to craftsmanship and the service sector.

One of the first decisions for rural households is how much of the production will be sold and how much will be kept for self-subsistence in order to avoid buying food in the following months. In periods of scarcity, costs for some staples (maize and beans) can rise significantly (250 per cent higher than the price in a regular year). This occurred in 2007–9. Household decision-making processes are influenced by factors such as debts and cash flows during the year. It was expected that an initial reaction of rural families would be to guarantee basic consumption of maize and beans. This could be achieved either by increasing planting areas and/or reducing marketed volumes. The second option would be related to opportunities to take advantage of good prices. However, expanding the area for cultivation increases demand for labour and inputs and probably means renting land. Increasing production areas also competes with other strategies such as integration with rural labour markets (yielding immediate income), or resources for cash crops, or commercial crop areas that are the main generator of liquid cash flow for the year.

Choices are influenced by the extent of the family's labour force and the nature of the production system for cash crops, especially coffee, peanuts, dairy and meat, which allows them sufficient income to buy food. Other options for obtaining cash income are non-agricultural activities such as urban labour markets, trade and remittances. These are primarily alternatives for households with higher incomes.

On average, the surveyed households devote 41 per cent of their production to consumption; however, there are differences between the territories, and in terms of income levels of households. The cattle-producing areas have higher than average self-consumption, with 53 per cent of production of grains, dairy and tubers being consumed in the household. Key determinants are the difficulty of accessing roads and low coverage of supermarkets and even traditional markets. The municipalities with greater integration of supermarkets and traditional markets reported lower levels of self-consumption (El Cua 32 per cent and El Viejo 24 per cent). In these municipalities, sales represent more than 65 per cent of production and self-consumption is mainly restricted to grains and dairy.

Three trends were identified in the household survey:

- Home self-consumption is based on crop production, and animal production tends to play a very limited role in this regard. Milk consumption is minimal and the tendency is to consume dairy products as cheese.
- In terms of self-consumption behaviour, the higher the household's income, in absolute terms (USD), the larger the amount of money is spent on self-consumption.
- However, in relative numbers (per cent), families with a limited income devote a large proportion of their production to self-consumption. Since their income is limited, every dollar in food consumption will have a higher weight (Figure 17.2).
- In areas with limited access to infrastructure and markets (La Libertad and Muy Muy), lower income households tend to consume more of their production, while in municipalities with access to markets and marketing channels (El Cua and El Viejo), the trend is different, since all households sell most of their production (Figure 17.2).

Because the survey was conducted in rural communities, it was expected that the level of food purchasing would be limited, since most rural families are food producers. However, it was found that 98 per cent of households bought food and it accounted for 67 per cent of the household total food consumption. The territories with greater integration of infrastructure and markets are those with a higher dependence on the market for access to food. The cattle production territories are the least dependent on purchasing. La Libertad is a municipality with a lower dependence on purchased food. Food production on-farm covers on average 50 per cent of consumption. Families with higher income levels tend to purchase more food, which is expected given that the land is used mainly for livestock production. In the municipality of Muy Muy, food production covers less than 40 per cent of consumption; 50 per cent of families must purchase rice and beans and 15 per cent of families must purchase the three basic staples in their diet: rice, beans and corn. On average, in the cattle production territories 97 per cent of households acquire food by purchasing, and purchased food accounts for 57.4 per cent of total food consumption.

Terrabona is the second most food insecure municipality because most households are net food consumers and they depend on labour markets. Fewer households produce food, with two key factors being the limited rainfall and the lack of fertile land for growing beans. In the poorest quintile, 38 per cent of households are net food buyers, and their livelihoods rely on rural labour markets and /or migration. Throughout the sample, food purchasing was found to be more important in the poorest quintile. The basic food basket is comprised of rice, cooking oil and sugar (over 90 per cent of households). The less frequent food basket consists of beans, maize, chicken and dairy. Thirty-one per cent of households purchase maize, beans and rice.

El Cua and El Viejo are territories with a strong presence of smallholders and they illustrate different situations. In the municipality of El Cua over 80 per cent of households produce food and coffee. However, in the poorest households, food production does not cover 30 per cent of total consumption, and 21 per cent of total households depend on the market for accessing basic grains (see Figure 17.3). The

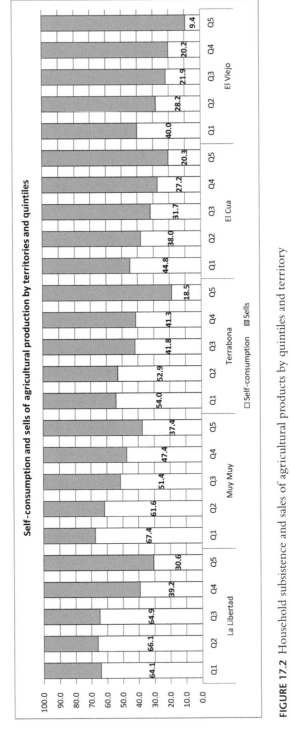

FIGURE 17.2 Household subsistence and sales of agricultural products by quintiles and territory

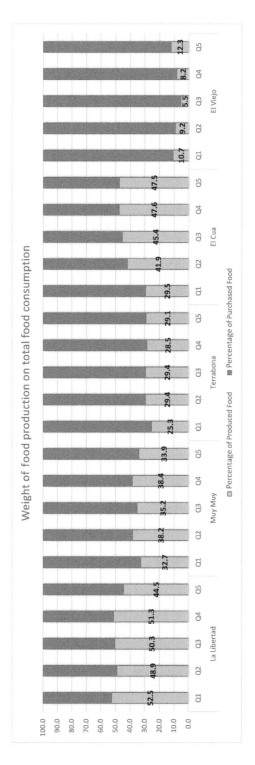

FIGURE 17.3 Nicaragua: weights of food produced and food purchased as a total of food consumption in five rural regions

municipality of El Viejo is the area most vulnerable to increases in food prices (for consumers), since 60 per cent of households do not produce food and 91 per cent of food is purchased at local markets. Beans, a key element in the diet, are a major component of food purchased. In the other territories beans were purchased by 16 and 49 per cent of families; in the case of El Viejo, this product is purchased by 95 per cent of households. The main items in the food basket include cooking oil, rice and beans. In the case of maize, El Viejo tends to be self-sufficient. Although it is a rice production area, 93 per cent of households have access to this product via purchase.

Concluding remarks

Rural families access food by two key alternatives: food production and food purchasing. It is commonly presumed that producers guarantee most of their food through production. However, our survey shows a different scenario; on average rural families buy more than 50 per cent of their food for consumption. When territories are analysed based on their access to infrastructure and markets, the survey shows that the more closely they are integrated into markets, the lower is the proportion of food which they have produced for total consumption.

In the case of food purchases, it was found that subsistence families (bottom income quintiles 1 and 2) purchase 68 per cent of their food, and have a high level of dependence on markets for their food security. This is related to small and non-fertile areas and livelihoods based on rural labour markets as key cash sources. Promoting food security for this group of families through agriculture faces the challenge that off-farm activities are the key income source for these families; thus food production is a low priority. Policies to control the Consumer Price Index for food and beverages, and/or those related to enhancing labour markets and wage levels, might be more effective ways for ensuring food access for around 26.6 per cent of rural households that live below the extreme poverty line, and even for the 63.3 per cent that live below the general poverty line.

When the dynamics of household self-consumption was analysed by specific territories, it was found that market integrated territories showed that initial perceptions of global integration were correct. Producers in municipalities with a high level of integration to infrastructure and markets tend to prioritize sales of food production rather than consumption. In this case, producers will use the scarce land on cash crops rather than self-consumption crops. They assume that cash crops will generate enough income for buying food at the harvest season.

In the other regions without regular access to infrastructure and markets, the basic pattern is that households with average incomes below the extreme poverty line tend to give priority to self-consumption, and as families increase their income, the proportion of sales is higher. In this case, since access to food for self-consumption is not regular, there is not stable food distribution; families tend to guarantee family needs. As income increases, the chance of travelling to the next urban area also increases, and with it access to a stable food market.

In areas with limited integration into dynamic markets, farm incomes are important for 97 per cent of interviewed households. Assets (land, livestock) and networks are key factors that determine income differentiation and integration into higher value chains in these areas. Poor households with limited access to resources, and with the ability to combine agriculture and livestock activities, can develop strategies to sell their products without endangering their food security. Livestock sales are a key element of food security, as those without animals face hunger in certain periods of the year.

Most of the households (87 per cent) grow grains but only 49 per cent of them sell some of the harvest; the rest opt to store part and use the remainder to meet the family's nutritional needs. The contribution of subsistence farming to income is a primary element, mainly in the context of high prices and shortages. Farm income ranges between an average 50 per cent and 60 per cent of total income, and shows a positive trend, in which the higher the agricultural income, the higher the household income. Non-agricultural employment reaches between 16 per cent and 30 per cent of total income, while remittances can represent between 6 per cent and 10 per cent of revenue. Thus, the possibility of entering the urban services labour market is important for improving household income.

Self-consumption is based on crop production rather than animal production. Milk consumption is minimal and the tendency is to consume dairy products as cheese. Self-consumption also represents an elastic behaviour, with respect to income levels in absolute terms. The higher the household's income, in absolute terms, the greater amount of self-consumption there is. In areas with limited access to infrastructure and markets, lower income households tend to depend on their own food production, while in municipalities with access to markets and marketing channels, households sell most of their production.

Notes

1 According to the national census (2011), 63.4 per cent of total farms produce maize. However, 37.5 per cent of maize producers own 1 ha or less, and this is the segment with the lowest production and yields in the country. In the case of beans, 52.2 per cent of total farms produce beans, and 64.5 per cent of bean producers own 1 ha or less. They have an average yield of 0.452 TM/ha, about 50 per cent of the national average (INIDE 2013).
2 Average 1999 to 2010.
3 Ministry for Promoting Industry and Trade
4 Metric Tons.
5 Prices for consumers have increased, but this did not lead to higher prices for producers because of high levels of intermediation and/or cartel dynamics.
6 It was developed for French Agricultural Research Centre for International Development (CIRAD) and implemented by the World Bank in the project Rural Struc Programme.

References

BCN (2013) *Reporte Anual de 2012,* Banco Central de Nicaragua.
De Franco, M. and Arias, D. (2010) *Are Food Markets in Central America Integrated with International Markets? An Analysis of Food Price Transmission in Honduras and Nicaragua,* Working Paper, World Bank.

ENABAS (2008) *Informe de acopio y comercialización, Febrero 2009,* Gobierno de Reconciliación y Unidad Nacional.

ENABAS (2009) *Plan Operaciones Institucional de ENABAS* (Contribuciones al Desarrollo del Plan del Sector Agroalimentario), Gobierno de Reconciliación y Unidad Nacional.

FUNICA (2012) *Estado actual, oportunidades y propuestas de acción del sector agropecuario y forestal en Nicaragua,* Managua: Fundación para el Desarrollo Tecnológico Agropecuario y Forestal de Nicaragua (FUNICA).

Grigsby, A. and Pérez, F. J. (2007) *RuralStruc Program. Structural Implications of Economic Liberalization on Agriculture and Rural Development in Nicaragua. First Phase: National Synthesis,* Managua: MAG-FOR, World Bank and Nitlapan.

Grigsby, A. and Pérez, F. J. (2009) *Peasant Strategies to Respond to Globalization's Structural Changes. Second Phase: Households Analysis,* Managua: World Bank and Nitlapan.

INEC (2002) *III CENAGRO Resultados Finales,* Managua: Instituto Nacional de Estadísticas y Censos de Nicaragua (INEC).

INIDE (2013) *IV Censo Agrario, Resultados,* Managua: Instituto Nacional de Información de Desarrollo de Nicaragua (INIDE).

Michelson, H., Reardon, T. and Perez, F. (2012) 'Small farmers and big retail: trade-offs of supplying supermarkets in Nicaragua', *World Development,* 40(2): 342–54. Online. Available <www.sciencedirect.com/science/journal/0305750X> (accessed 12 March 2013).

MIFIC (2007a) *Acuerdo Inter-Ministerial,* Ministerio de Industria Fomento y Comercio (MIFIC) 048–2007.

MIFIC (2007b) *Autorizaciones para la importación de frijoles emitidas conforme el Acuerdo Inter-Ministerial,* Ministerio de Industria Fomento y Comercio (MIFIC) 048–2007.

Pérez, F. (2011) *Nicaragua: Without Structural Changes There'll Be No Sustainable Reduction of Rural Poverty,* UNAN-Leon Working Paper No. 2, Leon: Universidad Nacional Autonoma de Nicaragua.

Pérez, F., Barrios, J. and Pavon, K. (2010) *Responses to High World Food Prices: Country Study: Nicaragua,* Managua: ODI, DFID and Nitlapan.

WB (2013) *World Development Indicators 2013,* World Bank.

PART III
Conclusions

18

ARE NEW CHALLENGES TO FOOD SECURITY GENERATING A NEW SOCIAL CONTRACT?

Ian Christoplos and Adam Pain

An underlying theme that is common to most of the chapters in this volume relates to the variable and changing nature of the social contract for food security between the state and its citizens. This contract can be seen as responding to the four pillars of food security: availability, access, utilization and stability (FAO 2009). The nature of these contracts between states and citizens varies both spatially and temporally. Some countries continue to demonstrate strong commitments, particularly South Korea and Viet Nam, but the extent to which these responsibilities are seen to be anchored in domestic production has shifted over time, as taking advantage of the benefits of globalization is now thought to be a more effective route to food security than gaining national subsistence through domestic production. Other countries, such as Nepal, show little evidence of the development of such a social contract and, as the Sudan case study shows, food aid may have contributed to the failure to establish such a contract.

The efforts to promote a right to food, analysed by Eide, (Chapter 4) and the India case study, obviously represent an attempt to make these social contracts more universal. However, the uncertainties around the right to food and the closely related but at the same time rather different discourses and dynamics related to rights to water (Munk Ravnborg, Chapter 6 and the Zambia case study), combined with a *de facto* shift of attention to achieving food entitlements from on-farm production sourced food security to market based food security (Nicaragua, Viet Nam and Afghanistan), continue to muddy the waters. The successes and limitations of a 'capabilities' perspective on rights, wherein participation in markets is expected to create entitlements to food security, show a highly mixed picture.

In the perspective of these varied social contracts, rights imply responsibilities, and in order to fulfil responsibilities there must be investments in government (and in many cases private sector) capacities to deal with malnutrition, agricultural services, food safety, access to water and capacities to respond to climatic events. The

Viet Nam and Uganda case studies describe the shifting perspectives on the services that are required in relation to agriculture; the Zambia example describes the conflicts over water related services; and the Sierra Leone case the building of capacities to address malnutrition. In Afghanistan and Sudan the need for national capacities to act on food security has largely been set aside as the international community has taken the lead. In contrast, the South Korea example can be seen as suggesting an effort to take advantage of weak capacities abroad to safeguard food security at home. In a similar vein, Brüntrup (Chapter 5) makes clear that in many respects the issues around land acquisitions relate largely to the capacities (or lack thereof) to manage investments that carry with them great risks to food security.

Food security is in many respects the arena wherein the humanitarian imperative of providing services directly, be it through food aid, nutritional interventions or seed fairs (Sudan, Sierra Leone and Uganda), comes into conflict with commitments to aid effectiveness that mandate greater attention to developing national capacities to ensure food security. The cases here of where duties are being shouldered, such as Viet Nam, illustrate that synergies are possible when local and national authorities invest in both relief and capacities to adapt to volatility in food access and availability simultaneously.

It is very striking that, despite the broad formal recognition of the significant and impending threats to food security from climate change, there is very little evidence found of the necessary capacities being developed to address the chronic nature of this risk. Thus far, climate change related food security risk is primarily being dealt with through relief response after droughts and floods have occurred. There is some awareness of the longer term risks, but so far efforts to develop the capacities to respond to emerging climate change risks have been grossly insufficient.

Food price volatility and dramatic scenarios around demographics and climate change have nonetheless begun to reawaken political commitments to ensure availability. Such growing commitments are reflected in measures such as overseas land acquisition and new investments in research and agricultural services. In some respects there is also a significant new awareness that access does not automatically follow from supply, and that ensuring access to food will not be solved by production increase alone, as illustrated in the shifting policies in Nicaragua and Viet Nam.

By contrast, food utilization and stability remain in a grey zone, both in terms of the extent of the acceptance of a responsibility by the state and the means to enforce it. Food safety is no longer just a concern in developed countries, as the Viet Nam case illustrates. Nutrition has re-emerged on the development cooperation agenda, but it is not certain that this agenda has brought with it durable solutions, as described in the Sierra Leone case. The Indian paradox of growth without reduction of malnutrition rates and the general lack of evidence of significant moves from words to action imply doubt about whether chronic hunger is on the food security agenda. Stability has also re-entered the aid discourse, together with the catchword of resilience, but here again, evidence of actual changes in the social contract in relation to transitional food insecurity is hard to discern. The case studies in this volume show that agriculture and livelihoods are still largely addressed through a modernization agenda reflecting expectations of a linear path to growth in production, productivity and industrialization. Because of this, states pushing for aggregate production increase at the national level may overlook those who are most vulnerable in the face of seasonal and climatic variability.

A new but selective social contract?

It is apparent across the cases and themes analysed that the nature of the social contract between the state and its citizens is increasingly selective, i.e. some risks to food security remain high on the agenda and some do not. One example of this is the great attention paid to conflict-related food security factors in Afghanistan, at the expense of attention to how drought and market dynamics play a greater role in generating food insecurity. In some countries where national food security commitments remain strong, there are localities that are effectively 'forgotten', such as northern Uganda. Furthermore, some food insecure people may be included as part of the social contract, whereas others are excluded, as particularly illustrated in the chapters on India and Nepal that highlight links between inequality and food insecurity. As such, some aspects, populations and phases of the 'crises' described in the introduction to this volume are generating a response, whereas others are largely ignored because the perceived scope and conceptualizations of these crises again varies both temporally and spatially, as described in the chapters by Willenbockel and Koning.

These findings suggest that large gaps exist in the new food security agendas in different countries. Efforts to attain national food security may be moving ahead at a significant cost to the rural households that lack stable access to food. Measures to avoid urban food riots and ensure national availability may impinge on those who live outside the cities and outside the countries that can afford to invest in increased production. Investments in land acquisition and 'grabbing' of water rights for large-scale mechanized farming, hydro-power or mining may be detrimental to the food security of those who were dependent on these resources in the past.

The question of 'whose food security counts' relates to that of 'whose capacities count', at local, national and global levels. At the local level this is often about targeted services and the inclusivity of markets. The ways that local authorities, aid agencies, elites and others intervene to capture or promote services and access to markets demarcates their influence on the food security of local households. These markets may be for agricultural products, but may also be for land and labour. The selectivity of the social contract is being justified in part because of the prevalence of faith in the equity-enhancing 'magic' of these markets, not only to promote food production, but also regarding expectations that the market will generate alternative forms of livelihoods and ensure that services are available to support smallholders.

At a national level, the breadth of the social contract is often about prioritization between what are perceived as high versus low potential areas, either deliberately, with Viet Nam's focus on areas suitable for rice production, or implicitly with the market-driven model of agricultural development in Afghanistan effectively favouring the irrigated areas with market access to the neglect of the significant rainfed, more remote areas. 'Low potential', may also relate to areas prone to conflict, e.g. the failure of the Ugandan government to invest in re-establishing agriculture in the north, or the normalization of food aid as part of the food security system in large parts of Sudan. Food may be a hot topic, but it may also be significantly cooler when it is about the accessibility of food for marginalized populations, be they in the hinterlands, areas written off because of conflict, those farming in 'unviable' conditions or people of the wrong caste or ethnic group.

An aspect that brings together the local, national and international is that of capacities to manage or even drive land acquisition and water resource management. Some governments see it as part of their social contract to acquire land abroad to ensure that food is available within their national borders. Transboundary water conflicts are also becoming part of how nations are struggling to retain control of this key resource amid growing scarcity due to climatic and demographic change. Others are satisfied with following narratives about the benefits of globalization as a justification for privatizing (and selling) land and water assets as an inevitable component of modernization. Some national and local officials are simply rent seeking through the sale or rental of national resources.

There is widespread recognition that politically fragile states are often characterized by multiple disasters and chronic and enduring food insecurity that affects significant proportions of the population. Further food insecurity can be a cause of violence as well as an outcome of it and is increasingly affecting urban as well as rural populations. Such conditions are unlikely to disappear anytime soon (as described by Pain in Chapter 7) and as the Sudan case shows, it is possible to have significant economic growth in such contexts based on natural resources without positive effects on food security. There is a degree of recognition that a quick fix to institutional and capacity deficiencies in such contexts is not on the cards. Simplistic assumptions about how, for example, introduction of new seed varieties or temporary non-governmental organization investments in agricultural extension will lead to food security, and with this faith in the existence of an 'exit strategy' for aid financed services, have been discredited.

Innovation, resilience and path dependency

The conundrums related to the different perspectives and priorities regarding food security described in this volume in many ways relate to the various meanings assigned to the current catchwords of innovation and resilience. And indeed a concept that is less often mentioned, that of path dependency.

There is renewed faith in technological innovation, driven by science, markets or even farmers themselves as a way to arrive at 'win-win' solutions that combine production growth with equity, and to combine greenness with growth. Declarations of the imminent arrival of 'new green revolutions' are increasingly heard.

The equity end of these equations, where mentioned, is often associated with the second catchword of resilience, as some acknowledge that smallholder productivity is inevitably and increasingly variable and unpredictable (due in part to climate change), and that new forms of innovation are therefore required. Crises in both production and consumption are being acknowledged as something that will not go away and are indeed increasingly recurrent.

However it is also recognized that the institutions and agencies that are expected to be the bearers of innovation and resilience are themselves subject to path dependency. The limits to the social contract noted above suggest that the risks of innovation may be perceived as greater than the security of continuing along well-trodden paths. Other signs of path dependence can be seen in the weak response to increased awareness about climate change and the failures to find alternatives to standard agricultural modernization narratives despite glaring evidence that linear growth can no longer be taken for granted.

Crises may constitute critical junctures that trigger innovation. The many descriptions of institutional change in this volume show that innovation is occurring, but highlight how the routes that are taken are rarely congruent with the normative claims being made. This is because the institutions related to research, extension, marketing, input provision, etc. are stuck in an uncertain space where they are expected to be driven by the roll-out of a new modernization agenda, while the clients of these services are struggling to maintain nutritional levels and find jobs when their shrinking farms are no longer sufficient.

Back to the future?

Half a century ago faith in innovation and the seemingly inevitable path to modernization meant that food security was often seen to be a temporary concern. Industrialization would absorb (and feed) smallholder farmers being pushed out of rural areas and science would ensure that the new, efficient mega-farms would guarantee food availability. Weberian bureaucracies would step in to overcome the 'cultural' factors that led to discrimination, exploitation and hunger. These optimistic and technocratic assumptions largely faded in the years that followed. More recently, pressures to address converging crises have led to a return to faith in 'innovation' and 'resilience' as touchstones of a renewed faith in modernity as the solution for food insecurity. States are rediscovering their preference for mega-farms as a simple and quick fix for applying science to feeding (and fuelling) the growing and politically volatile urban majority.

But the cases in this volume have shown that the ascendency of this new/old paradigm is disputed. There is an awareness that such a selective application of the social contract will not be accepted, and counter-narratives around social protection must encompass the rural poor and those who have shrinking opportunities to buy the food produced on these mega-farms. The meaning of food security is likely to remain disputed for the foreseeable future.

Reference

FAO (2009) *Reform of the Committee on World Food Security: Final Version,* CFS:2009/2 Rev.2, Rome: Food and Agriculture Organization. Online. Available <www.fao.org/fileadmin/templates/cfs/Docs0910/ReformDoc/CFS_2009_2_Rev_2_E_K7197.pdf> (accessed 25 March 2014).

INDEX

Note: page numbers followed by *t* or *f* refer to tables and figures respectively.